Lecture Notes in Artificial Intelligence   10308

Subseries of Lecture Notes in Computer Science

More information about this series at http://www.springer.com/series/1244

Karell Bertet · Daniel Borchmann
Peggy Cellier · Sébastien Ferré (Eds.)

Formal
Concept Analysis

14th International Conference, ICFCA 2017
Rennes, France, June 13–16, 2017
Proceedings

 Springer

Editors
Karell Bertet
Université de La Rochelle
La Rochelle
France

Daniel Borchmann (iD)
Institute of Theoretical Computer Science
Technische Universität Dresden
Dresden
Germany

Peggy Cellier
Université de Rennes
Rennes
France

Sébastien Ferré
Université de Rennes
Rennes
France

ISSN 0302-9743   ISSN 1611-3349   (electronic)
Lecture Notes in Artificial Intelligence
ISBN 978-3-319-59270-1   ISBN 978-3-319-59271-8   (eBook)
DOI 10.1007/978-3-319-59271-8

Library of Congress Control Number: 2017940844

LNCS Sublibrary: SL7 – Artificial Intelligence

Printed on acid-free paper

This Springer imprint is published by Springer Nature
The registered company is Springer International Publishing AG
The registered company address is: Gewerbestrasse 11, 6330 Cham, Switzerland

Preface

This volume features the accepted publications of the 14th International Conference on Formal Concept Analysis (ICFCA 2017), held during June 13–16, 2017, at IRISA, Rennes. The ICFCA conference series is one of the major venues for researchers from the field of formal concept analysis (FCA) and related areas to present and discuss their recent work with colleagues from all over the world. Since its first edition in 2003 in Darmstadt, the ICFCA conference series has been held in Europe, Australia, America, and Africa.

The field of FCA originated in the 1980s in Darmstadt as a subfield of mathematical order theory, with prior developments in other research groups. Its original motivation was to consider *complete lattices* as *lattices of concepts*, drawing motivation from philosophy and mathematics alike. FCA has since then developed into a wide research area with applications far beyond its original motivation, for example, in logic, data mining, learning, and psychology.

The FCA community is mourning the passing of Rudolf Wille on January 22, 2017, in Bickenbach, Germany. As one of the leading researchers throughout the history of FCA, he was responsible for inventing and shaping many of the fundamental notions in this area. Indeed, the publication of his article "Restructuring Lattice Theory: An Approach Based on Hierarchies of Concepts" is seen by many as the starting point of FCA as an independent direction of research. He was head of the FCA research group in Darmstadt from 1983 until his retirement in 2003, and remained an active researcher and contributor thereafter. In 2003, he was among the founding members of the ICFCA conference series. In memory of Rudolf Wille's contribution to the foundations of FCA, and in the tradition of previous conference proceedings, this volume contains a historical paper from the early days of FCA (1986), namely "Implikationen und Abhängigkeiten zwischen Merkmalen" by Bernhard Ganter and Rudolf Wille, together with a translation by Daniel Borchmann and Juliane Prochaska, authorized by Bernhard Ganter.

For the proceedings of this year's ICFCA, out of 37 submitted papers, 13 where chosen to be published in this volume, resulting in an acceptance rate of around 35%. Additionally, four papers were judged mature enough to be discussed at the conference and are included in a supplementary volume titled "Contributions to ICFCA 2017," published by the University of Rennes 1.

In addition to the regular contributions, this volume also contains the abstracts of the four invited talks at ICFCA 2017 as well as an invited contribution titled "An Invitation to Knowledge Space Theory." In this article, the authors again want to draw attention to the close links between FCA and *knowledge space theory*, a successful theory from psychology on how to learn.

This proceedings volume would not have been possible without the valuable work of the authors, the members of the Program Committee, and the members of the Editorial Board. Many thanks also go to the external reviewers, who provided valuable feedback across the board. We also want to express our gratitude to the team of local organizers, who made sure that the conference ran smoothly and was a pleasant experience for all its participants. We also thank Springer for sponsoring the Best Young Research Award. Finally, we would like to thank the EasyChair team for providing their system to organize the conference.

June 2017   Karell Bertet
                                                        Daniel Borchmann
                                                           Peggy Cellier
                                                         Sebastien Ferre

Organization

Executive Committee

Peggy Cellier   IRISA, INSA Rennes, France
Sébastien Ferré   IRISA, University Rennes 1, France

Program and Conference Proceedings

Program Chairs

Karell Bertet   L3I, Université de La Rochelle, France
Daniel Borchmann   Technische Universität Dresden, Germany

Editorial Board

Jaume Baixeries	Polytechnic University of Catalonia, Catalonia
Peggy Cellier	IRISA lab, Rennes, France
Florent Domenach	Akita International University, Japan
Peter Eklund	IT University of Copenhagen, Denmark
Sebastien Ferré	Université de Rennes 1, France
Bernhard Ganter	Technische Universität Dresden, Germany
Cynthia Vera Glodeanu	Germany
Mehdi Kaytoue	LIRIS, INSA-Lyon, France
Sergei Kuznetsov	Higher School of Economics, Moscow, Russia
Leonard Kwuida	Bern University of Applied Sciences
Rokia Missaoui	Université du Québec en Outaouais (UQO), Canada
Amedeo Napoli	LORIA, Nancy, France
Sergei Obiedkov	Higher School of Economics, Moscow, Russia
Manuel Ojeda-Aciego	University of Malaga, Spain
Uta Priss	Ostfalia University of Applied Sciences, Wolfenbüttel, Germany
Sebastian Rudolph	Technische Universität Dresden, Germany
Christian Sacarea	Babes-Bolyai University of Cluj-Napoca, Romania
Stefan E. Schmidt	Technische Universität Dresden, Germany
Gerd Stumme	Universität Kassel, Germany
Petko Valtchev	Université du Québec à Montréal, Canada
Karl Erich Wolff	University of Applied Sciences, Darmstadt, Germany

Honorary Members

Vincent Duquenne   ECP6-CNRS, Université Paris 6, France
Rudolf Wille   Technische Universität Darmstadt, Germany
                     (deceased)

Program Committee

Simon Andrews	Sheffield Hallam University, United Kingdom
François Brucker	Centrale Marseille, France
Claudio Carpineto	Fondazione Ugo Bordoni, Italy
Pablo Cordero	University of Málaga, Spain
Jean Diatta	Université de La Réunion, France
Felix Distel	d-fine GmbH, Germany
Christophe Demko	L3I lab, Université de La Rochelle, France
Stephan Doerfel	MICROMATA GmbH, Germany
Alain Gély	Université Paul Verlaine, Metz, France
Robert Godin	Université du Québec à Montréal, Canada
Marianne Huchard	LIRMM, Université Montpellier, France
Dmitry Ignatov	Higher School of Economics, Moscow, Russia
Robert Jäschke	University of Sheffield, England
Michal Krupka	Palacky University, Olomouc, Czech Republic
Marzena Kryszkiewicz	Warsaw University of Technology, Poland
Wilfried Lex	Universität Clausthal, Germany
Jesús Medina	Universidad de Cádiz, Spain
Engelbert Mephu Nguifo	LIMOS, Université Blaise Pascal Clermont-Ferrand, France
Lhouari Nourine	LIMOS, Université Blaise Pascal Clermont-Ferrand, France
Jan Outrata	Palacky University of Olomouc, Czech Republic
Jean-Marc Petit	LIRIS, INSA Lyon, France
Sandor Radeleczki	University of Miskolc, Hungary
Barış Sertkaya	Frankfurt University of Applied Sciences, Germany
Laszlo Szathmary	University of Debrecen, Hungary
Andreja Tepavčević	University of Novi Sad, Serbia
Jean-François Viaud	L3I lab, Université de La Rochelle, France

Additional Reviewers

Sabeur Aridhi	University of Lorraine, France
Alexandre Bazin	LIMOS, Université Blaise Pascal Clermont-Ferrand, France
Hinde Lilia Bouziane	LIRMM, Université Montpellier, France
Tom Hanika	Universität Kassel, Germany
Laurent Beaudou	LIMOS, Université Blaise Pascal Clermont-Ferrand, France
David Lobo	Universidad de Cádiz, Spain
Eloisa Ramírez-Poussa	Universidad de Cádiz, Spain

Invited Talks

Analogy Between Concepts

Laurent Miclet[1], Nelly Barbot[2], and Henri Prade[3]

[1] IRISA, Rennes, France
[2] IRISA-Lannion, Lannion, France
[3] IRIT Toulouse, Toulouse, France

Analogical reasoning exploits parallels between situations. It enables us to state analogies for explanation purposes, to draw plausible conclusions, or to create new devices by transposing old ones in new contexts. As such, reasoning by analogy plays an important role in human thinking, as it is widely acknowledged. For this reason, it has been studied for a long time, in philosophy, linguistics, cognitive psychology and artificial intelligence, under various approaches.

The classical view of analogy describes the parallel between two situations, which are described in terms of objects, properties of the objects, and relations linking the objects.

An analogical situation may be also expressed by *proportions*. In that respect, a key pattern is a statement of the form '*A* is to *B* as *C* is to *D*', as in the examples "a calf is to a bull as a foal is to a stallion", or "gills are to fishes as lungs to mammals".

In the first example, the four items *A, B, C, D* belong to *the same* category, and we speak of *analogical proportion*. In the second example, the four items belong to two *different* categories, here *A* and *C* are organs and *B* and *D* classes of animals. In this second type of analogical statement, we speak of *relational analogy* or *relational proportion*.

It is only recently that these proportions have been systematically studied in terms of mathematic properties and of applications in AI tasks. As important examples, an exhaustive study of the "logical" proportions (in propositional logic) has been produced by H. Prade and G. Richard; proportions between formal structures, including words over finite alphabets, have been studied by Y. Lepage, N. Stroppa and F. Yvon; the use of proportions for Pattern Recognition has been proved useful by S. Bayoud and L. Miclet. In this spirit, we present in this talk our results on proportions in lattices, with a focus on concept lattices.

Hence, the goal of this talk is

- to give a formalization of the analogical proportion between four elements of a general lattice,
- to see how it applies in Formal Concept Lattices,
- and to give operational definitions of an analogical proportion between formal concepts and of a relational proportion in a formal context.

More precisely, concerning lattices of formal concepts, we will describe how an analogical proportion can be defined between concepts and how such proportions are deeply related to subcontexts of a special structure, that we call *analogical complexes*.

We will also illustrate the fact that the set of these complexes is itself structured as a lattice. In that way, we will demonstrate how analogical proportions between concepts can be constructed from the formal context, without the construction of the whole lattice of concepts.

We will also give a acceptable definition of what can be a relational proportion in terms of formal concept analysis, and illustrate it with some linguistic examples.

The talk will be mainly illustrative. No purely formal results nor demonstrations will be given, but mainly examples and illustrations of new notions and of the running of algorithms. An exhaustive article, which is the base of this talk, is presently under review for a journal.

References

1. Prade, H., Richard, G.: From analogical proportion to logical proportions. Log. Univers. **7**, 441–505 (2013)
2. Lepage, Y.: De l'analogie rendant compte de la commutation en linguistique. HDR, Grenoble (2001). http://www.slt.atr.jp/lepage/pdf/dhdryl.pdf,
3. Miclet, L., Nicolas, J.: From formal concepts to analogical complexes. In: Proceedings of the 12th International Conference on Concept Lattices and their Applications, Clermont-Ferrand, pp. 159–170 (2015)
4. Stroppa, N., Yvon, F.: Formal models of analogical proportions. Technical report 2006D008, École Nationale Supérieure des Télécommunications, Paris, France (2006)
5. Miclet, L., Bayoudh, S., Delhay, A.: Analogical dissimilarity: definition, algorithms and two experiments in machine learning. JAIR **32**, 793–824 (2008)

Facilitating Exploration of Knowledge Graphs Through Flexible Queries and Knowledge Anchors

Alexandra Poulovassilis

Knowledge Lab, Birkbeck, University of London, London, UK
ap@dcs.bbk.ac.uk

Semantic web and information extraction technologies are enabling the creation of vast information and knowledge repositories, in the form of knowledge graphs comprising entities and the relationships between them. As the volume of such graph-structured data continues to grow, it has the potential to enable users' knowledge expansion in application areas such as web information retrieval, formal and informal learning, scientific research, health informatics, entertainment, and cultural heritage. However, users are unlikely to be familiar with the complex structures and vast content of such datasets and hence need to be assisted by tools that support interactive exploration and flexible querying.

Recent work has proposed techniques for automatic approximation and relaxation of users' queries over knowledge graphs, allowing query answers to be incrementally returned in order of their distance from the original form of the query. In this context, approximating a query means applying an edit operation to the query so that it can return possibly different answers, while relaxing a query means applying a relaxation operation to it so that it can return possibly more answers. Edit operations include insertion, deletion or substitution of a property label, while relaxation operations include replacing a class by a superclass, or a property by a superproperty.

The benefits of supporting such flexible query processing over knowledge graphs include: (i) correcting users' erroneous queries; (ii) finding additional relevant answers that the user may be unaware of; and (iii) generating new queries which may return unexpected results and bring new insights. However, although this kind of flexible querying can increase a user's understanding of the knowledge graph and underlying knowledge domain, it can return a large number of query results, all at the same distance away from the user's original query. Therefore, a key challenge is how to facilitate users' meaning making from flexible query results.

Meaning making is related to users' domain knowledge and their ability to make sense of the entities that they encounter during their interactions with the knowledge graph. Empirical studies have suggested that paths which start with familiar entities (knowledge anchors) and then add new, possibly unfamiliar, entities can be beneficial for making sense of complex knowledge graphs. Recent work has proposed an approach to identifying knowledge anchors that adopts the cognitive science concept of basic-level objects in a domain taxonomy, with the development of a formal framework for deriving a set of knowledge anchors from a knowledge graph.

In this talk we discuss how a combination of these two directions of work - namely, flexible querying of graph-structured data, and identification of knowledge anchors in a knowledge graph - can be used to support users in incrementally querying, exploring and learning from large, complex knowledge graphs. Our hybrid approach combines flexible graph querying and knowledge anchors by including in the query results paths to the nearest knowledge anchor. This makes more evident the relationships between the entities returned within the query results, and allows the user to explore increasingly larger fragments of the domain taxonomy.

Patterns, Sets of Patterns, and Pattern Compositions

Jilles Vreeken

Cluster of Excellence on Multi-Modal Computing and Interaction,
Max Planck Institute for Informatics, Saarland University,
Saarbrücken, Germany
`jilles@mpi-inf.mpg.de`

The goal of exploratory data analysis – or, data mining – is making sense of data. We develop theory and algorithms that help you understand your data better, with the lofty goal that this helps formulating (better) hypotheses. More in particular, our methods give detailed insight in how data is structured: characterising distributions in easily understandable terms, showing the most informative patterns, associations, correlations, etc.

Patterns, such as formal concepts, can give valuable insight in data. Mining all potentially interesting patterns is a useless excercise, however: the result is cumbersome, sensitive to noise, and highly redundant. Mining a small set of patterns, that together describes the data well, leads to much more useful results. Databases, however, typically consist of different parts, or, components. Each such component is best characterised by a different set of patterns. Young parents, for example, exhibit different buying behaviour than elderly couples. Both, however, buy bread and milk. A pattern composition models exactly this. It jointly characterises the similarities and differences between such components of a database, without redundancy or noise, by including only patterns that are descriptive for the data, and assigning those patterns only to the relevant components of the data. Knowing what a pattern composition is, this leads to the question, how can we discover these from data?

Semantic Web: Big Data, Some Knowledge and a Bit of Reasoning

Marie-Christine Rousset[1,2], Manuel Atencia[1], Jérôme David[1],
Fabrice Jouanot[1], Olivier Palombi[3,4], and Federico Ulliana[5]

[1] Université Grenoble Alpes, CNRS, Inria, LIG, 38000 Grenoble, France
Marie-Christine.Rousset@imag.fr
[2] Institut Universitaire de France, 75005 Paris, France
[3] Université Grenoble Alpes, Inria, CNRS, LJK, 38000 Grenoble, France
[4] Université Grenoble Alpes, Laboratoire d'Anatomie des Alpes Françaises
(LADAF), 38000 Grenoble, France
[5] Université de Montpellier, CNRS, Inria, LIRMM, 34000 Montpellier, France

Linked Data provides access to huge, continuously growing amounts of open data and ontologies in RDF format that describe entities, links and properties on those entities. Equipping Linked Data with reasoning paves the way to make the Semantic Web a reality. In this presentation, I will describe a unifying framework for RDF ontologies and databases that we call deductive RDF triplestores. It consists in equipping RDF triplestores with inference rules. This rule language allows to capture in a uniform manner OWL constraints that are useful in practice, such as property transitivity or symmetry, but also domain-specific rules with practical relevance for users in many domains of interest. I will illustrate the expressivity of this framework for Semantic Web applications and its genericity for developing inference algorithms with good behaviour in data complexity. In particular, we will show how it allows to model the problem of data linkage as a reasoning problem on possibly decentralized data. We will also explain how it makes possible to efficiently extract expressive modules from Semantic Web ontologies and databases with formal guarantees, whilst effectively controlling their succinctness.

Experiments conducted on real-world datasets have demonstrated the feasibility of this approach and its usefulness in practice for data integration and information extraction.

Contents

Invited Contribution

An Invitation to Knowledge Space Theory   3
   Bernhard Ganter, Michael Bedek, Jürgen Heller, and Reinhard Suck

Historical Paper

Implications and Dependencies Between Attributes 23
   Bernhard Ganter, Rudolf Wille, Daniel Borchmann,
   and Juliane Prochaska

Regular Contributions

The Implication Logic of (n, k)-Extremal Lattices 39
   Alexandre Albano

Making Use of Empty Intersections to Improve the Performance
of CbO-Type Algorithms 56
   Simon Andrews

On the Usability of Probably Approximately Correct Implication Bases 72
   Daniel Borchmann, Tom Hanika, and Sergei Obiedkov

FCA in a Logical Programming Setting for Visualization-Oriented
Graph Compression 89
   Lucas Bourneuf and Jacques Nicolas

A Proposition for Sequence Mining Using Pattern Structures 106
   Victor Codocedo, Guillaume Bosc, Mehdi Kaytoue,
   Jean-François Boulicaut, and Amedeo Napoli

An Investigation of User Behavior in Educational Platforms Using
Temporal Concept Analysis 122
   Sanda-Maria Dragoş, Christian Săcărea, and Diana-Florina Şotropa

Hierarchies of Weighted Closed Partially-Ordered Patterns for Enhancing
Sequential Data Analysis 138
   Cristina Nica, Agnès Braud, and Florence Le Ber

First Notes on Maximum Entropy Entailment for Quantified Implications . . . 155
   Francesco Kriegel

XVIII   Contents

Implications over Probabilistic Attributes   168
  Francesco Kriegel

On Overfitting of Classifiers Making a Lattice   184
  Tatiana Makhalova and Sergei O. Kuznetsov

Learning Thresholds in Formal Concept Analysis   198
  Uta Priss

The Linear Algebra in Extended Formal Concept Analysis
Over Idempotent Semifields   211
  Francisco José Valverde-Albacete and Carmen Peláez-Moreno

Distributed and Parallel Computation of the Canonical Direct Basis   228
  Jean-François Viaud, Karell Bertet, Rokia Missaoui,
  and Christophe Demko

Author Index   243

Invited Contribution

An Invitation to Knowledge Space Theory

Bernhard Ganter[1(✉)], Michael Bedek[2], Jürgen Heller[3], and Reinhard Suck[4]

[1] Technische Universität Dresden, Dresden, Germany
bernhard.ganter@tu-dresden.de
[2] Technische Universität Graz, Graz, Austria
[3] Universität Tübingen, Tübingen, Germany
[4] Universität Osnabrück, Osnabrück, Germany

Abstract. It has been mentioned on many occasions that Formal Concept Analysis and KST, the theory of Knowledge Spaces, introduced by J.-P. Doignon and J.-C. Falmagne, are closely related in theory, but rather different in practice. It was suggested that the FCA community should learn from and contribute to KST. In a recent workshop held at Graz University of Technology, researchers from both areas started to combine their views and tried to find a common language. This article is a partial result of their effort. It invites FCA researchers to understand some ideas of KST by presenting them in the language of formal contexts and formal concepts.

1 Introduction

Knowledge Space Theory (KST) [6] and *Formal Concept Analysis (FCA)* [21] share a range of similarities: Both approaches have been invented in the early 1980s and since then have experienced an astonishing development from a theoretical as well as a practical point of view. Since the very beginning, researchers from both fields were aware of each other, and it was clear that the approaches are also similar in their mathematical foundations (A first translation was worked out by Rusch and Wille [18]). However, KST and FCA developed and grew primarily in parallel, perhaps due to the different fields of application. While FCA always aimed at a broad spectrum of applications, KST from its beginnings has focused on the topic of *learning* and was impressively successful there.

This article is another invitation to the FCA community not only to be aware of KST, but to learn from its success and to contribute, whenever possible, to its theoretical basis. We recommend to study the KST literature, in particular the two monographs by Doignon and Falmagne [6,8]. Here we present some of the basic ideas in FCA language, hoping to spark the interest of FCA researchers.

In a nutshell, KST aims at an adaptive, and thus efficient, assessment of a persons' knowledge state with respect to a given knowledge domain. Such a domain is given by a (finite) set Q of questions, and a knowledge state is the set of problems the person is capable to master. Details will follow below. We sometimes hear from colleagues in pedagogical research that such an approach is much too simple to model a learning process. One must however admit that KST

© Springer International Publishing AG 2017
K. Bertet et al. (Eds.): ICFCA 2017, LNAI 10308, pp. 3–19, 2017.
DOI: 10.1007/978-3-319-59271-8_1

is remarkably and measurably successful in such applications. The professional ALEKS software (**A**ssessment and **LE**arning in **K**nowledge **S**paces) is an online tutoring and assessment program with many users that includes course material in mathematics, chemistry, introductory statistics, and business.

There are extensions of KST, for example *Competence-based Knowledge Space Theory (CbKST)* [1,16,17], which aims at an adaptive assessment of a person's skills and competencies that enable him or her to master certain problems. Based on that, CbKST suggests pedagogically reasonable learning paths. Natural applications of (Cb)KST are intelligent tutoring systems, as well as other kinds of technology-enhanced learning applications (e.g., game-based learning).

A main theoretical difference between the two fields is that FCA works with *closure systems*, that is, with families of sets that are closed under intersections, while knowledge spaces are closed under the *union of sets*. The translation seems obvious and is indeed not difficult.

2   Knowledge States and Formal Concepts

We investigate a *knowledge domain*, which is a finite set $\mathcal{Q}$ of **questions** (also called **tasks** or **test items**), which learners may or may not be able to master (see Example 6 for an instance). Each learner has an individual **knowledge state**, characterized by the collection of those questions in $\mathcal{Q}$ which that learner is able to solve. Such **learner-task data** is conveniently recorded in form of a formal context

$$(\mathcal{L}, \mathcal{Q}, \Box),$$

where the $\Box$ symbol denotes the "masters" relation between learners and questions. Symbolically, we abbreviate the statement that "learner l masters question q" to

$$l \ \Box \ q.$$

But beware! The table should not be understood as *empirical* data, which typically would contain *careless errors and lucky guesses* of the learners. Instead, each row of this data is meant to record the respective learner's true knowledge state, formally

$$l^{\Box} := \{q \in \mathcal{Q} \mid l \ \Box \ q\}$$

for each learner $l \in \mathcal{L}$. With a similar notation, we write for each question $q \in \mathcal{Q}$

$$q^{\Box} := \{l \in \mathcal{L} \mid l \ \Box \ q\}$$

to denote the set of those learners who master that question.

Example 1. *Figure 1 shows a very small example of a learner-task context. It lists nine learners $l_1, \ldots, l_9$, and eight tasks $a, \ldots, h$. Learner l_4 masters only tasks e, f, and g, so that the knowledge state of l_4 is*

$$l_4^{\Box} = \{e, f, g\}.$$

Problem g is mastered by learners l_3, l_4, l_6, l_8, and l_9, abbreviated

$$g^{\Box} = \{l_3, l_4, l_6, l_8, l_9\}.$$

□	a	b	c	d	e	f	g	h
l_1								
l_2	×							
l_3	×	×	×	×	×		×	×
l_4						×	×	×
l_5		×	×	×	×			
l_6		×	×	×	×	×	×	
l_7		×	×			×		
l_8	×	×	×	×	×	×	×	
l_9	×	×	×	×	×	×	×	×

Fig. 1. A learner-task context

Usually not every subset of Q qualifies as a possible knowledge state, since some of the tasks may be *prerequisites* for others. For example, it seems unlikely that a learner can divide, but not add.

The set K of all feasible knowledge states over Q is called a **knowledge structure** and is often denoted as (Q, K). Although often very large, any knowledge structure (Q, K) can be written down in the same form as learner-task data, namely as

$$(K, Q, \ni).$$

General knowledge structures are difficult to handle, since they are sets of sets and can easily be of enormous size. Reasonable assumptions about structural properties of knowledge structures are therefore welcome, because they may simplify their use. One such assumption relies on the idea that each knowledge state must be obtainable through a step-wise *learning* process, which respects the prerequisites, and that the knowledge states are precisely all subsets of Q that can be learned. A consequence of this assumption is that only *missing* prerequisites can be obstacles to learning, but the presence of additional knowledge never is. This in turn implies that in theory every learner could acquire the knowledge of each other learner, and that therefore the set union of any two knowledge states is again a knowledge state. A knowledge structure which is closed under arbitrary set unions is called a **knowledge space**.

The assumption that knowledge spaces are closed under unions of states may surprise a little. There are several possible justifications for this. One is that, since knowledge must be acquired step by step, there must be a graded *learning path* to every knowledge state. This leads to an even more specialized notion, that of a *learning space* (or *antimatroid* or *well-graded knowledge space*). We shall not treat this here and refer the reader to Falmagne and Doignon [8] (but we shall mention a characterization of these spaces in Sect. 8). A more simple-minded approach is to consider a knowledge space as consisting of those task sets which can be solved by *groups of learners*, assuming that a group solves a problem if at least one group member does. This obviously leads to union-closed families of states.

The knowledge space *generated* by given knowledge states consists of all set unions of these states, including the empty union, $\emptyset$. The knowledge space generated by the states in given learner-task data $(\mathcal{L}, \mathcal{Q}, \square)$, such as in Fig. 1, is denoted by

$$\mathcal{K}(\mathcal{L}, \mathcal{Q}, \square).$$

In Formal Concept Analysis terms, we find that this knowledge space consists of the intent-complements of $(\mathcal{L}, \mathcal{Q}, \boxslash)$, i.e.,

$$\mathcal{K}(\mathcal{L}, \mathcal{Q}, \square) = \{\mathcal{Q} \setminus B \mid (B', B) \in \underline{\mathfrak{B}}(\mathcal{L}, \mathcal{Q}, \boxslash)\}.$$

Example 2. *The knowledge space generated by the learner-task data in Fig. 1 has 16 elements. A diagram is shown in Fig. 2. We give an examples of how to read this diagram:*

- *Consider the node labeled with d. It represents the knowledge state $\{a, b, c, e, f, g\}$, which is the union of the knowledge states of the learners l_1, l_2, l_4, and l_7. The learners are the ones that can be reached via downward paths from the node. The tasks are the ones which cannot be reached via upward paths from the node labeled d.*

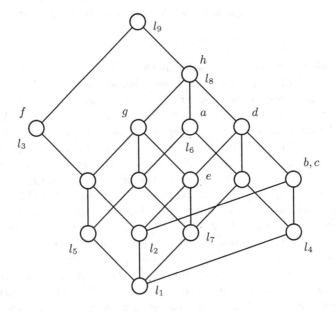

Fig. 2. The knowledge space generated by the learner-task data in Fig. 1, represented as the concept lattice of the complementary context $\underline{\mathfrak{B}}(\mathcal{L}, \mathcal{Q}, \boxslash)$. The concept intents are the complements of the knowledge states. The extents consists of those learners, whose knowledge state is contained in the one represented by the respective node.

Each knowledge space has a unique **basis** $\mathcal{B}$. It consists of the $\cup$-irreducible states, i.e., of those states that cannot be obtained as unions of other states.

It can be obtained from $(\mathcal{K}, \mathcal{Q}, \ni)$ by a process which we call **coreduction**, by removing all rows that can be written as set unions of other rows. This corresponds to (object-) reduction in Formal Concept Analysis, except that set union is used instead of set intersection. In Formal Concept Analysis terms one first goes to the complementary context

$$(\mathcal{K}, \mathcal{Q}, \not\ni),$$

object-reduces it and complements the outcome to obtain the basis.

Example 3. *The learner-task data in Fig. 1 is not coreduced because, for example, the knowledge state of learner l_6 is the set union of the states of learners l_4 and l_5:*

$$\{b, c, d, e, f, g\} = l_6^{\square} = l_4^{\square} \cup l_5^{\square} = \{e, f, g\} \cup \{b, c, d, e\}.$$

The basis of the knowledge space in Fig. 2 consists of five states, see Fig. 3.

$\square$	a	b	c	d	e	f	g	h
l_2	×							
l_3	×	×	×	×	×		×	×
l_4						×	×	×
l_5		×	×	×	×			
l_7		×	×			×		

Fig. 3. The basis of the knowledge space generated by the learner-task context in Fig. 1 consists of the complements of the object intents of the object-reduced formal context.

The basis uniquely determines the knowledge space. There are fast algorithms for generating a knowledge space from its basis. For example one can apply the standard NEXT INTENT algorithm to the complementary context and obtain the knowledge states as the complements of the generated intents. Actually, these algorithms can do more. For example, one often has only *incomplete information* about a learner's knowledge state. Some set $P \subseteq \mathcal{Q}$ may be known of questions which that learner has passed, and also some set F of questions which were failed. It then is clear that the learner's state must be some $S \in \mathcal{K}$ satisfying $P \subseteq S$ and $S \cap F = \emptyset$. The above-mentioned algorithm can indeed generate exactly these states, given a basis together with the sets P and F.

But complexity remains an issue, even when fast algorithms are available. This is due to the huge number of possibilities. Consider, for example, the case that only $|\mathcal{Q}| = 7$ questions are investigated. It was shown by Colomb et al. [4] that there are as many as 14 087 648 235 707 352 472 different knowledge spaces over seven questions. The largest one of them has 128 ($= 2^7$) states and a basis of size seven. It contains all other knowledge spaces over the same seven questions. Among them is one that has only 65 states, but a basis of size 35.

3   Prerequisites and Implications

A knowledge space can be specified by explicitly listing up sufficiently many of its knowledge states, so that the space then can be generated from these via set unions. However, this may be rather tedious and not intuitive. Another approach is to use expert judgments on the dependencies between questions. Such a judgment could say that question q_2 is more difficult than q_1 in the sense that *any learner who masters q_2 also masters q_1*. This is abbreviated to $q_1 \leq q_2$, and a natural assumption is that this relation $\leq$ is reflexive and transitive. Such relations are called **quasi-orders** or, synonymously, **preorders**.

Example 4. *The quasi-order for the data in Fig. 1 is given by the diagram in Fig. 4.*

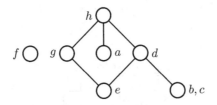

Fig. 4. The quasi-order for the learner-task data in Fig. 1. Whoever masters d also masters b, c, e, but not necessarily a, f, g or h.

A subset of $\mathcal{Q}$ **disrespects** a judgment $q_1 \leq q_2$ if it contains q_2, but not q_1. Otherwise, it **respects** it. Given a collection $\mathcal{J}$ of such judgments, we may consider the set

$$\mathcal{K} := \{S \subseteq \mathcal{Q} \mid S \text{ respects all judgments in } \mathcal{J}\}.$$

It is easy to see that this is a knowledge space: if two sets both respect $q_1 \leq q_2$, then so does their set union. Thus judgments can be used to define knowledge spaces. However, the knowledge spaces obtained in this way are rather special, since they are also closed under set intersections. They are called **quasi-ordinal** knowledge spaces, because their states are precisely the "downsets" or "order ideals" of the quasi-orders $\leq$.

Most spaces are not quasi-ordinal. Of the 14 087 648 235 707 352 472 different knowledge spaces over seven questions only 9 535 241 are quasi-ordinal. To describe all knowledge spaces we therefore need a more general form of expert judgment. It can be shown that judgments of the form

each learner who masters question q also masters at least one of the questions $q_1, \ldots, q_n$

are sufficient for specifying arbitrary knowledge spaces. This is no surprise since these expressions correspond to *implications* between attributes of the complementary context: they can be reformulated in a logically equivalent form *if a learner masters none of the questions $q_1, \ldots, q_n$, then that learner cannot master q.*

Symbolically, we abbreviate this to

$$\neg A \rightarrow \neg q \qquad \text{(where } A := \{q_1, \ldots, q_n\}\text{)}.$$

Mixed approaches are possible using the techniques of **attribute exploration** [11]. Typically information is collected in form of some judgments $A_1 \rightarrow q_1$, $A_2 \rightarrow q_2$, ...and of some "observations" (P_1, F_1), (P_2, F_2), ..., as described above. The exploration algorithm then finds, if possible, the *simplest piece of information* missing for a complete description of a knowledge space, and this can be given to an expert as a query.

Example 5. *The knowledge space generated by the learner-task data in Fig. 1 is not quasi-ordinal. It can be checked that its quasi-order, depicted in Fig. 4, is respected by 34 sets, which means that 34 is the number of states in a quasi-ordinal knowledge space with this quasi-order. But as already mentioned, the number of set unions of the states in Fig. 1 is only 16. An example of a "non-ordinal" rule that holds in Fig. 1 is that each learner who masters g also masters one of f and h, i.e., that*

$$\neg\{f, h\} \rightarrow \neg g.$$

4   A Quasi-Ordinal Example

Typically the problems in a knowledge domain $\mathcal{Q}$ are assumed to be dichotomous. Responses are coded as either correct (solved), or incorrect (not solved).

Example 6. *The knowledge domain $\mathcal{Q} = \{a, b, c, d, e, f\}$ defined in Table 1 is taken from the beginning algebra items in the ALEKS system. Below the table we show a prerequisite quasi-order $(\mathcal{Q}, \leq)$ (which in this case actually is an order), as it is used by ALEKS.*

So ALEKS assumes that a learner which masters Question f also masters all the other questions, while any learner mastering c also masters a, but not necessarily the other ones. The sets respecting this prerequisite order are precisely the downsets (order ideals). So these are the knowledge states if no other prerequisite conditions are imposed. It is well known that the family of order-ideals of a quasi-order is closed under set union and set intersection and therefore forms a distributive lattice. In fact, Birkhoff's representation theorem (Theorem 39 in [12]) shows that every finite distributive lattice is obtained in this way.

It is well known that the lattice of order ideals of a quasi-order $(\mathcal{Q}, \leq)$ is isomorphic to the concept lattice of the corresponding *contra-ordinal scale* $(\mathcal{Q}, \mathcal{Q}, \not\geq)$ (see [12], p. 49). Figure 6 shows this scale (derived from Fig. 5) and its concept lattice.

The *learning paths* of a knowledge space are defined as the maximal chains in the lattice of knowledge states. Figure 7 shows the learning paths for the

Table 1. An example of a knowledge domain from the ALEKS system.

	Problem type	Instance
a	Word problem on proportions	A car travels on the freeway at an average speed of 52 miles per hour. How many miles does it travel in 5 h and 30 min?
b	Plotting a point in the coordinate plane	Using the pencil, mark the point at coordinates $(1, 3)$
c	Multiplication of monomials	Perform the following multiplication: $4x^4y^4 \cdot 2x \cdot 5y^2$ and simplify your answer as much as possible
d	Greatest common factor of two monomials	Find the greatest common factor of expressions $14t^6y$ and $4tu^5y^8$. Simplify your answer as much as possible
e	Graphing the line through a given point with a given slope	Graph the line with slope -7 passing through the point $(-3, -2)$
f	Writing the equation of the line through a given point and perpendicular to a given line	Write an equation for the line that passes through the point $(-5, 3)$ and is perpendicular to the line $8x + 5y = 11$

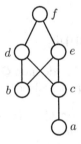

Fig. 5. The prerequisite (quasi-)order for the knowledge domain shown in Table 1

"beginning algebra" domain, depicted as a tree (right hand side). On the left, it is indicated how they can be counted for any given knowledge space. In the case of quasi-ordinal knowledge spaces, the learning paths correspond to the linear extensions of the underlying quasi-order. In general, the number of learning paths can be considerably smaller. The knowledge space shown in Fig. 2 has 18 learning paths, while the ordered set of its questions (shown in Fig. 4) has 198 linear extensions.

5   Probabilistic Knowledge Structures

The probabilistic theory of knowledge spaces is much more elaborate than what is available for FCA. This is due to the applications: it is desirable to guess a

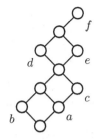

Fig. 6. The contra-ordinal scale and the knowledge space for the data in Table 1. The 10 knowledge states can be read from the reduced labeling as usual in Formal Concept Analysis. For example, the knowledge state corresponding to the "waist" node in the middle is $\{a, b, c\}$, since these are the labels at downward paths starting from this node.

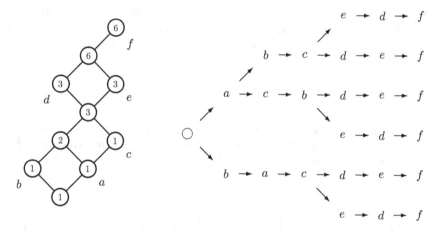

Fig. 7. The learning paths for the "beginning algebra" domain. Each number in the lattice diagram indicates how many learning paths lead to the respective knowledge state. They are obtained recursively by assigning 1 to the least element, and to each other element the sum of the numbers assigned to its lower neighbours. The right diagram shows the paths as a tree.

learner's knowledge state from only a few observations. Once the knowledge state is known, one can determine its "outer fringe", consisting of those problems that the learner "is ready to learn" (i.e., for which all prerequisites are already there), and this can be used with a tutorial system. The parameters of the stochastic models can be adjusted to observed learners' behaviour. Tutorial systems based on probabilistic KST are therefore able to improve upon usage. For a systematic presentation we refer the reader to Falmagne and Doignon [8]. Here we sketch some first ideas.

In practical applications we cannot assume that a student's response to an item is correct if and only if the student masters it (i.e. if the item is an element of the respective knowledge state). There are two types of response errors, namely

- careless errors, i.e. the response is *incorrect* although the item is an element of the learner's knowledge state (and could potentially be *mastered*),
- lucky guesses, i.e. the response is *correct* although the item is not an element of the learner's knowledge state (and could regularly *not* be *mastered*).

In any case, we need to dissociate knowledge states and response patterns. Let R denote a response pattern, which is a subset of Q, and $\mathcal{R} = 2^Q$ the set of all response patterns.

   In this approach the knowledge state K is a latent construct, and the response pattern R is a manifest indicator to the knowledge state. The conclusion from the observable response pattern to the unobservable knowledge state can only be of a stochastic nature. This requires to introduce a probabilistic framework capturing the facts that knowledge states will not occur with equal probability, and that response errors will occur with a certain probabilities.

   A *probabilistic knowledge structure* $(Q, \mathcal{K}, P)$ is defined by specifying a (marginal) distribution $P_\mathcal{K}$ on the states of $\mathcal{K}$, and the conditional probabilities $P(R \mid K)$ for all $R \in \mathcal{R}$ and $K \in \mathcal{K}$. The marginal distribution $P_\mathcal{R}$ on $\mathcal{R}$ then is predicted by

$$P_\mathcal{R}(R) = \sum_{K \in \mathcal{K}} P(R \mid K) \cdot P_\mathcal{K}(K).$$

The probabilistic knowledge structure that received most attention is the *basic local independence model* (BLIM), which satisfies the following condition: For each $q \in Q$ there are real constants $0 \leq \beta_q < 1$ and $0 \leq \eta_q < 1$ such that for all $R \in \mathcal{R}$ and $K \in \mathcal{K}$

$$P(R \mid K) = \left(\prod_{q \in K \setminus R} \beta_q \right) \cdot \left(\prod_{q \in K \cap R} (1 - \beta_q) \right) \cdot \left(\prod_{q \in R \setminus K} \eta_q \right) \cdot \left(\prod_{q \in Q \setminus (R \cup K)} (1 - \eta_q) \right).$$

   The constants β_q and η_q are interpreted as the probabilities of a careless error and a lucky guess, respectively, on item q.

   This condition entails two important properties. First, it assumes local stochastic independence. Given the knowledge state K, the responses are stochastically independent across the items. This means that any dependence between the responses to different items is mediated by the knowledge state K. Second, the response errors only depend on the respective item, but not on the knowledge state (beyond the fact that the item may be an element of the state, or not). This is a strong constraint which reduces the number of free parameters drastically and renders the BLIM testable. Procedures for estimating its parameters were proposed by Heller and Wickelmaier [14].

6   Competence-Based Knowledge Space Theory

The notion of a knowledge structure was formulated explicitly without reference to cognitive or mental concepts, like "latent traits", or specific problem components. Concepts like this were only introduced later on. There are parallel, but only partially overlapping approaches: Falmagne et al. [9], Doignon [5], Doignon and Falmagne [6], Albert and Held [1], as well as Korossy [15], and finally Düntsch and Gediga [7] and Gediga and Düntsch [13]. Korossy's approach is in a sense more general, as it allows for having a structure on the underlying cognitive abilities (a so-called *competence structure*).

We start by repeating an instructive example that was given by Falmagne and Doignon ([8], Example 6.1.1):

Example 7. *The question which could be included in a test of proficiency in the UNIX operating system is*

> *How many lines of the file* **lilac** *contain the word 'purple'? (Only one command line is allowed.)*

The skills required for solving this question include the use of basic UNIX commands, such as

$$S := \{grep, \quad wc, \quad cat, \quad more, | \, , \ldots\}.$$

There are several possible correct solutions, each one usually involving several of the basic commands, for example $\{cat, \quad grep, \quad | \, , wc\}$.

Example 7 suggests the following general approach: it is assumed that the ability to answer questions depends on the presence of certain *skills*, more precisely, on combinations of skills. Competence based Knowledge Space Theory (CbKST) is an extension of the KST approach explaining a learner's knowledge state by the competencies and skills that learner has. It is assumed that for each question there are one or several competencies that enable mastering the question, and that each of these competencies is a combination of skills. More precisely, one assumes for a given set $\mathcal{Q}$ of questions a set $\mathcal{C}$ of **competencies**, each of which is a subset of the set S of **skills**. For each question $q \in \mathcal{Q}$ there is a set $\mathcal{C}_q \subseteq \mathcal{C}$ of those competencies that suffice for mastering q. In FCA language one might write this in the form of a formal context

$$(\mathcal{C}, \mathcal{Q}, \models),$$

where the $\models$-relation expresses that a competency suffices for mastering a question:

$$C \models q :\Leftrightarrow \text{competency } C \text{ suffices for mastering question } q.$$

It is assumed that each learner has some, but not necessary all of these competencies. Abbreviating by $l \circ C$ the fact that learner l has competency C, we obtain a learner-competency context

$$(\mathcal{L}, \mathcal{C}, \circ).$$

The crucial model assumption then is that *a learner masters a question if and only if (s)he has a competency that suffices for mastering the question.* This can formally be expressed as

$$l \, \square \, q \Leftrightarrow \exists_{C \in \mathcal{C}} \, l \circ C \text{ and } C \models q,$$

which mathematically can be understood as a **Boolean product**

$$(\mathcal{L}, \mathcal{Q}, \Box) = (\mathcal{L}, \mathcal{C}, \circ) \cdot (\mathcal{C}, \mathcal{Q}, \models)$$

of formal contexts. Of course, the three formal contexts in this equation are usually not given, and it is a goal of *Learning Analytics* to provide reasonable models. Doignon and Falmagne write that their example (Example 7 above)

> ... *may lead one to believe that the skills associated with a particular domain could always be identified easily. In fact, it is by no means obvious how such an identification might proceed in general.*

Ganter and Glodeanu [10] have suggested that FCA could contribute to this important problem, and have made first steps into this direction.

Things get even more complicated when skills come into play. As said above, each competency corresponds to a skill combination, so that, sloppily spoken, the ability of mastering a question corresponds to a disjunction of conjunctions of skills. However, for reasons of space we shall not discuss the role of skills here in detail. Instead, we sketch the special case that skills and competencies coincide.

7   The Disjunctive Skill Model

In the disjunctive model each competency consists of a single skill, so that competencies and skills essentially may be identified. A disjunctive model thus is essentially described by the formal context

$$(\mathcal{C}, \mathcal{Q}, \models),$$

which is assumed to fulfill the Boolean matrix equation

$$(\mathcal{L}, \mathcal{Q}, \Box) = (\mathcal{L}, \mathcal{C}, \circ) \cdot (\mathcal{C}, \mathcal{Q}, \models).$$

A first research question is to identify competencies for given learner-task data $(\mathcal{L}, \mathcal{Q}, \Box)$. Is it possible to determine the minimum size of an appropriate set $\mathcal{C}$ of competencies? This question has two different answers.

Recall that the knowledge space $\mathcal{K}$ corresponding to $(\mathcal{L}, \mathcal{Q}, \Box)$ is generated by the object intents:

$$\mathcal{K} = \{\bigcup_{l \in L} l^{\Box} \mid L \subseteq \mathcal{L}\}.$$

Falmagne and Doignon [8] request that this knowledge space $\mathcal{K}$ is *delineated* by $(\mathcal{C}, \mathcal{Q}, \models)$, meaning that the knowledge states are precisely the unions of object intents. Written in terms of Boolean products, this reads as

$$\mathcal{K} = \{t \cdot (\mathcal{C}, \mathcal{Q}, \models) \mid t \in 2^{\mathcal{C}})\}.$$

More comprehensively this means that to every combination $T \subseteq \mathcal{C}$ of competencies, the set of questions that can be mastered having these competencies is a knowledge state. This set is

$$\bigcap_{C \in T} C^{\vDash},$$

which corresponds to

$$t \cdot (\mathcal{C}, \mathcal{Q}, \vDash),$$

where t is the Boolean vector for T.

Falmagne and Doignon show ([8], Theorem 6.3.8) that every finite knowledge space is delineated by some *minimal* skill map, which is unique up to isomorphism. In particular, they show that the minimal possible number of competencies equals the size of the basis of the knowledge space, which in turn is equal to the number of attributes of the complementary context of $(\mathcal{L}, \mathcal{Q}, \square)$. As an example, consider the knowledge space in Fig. 2, which corresponds to the learner-task data in Fig. 1. The unique minimal skill map is given, in form of a formal context, in Fig. 3, with the learners l_2, l_3, l_4, l_5, l_7 representing the competencies. Learner l_6 in Fig. 1 has the combined competencies of the learners l_4, l_5, and l_7, and so on.

The condition that $(\mathcal{C}, \mathcal{Q}, \vDash)$ delineates the knowledge space is, however, quite restrictive. It assumes that *every* combination of competencies may occur. In practice one may expect that some competencies are more advanced than others, so that the set of admissible competency combinations itself is structured, similarly to the questions. With this assumption, solutions with fewer competencies are possible, and uniqueness is lost. The formal problem statement under these conditions is to solve the Boolean matrix equation

$$(\mathcal{L}, \mathcal{Q}, \square) = (\mathcal{L}, \mathcal{C}, \circ) \cdot (\mathcal{C}, \mathcal{Q}, \vDash)$$

for given $(\mathcal{L}, \mathcal{Q}, \square)$, but unknown relations $\circ$ and $\vDash$, with a smallest possible set $\mathcal{C}$.

Again, the solution is known, but it may be hard to determine. It turns out that the smallest possible number of competencies equals the *2-dimension* of the knowledge space, which is the number of atoms of the smallest Boolean algebra that contains an order-embedding of the knowledge space. Bělohlávek and Vychodil [3] observed that for every competency C in a solution the set $C^{\circ} \times C^{\vDash}$ must be a "rectangular" subset of the relation $\square \subseteq \mathcal{L} \times \mathcal{Q}$, and that a solution is obtained if and only if these rectangles cover the relation, i.e., iff

$$\square = \bigcup_{C \in \mathcal{C}} C^{\circ} \times C^{\vDash}.$$

There is a more intuitive formulation, using the fact that every rectangle is contained in a maximal one. *The smallest possible number of competencies for a given formal context equals the smallest number of formal concepts which together cover all incidences of the incidence relation.* According to [12], Theorem 47, this number equals the 2-dimension of the complementary context, i.e., the 2-dimension of the knowledge space. The problem of determining, for given $\mathcal{K}$ and n, if $\dim_2(\mathcal{K}) \leq n$, is known to be $\mathcal{NP}$-complete.

8   Skills First

Explaining a knowledge space by a small number of skills is only one of several interesting challenges. Another task is finding a set of questions for testing given skills. Here we can make use of a result by R. Suck, which also has an interpretation for concept lattices.

Theorem 1 (Suck [20]). *Given an ordered set* $(S, \leq)$, *a finite set* $\mathcal{Q}$, *and an order embedding* $\varphi : (S, \leq) \to (\mathfrak{P}(\mathcal{Q}), \subseteq)$, *the set*

$$\{\varphi(s) \mid s \in S\}$$

is the basis of a knowledge space if and only if φ *satisfies*

$$|\varphi(s)| > \left| \bigcup_{t < s} \varphi(t) \right|. \tag{1}$$

Corollary 1. *Given an ordered set* $(G, \leq)$, *a finite set* M, *and an order embedding* $\varphi : (G, \leq) \to (\mathfrak{P}(M), \subseteq)$, *the formal context* (G, M, I) *defined by*

$$g \, I \, m :\Leftrightarrow m \notin \varphi(g)$$

is object-reduced if and only if φ *satisfies Condition (1) of Theorem 1.*

We prove the corollary:

Proof. By definition of the incidence relation we have $g' = M \setminus \varphi(g)$ for all $g \in G$. An object $g \in G$ could be reducible only if there was some subset $X \subseteq G$, $g \notin X$, with

$$g' = \bigcap_{x \in X} x', \tag{2}$$

which implies $g' \subset x'$ and therefore $\varphi(x) \subset \varphi(g)$ for all $x \in X$. Since φ is an order embedding, we conclude $x < g$ for all $x \in X$. Equation (2) transforms to

$$M \setminus \varphi(g) = \bigcap_{x \in X} (M \setminus \varphi(x)) = M \setminus \bigcup_{x \in X} \varphi(x),$$

which implies

$$\varphi(g) = \bigcup_{x \in X} \varphi(x).$$

But that contradicts Condition (1).

Suck's theorem provides us with an easy method for constructing formal contexts with a preset order of the object concepts. One way to achieve this is given by the contra-ordinal scale $(S, S, \not\geq)$ (see Fig. 6 for an example). In that case, the concept lattice is distributive and Condition (1) is fulfilled in the most parsimonious way:

$$|\varphi(s)| = \left| \bigcup_{t < s} \varphi(t) \right| + 1. \tag{3}$$

Theorem 4 in Suck [19] shows that this condition is characteristic for the well-graded knowledge spaces which were mentioned in Sect. 2.

Theorem 1 gives an answer to the following questions: Suppose we start with a set S of skills (or competencies, since we are in the disjunctive case), together with an order relation $\leq$ on S, where $s_1 \leq s_2$ expresses that *learners with skill s_2 master all questions which learners with skill s_1 master*. How can a set Q of questions be designed that tests all these skills?

The answer given by Theorem 1 is that for every skill $s \in S$ a set $\varphi(s)$ of questions must be given in such a way, that two conditions are fulfilled: φ must be an order embedding, and Condition (1) must be satisfied.

9   Conclusions and Outlook

Put simply, the main aim of this paper was to describe Knowledge Space Theory in the terminology of Formal Concept Analysis. Apart from theoretical insights on mathematical overlaps and differences between both theories, the main advantage of such a "translation" is that it may facilitate further developments of applications. We see possibilities in the field of Technology-Enhanced Learning (TEL):

As mentioned above, knowledge- and competence spaces can become very large. This may be problematic for applications which make use of (Cb)KST for learner modelling purposes, due to the required computational power. Time consuming tasks are the adaptive assessment of the learner's knowledge state or personalized recommendations of suitable learning paths. As an on-vogue example, Game-based Learning applications need to be smoothly running, because otherwise the envisaged flow experience of the learners might diminish. This requires ongoing updates of the learner model in real time, based on the learner's actions and interactions with the TEL-application. The FCA community developed a bundle of remarkably fast algorithms which could be exploited.

But also more "traditional" TEL applications could benefit from applying FCA to learner modelling. The recently finished European research project LEA's BOX (**L**earning **A**nalytics Tool**box**, see http://leas-box.eu/), aimed at providing a Web platform for teachers and students for activity tracking, domain modelling, student modelling as well as visualization of educational data. (Cb)KST and FCA have been applied for domain and student modelling.

Visualizations of concept lattices resulting from learner-task data as well as learner-competence matrices have been used by teachers to get an overview of the strengths and weaknesses of the whole classroom. After a short training on how to interpret and "read" the concept lattices, teachers were able to make use of interactive lattices by getting insights on several pedagogically relevant questions (Bedek et al. [2]). Consider a lattice similar to the one shown in Fig. 1. Pedagogically relevant questions which can be answered by reading the lattice are for example: "What are the best and weakest students?", "which students could be tutors for others in a collaborative, peer-learning scenario?", or "how coherent is the classroom performance" (As a "rule of thumb", a concept lattice with a large number of formal concepts is an indication for a high diversity among the students' performance- and competence states). Finally, in some cases it might be of great interest for a teacher to observe the learning progress over a longer period of time by comparing the lattices generated at different points in time with the empirical learner-competence matrices. In the perfect case, all students should finally end up (e.g. at the end of the semester) with the knowledge state Q.

The attempts of the LEA's BOX project to introduce such kind of visualizations into the classroom are just the starting point. So far, rather small learner-competence matrices have been used (i.e. around 30 students and 10 competencies). Further research is required to provide more intuitive and user-friendly interactive visualizations—even with larger lattices. In addition to that, empirical investigations on how such interactive visualizations are used in the classroom by teachers and students should and will be carried out in the near future.

Acknowledgement. This article is a partial result of an effort which has been facilitated by a workshop at Graz University of Technology. We would like to thank Dietrich Albert for hosting this workshop. It has been partly funded by the European Commission (EC) (7th Framework Programme contract no. 619762, LEA's BOX). This document does not represent the opinion of the EC and the EC is not responsible for any use that might be made of its content.

References

1. Albert, D., Held, T.: Establishing knowledge spaces by systematical problem construction. In: Albert, D. (ed.) Knowledge Structures, pp. 78–112. Springer, New York (1994)
2. Bedek, M.A., Kickmeier-Rust, M.D., Albert, D.: Formal concept analysis for modelling students in a technology-enhanced learning setting. In: ARTEL@ EC-TEL, pp. 27–33 (2015)
3. Bělohlávek, R., Vychodil, V.: Discovery of optimal factors in binary data via a novel method of matrix decomposition. J. Comput. Syst. Sci. **76**(1), 3–20 (2010)
4. Colomb, P., Irlande, A., Raynaud, O.: Counting of moore families for n=7. In: Kwuida, L., Sertkaya, B. (eds.) ICFCA 2010. LNCS, vol. 5986, pp. 72–87. Springer, Heidelberg (2010). doi:10.1007/978-3-642-11928-6_6
5. Doignon, J.-P.: Knowledge spaces and skill assignments. In: Fischer, G.H., Laming, D. (eds.) Contributions to Mathematical Psychology, Psychometrics, and Methodology, pp. 111–121. Springer, Heidelberg (1994)
6. Doignon, J.-P., Falmagne, J.-C.: Knowledge Spaces. Springer, Heidelberg (1999)

7. Düntsch, I., Gediga, G.: Skills and knowledge structures. Br. J. Math. Stat. Psychol. **48**(1), 9–27 (1995)
8. Falmagne, J.-C., Doignon, J.-P.: Learning Spaces. Springer, Heidelberg (2011)
9. Falmagne, J.-C., Koppen, M., Villano, M., Doignon, J.-P., Johannesen, L.: Introduction to knowledge spaces: how to build, test, and search them. Psychol. Rev. **97**(2), 201 (1990)
10. Ganter, B., Glodeanu, C.V.: Factors and skills. In: Glodeanu, C.V., Kaytoue, M., Sacarea, C. (eds.) ICFCA 2014. LNCS, vol. 8478, pp. 173–187. Springer, Cham (2014). doi:10.1007/978-3-319-07248-7_13
11. Ganter, B., Obiedkov, S.: Conceptual Exploration. Springer, Heidelberg (2016)
12. Ganter, B., Wille, R.: Formal Concept Analysis: Mathematical Foundations. Springer, Heidelberg (1999)
13. Gediga, G., Düntsch, I.: Skill set analysis in knowledge structures. Br. J. Math. Stat. Psychol. **55**(2), 361–384 (2002)
14. Heller, J., Wickelmaier, F.: Minimum discrepancy estimation in probabilistic knowledge structures. Electron. Notes Discrete Math. **42**, 49–56 (2013)
15. Korossy, K.: Modellierung von Wissen als Kompetenz und Performanz. Unpublished doctoral dissertation, see [15]. Universität Heidelberg, Germany (1993)
16. Korossy, K.: Extending the theory of knowledge spaces: a competence-performance approach. Zeitschrift für Psychol. **205**(1), 53–82 (1997)
17. Korossy, K., et al.: Modeling knowledge as competence and performance. In: Albert, D., Lukas, J. (eds.) Knowledge Spaces: Theories, Empirical Research, and Applications, pp. 103–132. Lawrence Erlbaum Associates, Mahwah (1999)
18. Rusch, A., Wille, R.: Knowledge spaces and formal concept analysis. In: Bock, H.H., Polasek, W. (eds.) Data Analysis and Information Systems: Statistical and Conceptual Approaches, pp. 427–436. Springer, Heidelberg (1996)
19. Suck, R.: Parsimonious set representations of orders, a generalization of the interval order concept, and knowledge spaces. Discrete Appl. Math. **127**(2), 373–386 (2003)
20. Suck, R.: Knowledge spaces regarded as set representations of skill structures. In: Dzhafarov, E., Perry, L. (eds.) Descriptive and Normative Approaches to Human Behavior, pp. 249–270. World Scientific, Singapore (2012)
21. Wille, R.: Restructuring lattice theory: an approach based on hierarchies of concepts. In: Rival, I. (ed.) Ordered Sets, pp. 445–470. Reidel, Dordrecht-Boston (1982)

Historical Paper

Implications and Dependencies
Between Attributes

Bernhard Ganter[1], Rudolf Wille[2], Daniel Borchmann[3]([⊠]) [iD],
and Juliane Prochaska[1]([⊠])

[1] Institute of Algebra, Technische Universität Dresden, Dresden, Germany
juliane.prochaska@gmx.de
[2] Fachbereich Mathematik, Technische Universität Darmstadt, Darmstadt, Germany
[3] Institute of Theoretical Computer Science, Technische Universität Dresden,
Dresden, Germany
daniel.borchmann@tu-dresden.de

Abstract. This work is a translation of "Implikationen und Abhäng-igkeiten zwischen Merkmalen" by Bernhard Ganter and Rudolf Wille, Technische Hochschule Darmstadt, Preprint-Number 1017, 1986. The manuscript has originally been published in "Die Klassifikation und ihr Umfeld", edited by P. O. Degens, H. J. Hermes, and O. Opitz, Indeks-Verlag, Frankfurt, 1986 (rights now with Ergon-Verlag).
This translation has been authorized by Bernhard Ganter.

1 Introduction

Implications and dependencies between attributes require primal interest when one wants to study the connection between objects and attributes in application domains. In the following, it shall be shown how implications and dependencies between attributes can be investigated within the framework of Formal Concept Analysis. It is assumed that the reader has knowledge about the basic notions of Formal Concept Analysis as described in [12,13]. Note that attributes of a context (G, M, I) shall also be denoted as single-valued attributes, in order to characterize them as a special case of many-valued attributes [12]. In the following, many aspects will only be addressed briefly and further references will be provided. It should be mentioned that there are many computer programs available for Formal Concept Analysis that can also be used to investigate implications and dependencies between attributes [6].

Translated by Daniel Borchmann and Juliane Prochaska.
Electronic supplementary material The online version of this chapter (doi:10.1007/978-3-319-59271-8_2) contains supplementary material, which is available to authorized users.

K. Bertet et al. (Eds.): ICFCA 2017, LNAI 10308, pp. 23–35, 2017.
DOI: 10.1007/978-3-319-59271-8_2

Typ	Beispiel	a	b	c	d	e
1		×				
2		×	×			
3	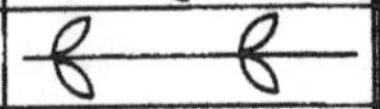	×		×		
4		×			×	
5		×				×
6		×	×		×	×
7		×		×	×	×

The attributes mean:

a: translation
b: glide reflection
c: horizontal reflection
d: vertical reflection
e: rotation by
    180 degrees

Fig. 1. Context of symmetry types of one-sided frieze patterns ("Beispiel" is "Example")

2 Implications Between Single-Valued Attributes

Let us first illustrate by means of an example what implications between attributes of a context are. For that purpose we choose the context from Fig. 1, containing the symmetry transformations of (one-sided) frieze patterns [9]. Objects of this context are the different types of symmetries, each comprising the frieze patterns with similar symmetry transformations; each type is represented by a corresponding pattern. Attributes of the context are possible transformations of frieze patterns. A cross in the table indicates which transformations are permitted by the corresponding symmetry type.

In this context the attributes "glide reflection" and "horizontal reflection" together imply the attribute "rotation by 180°", because each symmetry type with the first two attributes also has the attribute "rotation by 180°". Analogously, the attributes "vertical reflection" and "rotation by 180°" together imply the attribute "horizontal reflection". The attribute "translation" even is implied by every other attribute separately. The attribute "glide reflection" is not implied by the attributes "vertical reflection" and "horizontal reflection", because Symmetry Type 7 has the latter two attributes, but not "glide reflection" (remark: in the given context, the attribute "glide reflection" only denotes glide reflections with a translation by half the period of the corresponding frieze pattern).

In general an implication between sets A and B of attributes of a context (G, M, I) is defined as follows: A implies B (in symbols: $A \to B$), if $B \subseteq A''$ holds; in particular, an attribute m of the context is implied by A (in symbols: $A \to m$), if $m \in A''$ holds [12]. The implications can be read from the concept lattice $\underline{\mathfrak{B}}(G, M, I)$: A implies m if and only if

$$\mu m := (\{m\}', \{m\}'') \geq \bigwedge \{\mu n \mid n \in A\},$$

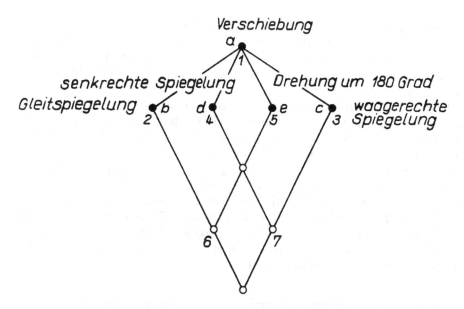

Fig. 2. Concept lattice of the context from Fig. 1 ("Gleitspiegelung" is "glide reflection", "senkrechte Spiegelung" is "vertical reflection", "Verschiebung" is "translation", "Drehung um 180 Grad" is "rotation by 180°", "waagerechte Spiegelung" is "horizontal reflection")

i.e., if in the line diagram of the concept lattice, the concept labeled with m is above or equal to the infimum of all those concepts labeled with some n from A. The implications mentioned above as well as others can thus be seen directly in the line diagram of the corresponding concept lattice as shown in Fig. 2.

How to determine and describe all attribute implications of a context (G, M, I)? Trivially, $A \to B$ holds if B is contained in A, or if $B \subseteq C$ and $A \to C$ is true, which is why it is not necessary to list such implications separately. Furthermore, $A_t \to B_t$ $(t \in T)$ directly entails $(\bigcup \{A_t \mid t \in T\}) \to (\bigcup \{B_t \mid t \in T\})$, and thus such derivable implications also do not need to be listed separately. This motivates the following definition: a set of attributes A of a context (G, M, I) is called a proper premise if

$$A \neq A'' \neq \bigcup \{(A \setminus \{n\})'' \mid n \in A\};$$

in particular, $\emptyset$ is a proper premise if $\emptyset'' \neq \emptyset$. $A \to B$ shall be called a proper implication if A is a proper premise and

$$B = A'' \setminus \bigcup \{(A \setminus \{n\})'' \mid n \in A\}.$$

To obtain a complete overview over all attribute implications of (G, M, I), it is sufficient to list all proper implications of (G, M, I), because from these implications the rest can be derived directly by the above mentioned rules. For the context from Fig. 1 one can read the following implications from Fig. 2: $\emptyset \to a$, $\{b,d\} \to e$, $\{b,e\} \to d$, $\{c,d\} \to e$, $\{c,e\} \to d$, $\{b,c\} \to \{d,e\}$. Implications like the last one in this list, where no objects satisfy the premise, will also be omitted in cases where such implications do not contribute to the understanding of the domain in question.

For contexts with fixed numbers of objects and attributes, the following rule of thumb roughly holds true: the more concepts, the fewer implications. Hence it is usually advisable to switch from the original context (G, M, I) to its complementary context $(G, M, (G \times M) \setminus I)$. For $m \in M$ and $A \subseteq M$ the following equivalences are true: $m \in A'' \iff \{m\} \subseteq A'' \iff A' \subseteq \{m\}' \iff \bigcap \{\{n\}' \mid n \in A\} \subseteq \{m\}' \iff G \setminus \{m\}' \subseteq \bigcup \{G \setminus \{n\}' \mid n \in A\}$. Therefore $A \to m$ is true in (G, M, I) if and only if in the complementary context every object having attribute m also has at least one attribute n from A. Figure 4 illustrates that this condition can easily be read from the line diagram of the concept lattice of the complementary context; the original context in Fig. 3 has 54 concepts, whereas the complementary context only has 17.

	Urlaubsort	a	b	c	d	e	f	g	h	i	j	k	a	b	c	d	e	f	g	h	i	j	k
1	Kassel	X	X	X	X	X	X	X	X	X	X												X
2	Bad Karlshafen		X	X	X	X	X	X	X		X	X	X								X		
3	Naumburg	X	X			X	X	X	X		X	X			X	X					X		
4	Emstal	X	X	X		X	X		X		X	X				X			X		X		
5	Reinhardshagen		X	X		X		X	X		X		X			X		X			X		X
6	Arolsen	X	X	X	X	X	X	X	X		X	X									X		
7	Diemelsee	X		X	X	X	X	X	X		X	X		X							X		
8	Willingen		X	X	X	X	X	X	X		X	X	X								X		
9	Bad Wildungen		X	X	X	X	X	X	X	X	X	X	X										
10	Waldeck	X		X	X	X	X	X	X		X	X		X							X		
11	Battenberg		X	X	X	X	X	X	X		X	X	X								X		
12	Vöhl	X		X	X	X	X	X	X		X	X		X							X		
13	Frankenau		X	X		X	X	X	X		X	X	X			X					X		
14	Bad Hersfeld		X	X	X	X	X	X	X		X	X	X								X		
15	Kirchheim	X		X	X	X	X	X	X		X	X		X							X		
16	Ronshausen		X	X	X	X	X	X	X		X		X								X		X
17	Rotenburg	X	X	X	X	X	X	X	X		X	X									X		
18	Knüllwald	X	X	X	X	X	X	X	X		X	X									X		
19	Melsungen		X	X		X	X	X	X		X		X			X					X		X
20	Neukirchen	X		X	X	X	X	X	X		X	X		X							X		
21	Zwesten		X	X	X	X	X	X	X		X	X	X								X		
22	Bad Sooden-Allendorf		X	X	X	X	X	X	X		X	X	X								X		
23	Witzenhausen	X	X	X	X	X	X	X	X		X	X									X		
24	Wanfried	X				X		X	X		X	X		X	X	X		X			X		
25	Ringgau		X	X	X	X	X	X	X		X		X								X		X

Fig. 3. Context of holiday resorts in the Hessian Highlands [10] and its complement. The attributes have the following meaning: a: open-air bath, b: heated open-air bath, c: public indoor pool, d: hotel indoor pool, e: bowling alley, f: riding, g: tennis, h: minigolf, i: golf, j: fishing, k: farm holidays; ("Urlaubsort" is "holiday resort")

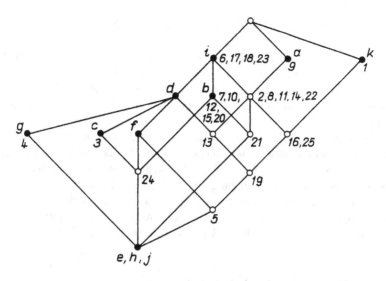

Fig. 4. Concept lattice of the completementary context in Fig. 3

For example, Fig. 4 shows that only the attributes d and f are implied by proper premises with more than one attribute, because these are the only attributes labeling concepts that are not labeled with objects.

3   Concept Lattices as Structures of Attribute Implications

As explained in the first section, attribute implications of a context (G, M, I) can be read from its concept lattice. Conversely, the concept lattice is uniquely determined by the attribute implications of (G, M, I): if $\mathcal{L}$ denotes the list of all proper implications of (G, M, I) and $A \subseteq M$, then

$$A'' = A \cup \bigcup \{Y \mid (X \rightarrow Y) \in \mathcal{L} \text{ where } X \subseteq A\}.$$

Hence the concept lattice of (G, M, I) can be considered as the structure of the attribute implications of (G, M, I) [14].

The question arises whether it is possible for each set $\mathcal{L}$ of pairs $(X, Y), X, Y \subseteq M$ to find a context (G, M, I), such that $X \rightarrow Y$ are attribute implications for all $(X, Y) \in \mathcal{L}$ and the concept lattice of (G, M, I) is uniquely determined by these attribute implications. Since concept intents are closed under attribute implications, it is not far-fetched to call a subset A of M to be $\mathcal{L}$-closed, if $(X, Y) \in \mathcal{L}$ and $X \subseteq A$ implies $Y \subseteq A$. The $\mathcal{L}$-closed subsets of M form a closure system $\mathcal{H}(\mathcal{L})$, consisting exactly of the concept intents of the context $(\mathcal{H}(\mathcal{L}), M, \ni)$ [4]. This yields an affirmative answer to the original question. The intents of $(\mathcal{H}(\mathcal{L}), M, \ni)$ can also be generated iteratively. For this we define for $A \subseteq M$ the set

$$A^{\mathcal{L}} := A \cup \bigcup \{Y \mid (X,Y) \in \mathcal{L} \text{ where } X \subseteq A\}.$$

Obviously, $A^{\mathcal{L}}$ is contained in the concept intent A'', but does not have to be equal to it. Forming the sets $A^{\mathcal{L}}$, $A^{\mathcal{LL}}$, $A^{\mathcal{LLL}}$, ... iteratively, one eventually arrives at a set $\mathcal{L}(A)$ satisfying $\mathcal{L}(A) = \mathcal{L}(A)^{\mathcal{L}}$, i.e., $\mathcal{L}(A) = A''$ in the context determined by $\mathcal{L}$ (if M is infinite, one may have to repeat this procedure transfinitely).

Given a context (G, M, I), how to find suitable lists of attribute implications, by which the concept lattice of the context is uniquely determined? As explained before, the proper implications of (G, M, I) constitute such a list. By means of a simple procedure, this list can be generated gradually: one successively tests all subsets of M in such a manner that a set is always considered later than all of its proper subsets. Consequently, one starts with the empty subset of M; one sets $\mathcal{L} := \emptyset$. For a subset A of M it is tested whether $A^{\mathcal{L}} = A''$ for

```
 1.  {b} => {a,c,f,q,s,t,v}
 2.  {c} => {f}
 3.  {d} => {a,b,c,f,g,h,k,l,m,q,r,s,t,u,v,w}
 4.  {e} => {n}
 5.  {f} => {c}
 6.  {a,c,f} => {b,q,s,t,v}
 7.  {g} => {l}
 8.  {h} => {g,k,l,m,q,r,v,w}
 9.  {i} => {e,g,h,j,k,l,m,n,o,p,q,r,v,w}
10.  {k} => {g,h,l,m,q,r,v,w}
11.  {l} => {g}
12.  {m} => {g,h,k,l,q,r,v,w}
13.  {n} => {e}
14.  {c,e,f,j,n} => {a,b,q,s,t,v}
15.  {e,g,l,n} => {p}
16.  {o} => {e,g,h,i,j,k,l,m,n,p,q,r,v,w}
17.  {g,l,p} => {e,n}
18.  {e,n,p} => {g,l}
19.  {e,g,j,l,n,p} => {h,i,k,m,o,q,r,v,w}
20.  {c,f,q} => {a,b,s,t,v}
21.  {g,l,q} => {h,k,m,r,v,w}
22.  {e,j,n,q} => {v}
23.  {j,p,q} => {e,g,h,i,k,l,m,n,o,r,v,w}
24.  {r} => {g,h,k,l,m,q,v,w}
25.  {a,s} => {b,c,f,q,t,v}
26.  {c,f,s} => {a,b,q,t,v}
27.  {e,j,n,s} => {a,b,c,f,q,t,v}
28.  {q,s} => {a,b,c,f,t,v}
29.  {t} => {a,b,c,f,q,s,v}
30.  {u} => {a,b,c,d,f,g,h,k,l,m,q,r,s,t,v,w}
31.  {v} => {q}
32.  {a,q,v} => {b,c,f,s,t}
33.  {e,n,q,v} => {j}
34.  {w} => {g,h,k,l,m,q,r,v}
35.  {g,h,j,k,l,m,q,r,v,w} => {e,i,n,o,p}
36.  {a,b,c,f,g,h,k,l,m,q,r,s,t,v,w} => {d,u}
```

Fig. 5. Minimal list of implications for Fig. 6

Viereck	a	b	c	d	e	f	g	h	i	j	k	l	m	n	o	p	q	r	s	t	u	v	w
1					X		X	X		X	X	X	X	X	X	X	X	X				X	X
2	X	X	X	X		X	X	X			X	X	X				X	X	X	X	X	X	X
3	X	X	X		X	X				X				X			X		X	X		X	
4			X		X	X	X				X	X		X		X							
5	X									X							X						
6							X			X	X	X							X				
7	X				X		X				X	X		X		X							
8							X			X	X	X											
9										X						X							
10	X						X			X	X	X											
11	X									X						X							
12	X	X	X			X										X	X		X	X		X	
13	X	X	X			X											X		X	X		X	
14					X		X				X	X		X		X			X				
15	X									X													
16	X																X						
17	X															X	X						
18										X						X			X				

Fig. 6. Context of properties of planar quadrangles, as given in [8] (A. Jung has participated in the creation of this context). The attributes mean the following: a: two neighboring sides have same length, b: sides are divided into pairs of neighboring sides, each of which have same length, c: sum of length of opposite sides is equal, d: sides have same length, e: sides are chords of a circle, f: sides touch a circle, g: two sides are parallel, h: opposite sides are parallel, i: opposite sides are perpendicular, j: one interior angle is perpendicular, k: neighboring angles add up to a straight angle, l: interior angles form pairs of neighboring angles, which add up to a straight angle, m: opposite angles are equal, n: opposite angles add up to a straight angle, o: interior angles are equal, p: diagonals have same length, q: one diagonal bisects the other, r: diagonals bisect each other, s: diagonals are perpendicular to each other, t: one diagonal bisects interior angles, u: each diagonal bisects interior angles, v: one diagonal partitions into congruent triangles, w: each diagonal partitions into congruent triangles. The quadrangles in the context have the following vertices:

1 : $(0,0), (2,0), (2,1), (0,1)$,   2 : $(0,0), (2,1), (3,3), (1,2)$,
3 : $(0,0), (3,9), (0,10), (-3,9)$, 4 : $(-4,0), (4,0), (1,4), (-1,4)$,
5 : $(0,0), (5,0), (8,16), (0,10)$, 6 : $(0,0), (4,0), (1,2), (0,2)$,
7 : $(0,0), (11,0), (8,4), (3,4)$,   8 : $(0,0), (6,0), (3,4), (0,4)$,
9 : $(-5,0), (3,-4), (5,0), (3.7309\ldots, 5.9733\ldots)$
10 : $(0,0), (8,0), (5,4), (0,4)$ 11 : $(0,0), (5,0), (7,1), (0,5)$,
12 : $(2,0), (0,1), (-2,0), (0,-3)$, 13 : $(0,0), (1,0), (2,2), (0,1)$,
14 : $(0,0), (2,2), (0,3), (-1,2)$ 15 : $(0,0), (5,5), (4,8), (-5,5)$,
16 : $(0,0), (1,-3), (2,4), (1,3)$, 17 : $(0,0), (11,4), (6,8), (1,4)$,
18 : $(0,0), (12,0), (5,12), (0,5)$.

("Viereck" is "quadrangle")

the currently known $\mathcal{L}$; in case of inequality, $\mathcal{L}$ is extended by the pair (A, B), where $B := A'' \setminus \bigcup \{Y \mid (X, Y) \in \mathcal{L}$ where $X \subseteq A\}$. One obtains a minimal list of attribute implications that determine the concept lattice of (G, M, I), if in the given procedure one tests $\mathcal{L}(A) = A''$ instead of $A^{\mathcal{L}} = A''$ [2]. In this procedure considerable amounts of computation time can be saved if one exploits the fact that among all subsets A with equal closure $\mathcal{L}(A)$ there is always a lexicographically first one [4,5].

Minimal lists of attribute implications are interesting when one wants to verify the correctness of a context by means of its implications. It is claimed for the context from Fig. 6 that it has exactly those implications that are valid between the listed attributes of planar quadrangles. This can be verified by using the above mentioned procedure to generate a minimal list of attribute implications that determine the concept lattice, and then proving these implications within the realm of planar Euclidean geometry. After this it is certain that the concept lattice in Fig. 7 reflects the implications between geometric properties. A variant of the procedure allows for incrementally generating the context from a fixed set of attributes in the domain of objects, such that the resulting context admits exactly the valid attribute implications in this domain [4,5]. We have implemented this as an interactive computer program: the computer asks for each newly found implication whether it should be included in the list, or whether instead the context could be extended by a new object that rejects the proposed implication. In this way, both a minimal list of implications and a context of objects, which can be considered as counterexamples for non-valid implications, are generated.

4   Dependencies Between Many-Values Attributes

In a many-valued context (G, M, W, I), a set of objects G, a set of attributes M, and a set of attribute manifestations or attribute values W, are connected by a ternary relation $I \subseteq G \times M \times W$ in such a way that $(g, m, v) \in I$ and $(g, m, w) \in I$ always implies $v = w$ [7,12]. For each object an attribute can thus have at most one value, for different objects, however, multiple values are possible, which is why we call such an attribute many-valued. If $(g, m, w) \in I$, then this is also denoted by writing $m(g) = w$, where m is considered as a partial function from G to W; the domain of m is then $\mathrm{Def}(m) := \{g \in G \mid (g, m, w) \in I$ for some $w \in W\}$.

The model of relational databases from database theory can be considered as a many-valued context, suggesting to define functional dependencies between many-valued attributes as in database theory [11]. In a many-valued context (G, M, W, I) an attribute m is called nominally dependent from a set of attributes A (in symbols: $A \xrightarrow{n} m$), if $\mathrm{Def}(m) \subseteq \mathrm{Def}(n)$ for all $n \in A$ and if for all $g, h \in \mathrm{Def}(m)$ with $m(g) \neq m(h)$ there exists at least one $n \in A$ where $n(g) \neq n(h)$. Thus one has $A \xrightarrow{n} m$, if and only if there exists a function f from the set of all tuples $(n(g) \mid n \in A)$ with $g \in \mathrm{Def}(m)$, such that $f(n(g) \mid n \in A) = m(g)$ for all

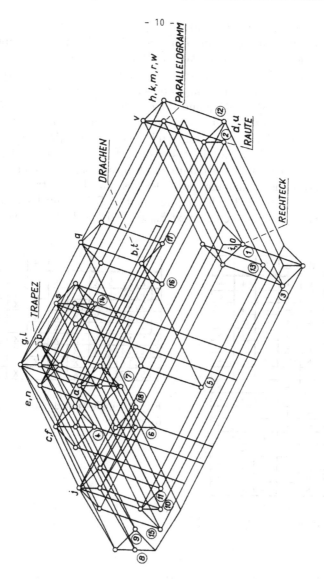

Fig. 7. Concept lattice of context in Fig. 6 ("Trapez" is "trapezoid", "Drachen" means "kite", "Parallelogramm" is "parallelogram", "Raute" is "rhombus", "Rechteck" is "rectangle")

$g \in \mathrm{Def}(m)$; from this it is apparent why nominal dependencies are also called functional dependencies.

To determine all nominal dependencies of a many-valued context (G, M, W, I) it is beneficial to assign to each $m \in M$ the relation $E(m) := \{(g, h) \in \mathrm{Def}(m) \times \mathrm{Def}(m) \mid m(g) = m(h)\}$ on G. Obviously, for $m \in M$ and $A \subseteq M$ one has $A \xrightarrow{n} m$ if and only if $\bigcap\{E(n) \mid n \in A\} \cap \mathrm{Def}(m) \times \mathrm{Def}(m) \subseteq E(m)$. This characterization suggests to compute nominal dependencies as attribute implications of a single-valued contexts. We therefore define for the many-valued context $\mathbb{K} := (G, M, W, I)$ the (single-valued) context $\mathbb{K}_n := (\mathfrak{P}_2(G), M \,\dot{\cup}\, \hat{M}, I_n)$, where $\mathfrak{P}_2(G)$ denotes the set of all two-elemental subsets of G, $\hat{M} := \{\hat{m} \mid m \in M$ where $\mathrm{Def}(m) \neq G\}$, and $\{g, h\} I_n m \iff (g, h) \in E(m)$ and $\{g, h\} I_n \hat{m} \iff (g, h) \in \mathrm{Def}(m) \times \mathrm{Def}(m)$. In $\mathbb{K}$ an attribute m is nominally dependent on A, if and only if m is implied by A or $A \cup \{\hat{m}\}$ (in case $\mathrm{Def}(m) \neq G$) in $\mathbb{K}_n$. If $\mathrm{Def}(m) = G$ for all attributes m of $\mathbb{K}$, then the nominal dependencies of $\mathbb{K}$ correspond exactly to the attribute implications of $\mathbb{K}_n$. The procedure thus outlined to determine all nominal dependencies shall be illustrated at the context of Chinese pots from [3] (also see [13], p. 41). This two-valued context $\mathbb{K}$ is given in Fig. 8, and the derived context $\mathbb{K}_n$ (in reduced form) in Fig. 9.

a	b	c	d	e	f	g	h
+	+	+	+	+	+	+	+
+	+	+	-	+	+	+	+
+	-	+	+	+	+	+	+
+	-	-	+	+	+	+	+
+	+	+	+	-	+	+	+
+	-	+	+	-	+	+	+
+	-	-	+	-	+	+	+
+	-	-	-	-	+	+	+
+	-	-	-	-	-	+	+
+	-	-	-	+	+	+	+
+	-	-	-	+	-	+	+
+	-	-	-	+	-	+	-
+	-	-	-	+	+	-	-
+	-	-	-	+	-	-	-
+	-	-	-	-	-	-	-
-	-	-	+	-	+	-	-
-	-	-	+	-	-	-	-

Fig. 8. Two-valued context $\mathbb{K}$ from [3]

	a	b	c	d	e	f	g	h
{1,2}	X	X	X		X	X	X	X
{1,3}	X		X	X	X	X	X	X
{3,4}	X	X		X	X	X	X	X
{1,5}	X	X	X	X		X	X	X
{8,9}	X	X	X	X	X		X	X
{11,12}	X	X	X	X	X	X	X	
{12,14}	X	X	X	X	X	X		X
{7,16}		X	X	X	X	X		
{15,17}		X	X		X	X	X	X

Fig. 9. Single-valued context $\mathbb{K}_n$ for $\mathbb{K}$ from Fig. 8

To visualize the attribute implications of $\mathbb{K}$, it is advisable to depict the concept lattice of the context complementary to $\mathbb{K}_n$, as explained in the first section, which has been done in Fig. 10. In Fig. 10 one can see directly that $\{d, g\} \to a$ and $\{d, h\} \to a$ are the only proper implications; this means that $\{d, g\} \xrightarrow{n} a$ and $\{d, h\} \xrightarrow{n} a$ are the only "proper" nominal dependencies of $\mathbb{K}$.

In some many-valued context not the nominal dependencies are of interest, but rather a form of attribute dependency that additionally takes into account the structure on the attribute values. This shall be discussed on the basis of the school grade context from Fig. 11. In this context, the attribute "English" depends nominally on { "Greek", "Mathematics", "Chemistry" }; however, as one can already see at the first eight students, in spite of better grades in Greek,

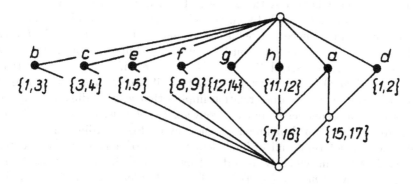

Fig. 10. Concept lattice of the complementary context for the context in Fig. 9

		Betragen	Fleiß	Aufmerksamk.	Ordnung	Deutsch	Geschichte	Sozialkunde	Latein	Griechisch	Englisch	Mathematik	Chemie	Biologie	Kunst	Leibeserz.
1	Anna	3	4	3	4	4	4	4	5	4	5	3	4	3	3	3
2	Berend	3	4	3	4	4	4	4	3	4	4	5	2	3	3	3
3	Christa	2	2	2	2	4	3	3	4	3	4	3	2	3	2	2
4	Dieter	1	1	1	1	1	1	1	2	2	2	1	1	1	3	3
5	Ernst	2	2	2	2	3	3	3	3	2	3	3	3	3	3	2
6	Fritz	2	1	2	2	2	2	2	2	2	2	2	2	3	3	1
7	Gerda	2	2	2	3	3	3	3	3	2	2	2	2	3	2	2
8	Horst	2	2	2	3	4	3	4	4	2	4	3	2	3	3	2
9	Ingolf	2	3	3	2	3	4	2	3	3	4	2	2	3	4	2
10	Jürgen	2	2	3	2	2	3	2	4	2	4	4	2	2	2	1
11	Karl	2	3	2	2	3	2	2	3	3	3	2	1	3	3	2
12	Linda	2	1	2	2	4	3	3	3	2	4	3	4	3	4	2	3
13	Manfred	2	2	2	2	2	3	4	1	1	2	3	2	2	2	3
14	Norbert	3	3	2	3	3	4	3	3	3	3	4	2	2	3	2
15	Olga	1	1	2	2	2	3	3	1	1	2	2	2	3	2	2
16	Paul	1	1	1	1	2	1	2	1	1	2	2	1	2	2	1
17	Quax	2	2	2	2	3	2	2	3	2	3	2	1	2	3	3
18	Rudolf	3	4	3	3	4	4	4	4	4	5	5	4	4	4	2
19	Stefan	1	1	1	1	1	1	1	1	1	2	2	1	1	1	2
20	Till	1	1	1	1	2	3	2	1	2	1	1	2	1	2	2
21	Uta	1	1	2	2	2	2	2	1	1	2	3	2	3	2	2
22	Volker	2	2	3	2	4	3	2	4	4	4	4	1	2	3	3
23	Walter	3	4	3	4	4	5	4	4	4	2	4	2	1	4	3
24	Xaver	1	1	1	1	2	2	1	2	1	2	2	2	3	3	2
25	Zora	2	2	2	2	2	4	4	3	3	3	3	4	4	2	2

Fig. 11. Context of school grades of a secondary school (1981) ("Betragen" is "Conduct", "Fleiß" is "Diligence", "Aufmerksamk(eit)" is "Attention", "Ordnung" is "Tidiness", "Deutsch" is "German", "Geschichte" is "History", "Sozialkunde" is "Social Studies", "Latein" is "Latin", "Griechisch" is "Greek", "Englisch" is "English", "Mathematik" is "Mathematics", "Chemie" is "Chemistry", "Biologie" is "Biology", "Kunst" is "Art", "Leibeserz(iehung)" is "Physical Education")

Mathematics, and Chemistry, the grade in English can turn out to be worse. Clearly, those dependencies between school grades which avoid this are particularly interesting. Hence, we define for a many-valued context (G, M, W, I) together with an order relation $\leq$ on W that an attribute m is ordinally dependent from an attribute set A (in symbols: $A \overset{o}{\to} m$), if $\mathrm{Def}(m) \subseteq \mathrm{Def}(n)$ for all $n \in A$ and $m(g) \leq m(h)$ for $g, h \in \mathrm{Def}(m)$ if $n(g) \leq n(h)$ for all $n \in A$. The anti-symmetry of the order relation implies that every ordinal dependency is also a nominal dependency, but the converse is not true. In the school grade context roughly every fifth nominal dependency is also an ordinal one.

In order to determine all ordinal dependencies of a many-valued context $\mathbb{K} := (G, M, W, I)$ in which W is ordered by $\leq$, we define for each $m \in M$ a relation on G by $D(m) := \{(g, h) \in \mathrm{Def}(m) \times \mathrm{Def}(m) \mid m(g) \leq m(h)\}$. For $m \in M$ and $A \subseteq M$ one obviously has $A \overset{o}{\to} m$, if and only if $\bigcap\{D(n) \mid n \in A\} \cap \mathrm{Def}(m) \times \mathrm{Def}(m) \subseteq D(m)$. Now let $\mathbb{K}_o := (G \times G, M \mathbin{\dot{\cup}} \hat{M}, I_o)$, where $(g, h) \, I_o \, m \iff (g, h) \in D(m)$ and $(g, h) \, I_o \, \hat{m} \iff (g, h) \in \mathrm{Def}(m) \times \mathrm{Def}(m)$. In $\mathbb{K}$ the attribute m is ordinally dependent on A if and only if m is implied in $\mathbb{K}_o$ by A or by $A \cup \{\hat{m}\}$ (in case $\mathrm{Def}(m) \neq G$). One has thus obtained a method to determine all ordinal dependencies. For the context of school grades, this yields that only the attributes "Conduct", "Diligence", "Attention", "German", and "English" ordinally depend on other attributes.

References

1. Duquenne, V.: Contextual implications between attributes and some properties for finite lattices. Groupe Ensembles Ordonnes, EHESS, Paris (1986, preprint)
2. Duquenne, V., Guigues, J.-L.: Informative implications derived from a table of binary data. Groupe Math. et Psychol. Univ. P.V., Paris (1984, Preprint)
3. Elissef, V.: Possibilities du scalogramme dans l'etude des bronzes chinois archaiques. Math. Sci. Hum. **11**, 1–10 (1965)
4. Ganter, B.: Two basic algorithms in concept analysis. THD-Preprint Nr. 831, Darmstadt (1984)
5. Ganter, B.: Algorithmen zur Formalen Begriffsanalyse. In: Ganter, B., Wille, R., Wolff, K.E. (Hrsg.) Beiträge zur Begriffsanalyse (to appear)
6. Ganter, B., Rindfrey, K., Skorsky, M.: Software for formal concept analysis. In: Gaul, W., Schader, M. (eds.) Classification as a Tool of Research, pp. 161–167. North-Holland, Amsterdam (1986)
7. Ganter, B., Stahl, J., Wille, R.: Conceptual measurement and many-valued contexts. In: Gaul, W., Schader, M. (eds.) Classification as a Tool of Research, pp. 169–176. North-Holland, Amsterdam (1986)
8. Gellert, W., Kästner, H., Neuber, S.: Fachlexikon ABC Mathematik. Verlag H. Deutsch, Thun und Frankfurt (1978)
9. Shubnikov, A.V., Koptsik, V.A.: Symmetry in Science and Art. Plenum Press, New York/London (1974)
10. Traurig, W.: Topographie Hessisches Bergland. In: Die Schöne Welt, p. 38, June 1986
11. Wedekind, H.: Datenbanksysteme I. B.I.-Wissenschaftsverlag, Mannheim (1974)

12. Wille, R.: Restructuring lattice theory: an approach based on hierarchies of concepts. In: Rival, I. (ed.) Ordered Sets, pp. 445–470. Reidel Publishing Company, Dordrecht-Boston (1982)

13. Wille, R.: Liniendiagramme hierarchischer Begriffssysteme. In: Bock, H.H. (Hrsg.) Anwendungen der Klassifikation: Datenanalyse und numerische Klassifikation, pp. 32–51. INDEKS Verlag, Frankfurt (1984)

14. Wille, R.: Bedeutungen von Begriffsverbänden. In: Ganter, B., Wille, R., Wolff, K.E. (eds.) Beiträge zur Begriffsanalyse (to appear)

Regular Contributions

The Implication Logic of (n, k)-Extremal Lattices

Alexandre Albano[✉] [iD]

Institut für Algebra, Technische Universität Dresden, Dresden, Germany
Alexandre_Luiz.Junqueira_Hadura_Albano@tu-dresden.de

Abstract. We characterize the canonical bases of lattices which attain the upper bound in Sauer-Shelah's lemma, i.e., the (n, k)-extremal lattices of [AC15]. A characteristic construction of such bases is presented. We make the case that this approach sheds light on important combinatorial properties. In particular, we give an explicit description of an (n, k)-extremal lattice with precisely $\binom{n}{k-1} + k - 2$ meet-irreducibles, together with its canonical basis and Whitney numbers.

1 Motivation and Objectives

The following result was discovered independently and almost simultaneously by Shelah [She72], Sauer [Sau72] and Vapnik and Chervonenkis [VC71]. It is an important result in at least logic, set theory and probability [Juk10] (see Subsect. 2.1 for the definition of *shattered set* and other basic notions).

Theorem 1 (Sauer, Shelah). *A family of subsets $\mathcal{F} \subseteq \mathcal{P}(\{1, 2, \ldots, n\})$ which does not shatter a set of size k can have at most $\sum_{i=0}^{k-1} \binom{n}{i}$ members.*

For the Formal Concept Analysis community, the particular case where $\mathcal{F}$ is not only a family of subsets, but actually a closure system, clearly sparks interest. Whenever $\mathcal{F}$ is a closure system, one may use the contextual perspective by considering its standard context. The advantage is that one may be able to "estimate" the size of a (concept) lattice as a function of a (formal) context $\mathbb{K}$ which gives rise to it. This is, indeed, possible because a closure system $\mathcal{F}$ shatters a set of size k if and only if that set corresponds to a subcontext of $\mathbb{K}$ which is a *contranominal scale* [AC15]. A contranominal scale having k objects and attributes is the context $([k], [k], \neq) =: \mathbb{N}^c(k)$, with $[k] := \{1, 2, \ldots, k\}$. The following is part of a stronger result which was shown in that same paper:

Theorem 2 [AC15]. *The concept lattice of any context $\mathbb{K}$ without $\mathbb{N}^c(k)$ as a subcontext can have at most $\sum_{i=0}^{k-1} \binom{n}{i}$ elements, where n is the number of objects of $\mathbb{K}$. This bound is sharp: for every n and k, there exists a context with n objects, no $\mathbb{N}^c(k)$ as a subcontext and exactly $\sum_{i=0}^{k-1} \binom{n}{i}$ concepts.*

The contexts which attest the sharpness mentioned in Theorem 2 are called (n, k)-*extremal contexts*. It turns out that instead of constructing these contexts

© Springer International Publishing AG 2017
K. Bertet et al. (Eds.): ICFCA 2017, LNAI 10308, pp. 39–55, 2017.
DOI: 10.1007/978-3-319-59271-8_3

directly, it is more convenient to construct their lattices. Fortunately, it is possible to change back and forth from the lattice to the contextual perspective, because a context $\mathbb{K}$ has $\mathbb{N}^c(k)$ as a subcontext if and only if $\mathfrak{B}(\mathbb{K})$ contains the boolean lattice $B(k)$ as a suborder [AC15]. Thus, the following definition comes naturally: a lattice is said to be (n, k)-*extremal* if it does not contain a boolean lattice with k atoms as a suborder, has exactly $\sum_{i=0}^{k-1} \binom{n}{i}$ elements and, amongst those, at most[1] n are join-irreducible (the least element is not irreducible according to our definitions). The positive natural numbers n and k are called *parameters*. Observe that if a context is (n, k)-extremal, then its lattice is (n, k)-extremal. Conversely, every (n, k)-extremal lattice is realizable by an (n, k)-extremal context.

The (n, k)-extremal lattices were characterized in [AC15]. A first contribution of the present paper is the description and construction of the canonical bases of those lattices, which was not done yet. As a second contribution, we show that this logical approach provides a first step into tackling an important problem, which is to be described below. For that purpose, we note that it remains elusive how many meet-irreducibles the (n, k)-extremal lattices possess. Moreover, observe that there is no mention to the number of attributes of $\mathbb{K}$ in Theorem 2. Therefore, the following problem arises very naturally:

Question. *Is it possible to prove an upper bound which is sharper than $\sum_{i=0}^{k-1} \binom{n}{i}$ by exploiting the number of attributes of $\mathbb{K}$? More generally, what are (n, m, k)-extremal lattices, where m stands for the maximum number of meet-irreducibles?*

We do not know how hard these questions are. Nevertheless, it seems that the alternative characterization of (n, k)-extremal lattices presented here makes their number of meet-irreducibles more evident. The reader is invited to use our approach to characterize the (n, m, k)-extremal lattices.

2   A Logical Perspective

The presented results are structured as follows. In this section, we introduce the notion of (n, k)-extremal sets of implications and provide an interesting example in Subsect. 2.3. In Sect. 3, we construct (Subsect. 3.1) and characterize (Subsect. 3.2) these objects. In the fourth section, it is first established that the introduced notion encompasses the canonical bases of extremal lattices. Then, we explain how important combinatorial properties can be read with the developed approach. We conclude in the fifth and last section.

2.1   Basic Definitions

The following definitions were already employed in [AC15] and are consistent with the terminology of [GW99]. The least and greatest elements of a lattice

---

[1] The restriction of having *at most* (and not exactly) n join-irreducibles is technical: the advantage is that $(n, 1)$-extremal lattices exist for any n. It makes no difference for the conclusions we draw.

L will be denoted, respectively, by 0_L and 1_L. An *atom* is an element covering 0_L, while a *coatom* is an element covered by 1_L. We denote by $J(L)$ and $M(L)$, respectively, the set of (completely) join-irreducible and meet-irreducible elements of a lattice L. Notice that, with these definitions, 0_L (1_L) is not join-irreducible (meet-irreducible). The *standard context* of a finite lattice L is $(J(L), M(L), \leq)$, where $\leq$ is the order of L. Given a finite lattice, its *length* is the number of elements in a maximum chain minus one. For a family of subsets $\mathcal{F} \subseteq \mathcal{P}(G)$ and $S \subseteq G$, the *trace* of $\mathcal{F}$ on S is $\{S \cap A \mid A \in \mathcal{F}\}$. We say that $\mathcal{F}$ *shatters* $S \subseteq G$ if its trace on S is $\mathcal{P}(S)$. The boolean lattice with k atoms is denoted $\mathrm{B}(k)$. If a lattice does not contain $\mathrm{B}(k)$ as a suborder, then it is $\mathrm{B}(k)$-*free*. For an element $l \in L$, we shall write $\downarrow l := \{x \in L \mid x \leq l\}$ and $\uparrow l := \{x \in L \mid x \geq l\}$.

2.2 Fundamental Results

In this subsection, we prove some elementary results. Most of them are very intuitive and some are already well known facts in the area.

We will employ a language for implications which is essentially the same as in [GW99]. However, our implications here are between objects. An implication $A \to B$ is a pair of subsets of G. The set A is called *premise* and B is called its *conclusion*. If $\mathcal{L}$ is a set of implications and every premise and every conclusion is a subset of G, then we say that G is an *implication set* or *set of implications over* G. The *ground set of* $\mathcal{L}$ is the union of all premises and conclusions and denoted $\Gamma(\mathcal{L})$. Occasionally, we refer to $\mathcal{L}$ as "set" (and leave implicit the expression "of implications").

A subset $T \subseteq G$ *respects an implication* $A \to B$ if $A \nsubseteq T$ or $B \subseteq T$. Such a subset *respects a set* $\mathcal{L}$ *of implications* if it respects every implication in $\mathcal{L}$. We may also say that T is a *respecting set*, in case $\mathcal{L}$ is implicitly understood. The collection of all subsets of G which respect $\mathcal{L}$ is denoted by $\mathfrak{H}(\mathcal{L})$. As it is well known, these form a closure system:

Proposition 1. *If $\mathcal{L}$ is a set of implications over G, then $\mathfrak{H}(\mathcal{L})$ is a closure system.*

The closure operator associated with an implication set can be described as follows [GW99]. For $S \subseteq G$, define $S^{\mathcal{L}} = S \cup \bigcup \{B \mid A \to B \in \mathcal{L}, S \supseteq A\}$. The application of $(\cdot)^{\mathcal{L}}$ onto S may be seen as the one-step *modus ponens* deduction of S. The set $S^{\mathcal{L}}$ is not, in general, closed (i.e. it does not respect every implication in $\mathcal{L}$). Instead, the closed sets are precisely the fixed points of this operator. The closure of an arbitrary $S \subseteq G$ will be denoted by $S^{\mathcal{L}\cdots\mathcal{L}}$. The sets G and $\mathcal{L}$ may be infinite; in such situation, the conventional transfinite constructions are employed.

As usual, the structure $\mathfrak{H}(\mathcal{L})$ may also be seen as a lattice: for an arbitrary family $(T_i)_i$ of respecting sets, its meet is given by intersection of all members and the supremum is the intersection of all sets which contain each T_i. This lattice has at most $|G|$ join-irreducible elements, as the next well known fact shows:

Proposition 2. *Let $\mathcal{L}$ be a set of implications over G. Then, every join-irreducible of $\mathfrak{H}(\mathcal{L})$ is the closure of a singleton $\{g\} \subseteq G$.*

Proof. Contraposition: let T be a set respecting $\mathcal{L}$ which is not the closure of a singleton. If T is the closure of the empty set, then certainly it is the least element of $\mathfrak{H}(\mathcal{L})$, therefore not join-irreducible and we are done. Otherwise, T is the closure of some non-empty set and T is therefore certainly non-empty. Every set which respects $\mathcal{L}$ and contains some element, say, $t \in G$, must also contain each element in its closure $\{t\}^{\mathcal{L}\cdots\mathcal{L}}$. Therefore, it is clear that $T = \cup_{t \in T}(\{t\}^{\mathcal{L}\cdots\mathcal{L}}) = \bigvee_{t \in T}\{t\}^{\mathcal{L}\cdots\mathcal{L}}$, which shows that T is not join-irreducible because of $T \neq \{t\}^{\mathcal{L}\cdots\mathcal{L}}$ for each $t \in T$.

From now on a lighter notation regarding braces will be adopted: $g^{\mathcal{L}}$ denotes $\{g\}^{\mathcal{L}}$ and implications like $\{g_1, g_2, \ldots, g_k\} \to \{h_1, h_2, \ldots, h_l\}$ will be simply written as $g_1 g_2 \ldots g_k \to h_1 h_2 \ldots h_l$.

Consider an arbitrary implication $P \to Q \in \mathcal{L}$. Obviously, one has that $P^{\mathcal{L}\mathcal{L}} \supseteq P^{\mathcal{L}} \supseteq P \cup Q$. If the containment $P^{\mathcal{L}\mathcal{L}} \supseteq P \cup Q$ holds with equality, then the second containment collapses as well and forces $P^{\mathcal{L}\mathcal{L}} = P^{\mathcal{L}}$. That is, in this case, $P^{\mathcal{L}}$ is a fixed point of the operator $(\cdot)^{\mathcal{L}}$, causing $P^{\mathcal{L}} = P \cup Q$ to be the closure of P. This will be of great utility in the present paper because it allows an exact determination of the closures of all premises. Hence, we define that an implication set $\mathcal{L}$ is *straight* if $P^{\mathcal{L}\mathcal{L}} = P \cup Q$ for each $P \to Q \in \mathcal{L}$. Being straight is not very restrictive, in the following sense: it will be demonstrated later in Lemma 6 that the canonical basis of a finite lattice is always straight.

Lemma 1. *A set of implications $\mathcal{L}$ is straight if and only if*

$$P \cup Q \supseteq R \Rightarrow P \cup Q \supseteq S \quad \text{(``condition for straightness'')}$$

holds for every $P \to Q, R \to S \in \mathcal{L}$.

Proof. Let $P \to Q \in \mathcal{L}$. Observe that $P^{\mathcal{L}} \supseteq P \cup Q$. Directly from the definition,

$$P^{\mathcal{L}\mathcal{L}} = P^{\mathcal{L}} \cup \bigcup \{S \mid R \to S, P^{\mathcal{L}} \supseteq R\}. \tag{1}$$

For one direction, suppose that $\mathcal{L}$ is straight. Then, $P^{\mathcal{L}\mathcal{L}} = P \cup Q$. Equation 1 forces that $P^{\mathcal{L}}$, as well as each set S inside the arbitrary union, to be contained in $P \cup Q$. Hence, we have $P^{\mathcal{L}} = P \cup Q$ and $S \subseteq P \cup Q$ for every implication $R \to S$ with $P \cup Q = P^{\mathcal{L}} \supseteq R$, i.e., the condition for straightness. Conversely, $P^{\mathcal{L}} = P \cup \bigcup \{S \mid R \to S, P \supseteq R\}$. Of course, $P \supseteq R$ implies $P \cup Q \supseteq R$, and the condition for straightness guarantees that each set S appearing in the arbitrary union must satisfy $S \subseteq P \cup Q$. Thus, $P^{\mathcal{L}} = P \cup Q$. Using Eq. 1 and making use of the condition again gives $P^{\mathcal{L}\mathcal{L}} = P \cup Q$.

An implication set is called *injective* if each pair of distinct premises has distinct closures. Consider a context $\mathbb{K} = (G, M, I)$. A set $S \subseteq G$ is called a *minimal generator with respect to $\mathbb{K}$* if $T'' \neq S''$ for every proper subset $T \subsetneq S$. The following is a weaker but consistent definition of a minimal generator from the logical perspective: a set $S \subseteq G$ is called a *minimal generator with respect to $\mathcal{L}$* if for each $P \rightarrow Q \in \mathcal{L}$, the implication $S \supseteq P \Rightarrow S \cap (Q \setminus P) = \emptyset$ holds. Observe that, in both notions, any subset of a minimal generator is once again a minimal generator.

Soon enough, we will link the absence of a minimal generator with respect to an implication set to the $B(k)$-freeness of its associated lattice. The main work in that direction is conducted by the following proposition:

Proposition 3. *Let $\mathcal{L}$ be a set of implications over G and $S \subseteq G$. If S is a minimal generator with respect to the standard context of $\mathfrak{H}(\mathcal{L})$, then S is a minimal generator with respect to $\mathcal{L}$.*

Proof. Set $L := \mathfrak{H}(\mathcal{L})$, and let $\mathbb{K}$ be the standard context of L. Denote by $''$ the closure operator of object sets of $\mathbb{K}$. We show the contraposition. Suppose that S is not a minimal generator with respect to $\mathcal{L}$. We may suppose $S \subseteq J(L)$, since otherwise the claim holds trivially. One can find $P \rightarrow Q \in \mathcal{L}$ with $S \supseteq P$ and $S \cap (Q \setminus P) \neq \emptyset$. Let $s \in S \cap (Q \setminus P)$. Note that $S \setminus \{s\} \supseteq P$. Any set T which respects $\mathcal{L}$ (equivalently, any extent T of $\mathbb{K}$) and contains $S \setminus \{s\}$ must have s as well, causing $(S \setminus \{s\})'' = S''$ and implying that S is not a minimal generator with respect to $\mathbb{K}$.

An implication set is said to be *r-regular* if every premise has exactly r elements. For an r-regular set $\mathcal{L}$, we say that $\mathcal{L}$ is *saturated* if no $(r+1)$-element subset of its ground set is a minimal generator.

To illustrate the properties "straight", "injective" and "saturated", we give examples below of regular sets possessing each two of the properties but not the third.

Examples: Consider the sets $\mathcal{L} = \{3 \rightarrow 21\}$, $\mathcal{M} = \{3 \rightarrow 21, 1 \rightarrow 2, 2 \rightarrow 1\}$ and $\mathcal{N} = \{3 \rightarrow 2, 2 \rightarrow 1\}$. The reader should have little or no trouble verifying that $\mathcal{L}$ is straight and injective but not saturated ("21" is a minimal generator with more than r elements), $\mathcal{M}$ is straight and saturated but not injective (we have $1^{\mathcal{M}} = 2^{\mathcal{M}}$ and this implies $1^{\mathcal{M}...\mathcal{M}} = 2^{\mathcal{M}...\mathcal{M}}$), and $\mathcal{N}$ is injective and saturated but not straight.

Whenever $\mathcal{L}$ is r-regular and saturated, a natural upper bound is imposed over the numbers of elements of the closure system $\mathfrak{H}(\mathcal{L})$, as the next proposition shows.

Proposition 4. *Let $\mathcal{L}$ be an r-regular, saturated set of implications over a finite set G. Then, the closure of any $S \subseteq G$ with $|S| \geq r + 1$ equals the closure of some premise P of $\mathcal{L}$ with $P \subseteq S$.*

Proof. Because S can not be a minimal generator, there exists an implication $P \rightarrow Q \in \mathcal{L}$ with $S \supseteq P$ and an element $s \in S \cap (Q \setminus P)$. Now, one clearly has that $S \setminus \{s\} \supseteq P$ which implies that S and $S \setminus \{s\} =: T$ have the same closure. If T has r elements, then $T \supseteq P$ together with $|P| = r$ (regularity of $\mathcal{L}$) force $P = T$ and we are done. If T has more than r elements, then one repeats this argument a necessary number of times, obtaining at each step another proper subset of S with the same closure.

We summarize the assertion present in Proposition 4 and a few other facts below:

Proposition 5. *The sets which respect an r-regular, saturated set of implications over a finite G are precisely the subsets of G with at most r − 1 elements, together with the r-element subsets which are not premises and the closures of each premise (which are given by P ∪ Q for each P → Q, in case straightness is satisfied).*

It was shown in [AC15] that if a context has no minimal generator with k elements, then its concept lattice is B(k)-free. Therefore, by combining this with Proposition 3, one has the following: if $\mathcal{L}$ is an implication set without a minimal generator having $r + 1$ elements, then the lattice $\mathfrak{H}(\mathcal{L})$ is B($r + 1$)-free. Recall that Proposition 2 upper bounds the number of join-irreducible elements of $\mathfrak{H}(\mathcal{L})$. Combining all those facts with Proposition 5 and supposing that injectivity is given, we summarize:

Proposition 6. *The sets which respect an injective, r-regular, saturated set of implications over an n-element set form an (n, r + 1)-extremal lattice.*

It would be natural to define that an implication set is extremal if it is injective, regular and saturated. We will, however, also require that it is straight: those mathematical objects are easier to characterize (as will be discussed after Theorem 3, making use of straightness helps to prove injectivity). Reaching this characterization is sufficient for the goals of this paper. Thus, we define that an implication set in an n-element set is *(n,k)-extremal* if it is injective, $(k − 1)$-regular, saturated and straight.

We now illustrate the just introduced notion for small values of k. For $k = 1$ and arbitrary n, it is clear that $\{\emptyset \to [n]\}$ is injective, 0-regular, saturated and straight. Therefore, its associated closure system is an $(n, 1)$-extremal lattice (i.e. it has only one element). On the left side of Fig. 1, the $(3, 2)$-extremal implication set $\{3 \to 21, 2 \to 1\}$ is displayed, together with its respecting sets (i.e. the associated lattice). Similarly, one can see on the right the same representation idea applied for the $(3, 3)$-extremal set $\{13 \to 2\}$. Note that, if an element has as label an implication $P \to Q$, it is to be understood that the element is the respecting set $P \cup Q$.

2.3 An Explicit Description of an Extremal Lattice with Many Meet-Irreducibles

One interesting task reveals itself: is it possible to explicitly describe extremal implication sets for arbitrary $n, k \geq 2$? This question can be promptly answered in the positive, as Theorem 3 shows. Nevertheless, it is still unclear (but will be revealed after Lemma 4) how one comes up with that set of implications. For a set G, the set of all of its r-element subsets is denoted $\binom{G}{r}$. At times, we refer to (n, k)-extremal lattices (sets) simply by *extremal lattices (sets)* if the parameters do not play an important role. The proof of Theorem 3 gives a hint as to why we

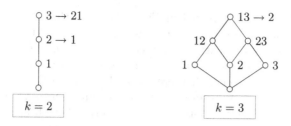

Fig. 1. Two small extremal lattices. A diagram with such a labelling allows one to depict $\mathfrak{H}(\mathcal{L})$ as well as $\mathcal{L}$ itself.

require that extremal sets be straight: such property makes the task of proving injectivity much simpler.

Theorem 3. *For any set* $\{g_1, g_2, \ldots, g_n\}$ *and* $k \geq 2$,

$$\bigcup_{i=1}^{n-k+1} \left\{ P \cup \{g_{i+1}\} \to \{g_1, \ldots, g_i\} \mid P \in \binom{\{g_{i+2}, \ldots, g_n\}}{k-2} \right\}$$

is (n,k)-*extremal.*

Proof. Denote by $\mathcal{L}_i$ each set appearing in the union above. Regularity is trivial. Regarding being saturated, consider a set $T \subseteq G := \{g_1, \ldots, g_n\}$ with k elements. By the pigeonhole principle, the intersection between T and $\{g_1, \ldots, g_{n-k+2}\}$ must contain at least two elements. Let g_i and g_j denote, respectively, the elements in said intersection with smallest and second to smallest indices. Then, $P := T \setminus \{g_i, g_j\}$ is a subset of $\{g_{j+1}, \ldots, g_n\}$ with $k-2$ elements and, without effort, one sees that $P \cup \{g_j\} \to \{g_1, \ldots, g_{j-1}\}$ has g_i in its conclusion and belongs to $\mathcal{L}_{j-1}$, which shows that T is not a minimal generator with respect to $\mathcal{L}$. Regarding straightness, let $P \to Q \in \mathcal{L}_i, R \to S \in \mathcal{L}_j$ for some i and j. If $i \geq j$, then $Q \supseteq S$, which implies $P \cup Q \supseteq S$ and the condition is satisfied. If $i < j$, then it holds that $P \not\supseteq R$ (both have exactly $k-1$ elements and are distinct, since the first contains g_{i+1}, whereas the latter does not). Moreover, observe that $Q \cap R = \emptyset$ and that this, together with $P \not\supseteq R$, yields $P \cup Q \not\supseteq R$ and the condition for straightness is satisfied. Considering injectivity, let P be a premise of $\mathcal{L}_i$ and Q be a premise of $\mathcal{L}_j$ with $P \neq Q$. Because of straightness, $P^{\mathcal{L}\ldots\mathcal{L}} = P \cup \{g_1, \ldots, g_i\}$ and $Q^{\mathcal{L}\ldots\mathcal{L}} = Q \cup \{g_1, \ldots, g_j\}$. If $i = j$, then by definition both P and Q have empty intersection with $\{g_1, \ldots, g_i\}$. Thus, $P \neq Q$ implies $P^{\mathcal{L}\ldots\mathcal{L}} \neq Q^{\mathcal{L}\ldots\mathcal{L}}$. If $i \neq j$, and without loss of generality, $i < j$, then $Q \cap \{g_1, \ldots, g_i\} = \emptyset$, which implies $P^{\mathcal{L}\ldots\mathcal{L}} = P \cup \{g_1, \ldots, g_i\} \not\supseteq Q$ and, on account of $Q^{\mathcal{L}\ldots\mathcal{L}} \supseteq Q$, the inequality $P^{\mathcal{L}\ldots\mathcal{L}} \neq Q^{\mathcal{L}\ldots\mathcal{L}}$ follows.

Corollary 1. *Extremal sets of implications, and therefore lattices, exist for every pair of parameters.*

Applying Theorem 3 with 1234 as ground set (with the natural order) and $k = 3$, one obtains the set of implications $\mathcal{L} = \{32 \to 1, 42 \to 1\} \cup \{43 \to 12\}$.

The associated lattice is depicted in Fig. 2. Observe that $\mathfrak{H}(\mathcal{L})$ has four join-irreducibles, no B(3) as a suborder and has precisely $1 + 4 + 6 = 11$ elements: in other words, it is an (4,3)-extremal lattice. Even though this particular lattice has 7 meet-irreducible elements (including those which are *doubly-irreducible*: i.e. both join and meet-irreducible), there exist (4,3)-extremal lattices with fewer meet-irreducibles: an easy example is an interordinal scale. Such example will be revisited inside our setting later, more precisely, after Theorem 5.

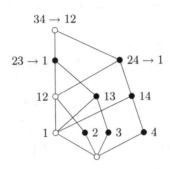

Fig. 2. A (4,3)-extremal lattice. Black circles represent meet (including doubly)-irreducibles.

In the following lemma, the ordered set $\{g_1, \ldots, g_i\}$ is the same as the one mentioned in Theorem 3.

Lemma 2. *Every set with at least $k-1$ elements and which respects the set of implications present in Theorem 3 has the form $Q \cup \{g_1, \ldots, g_i\}$ with $|Q| = k-2$.*

Proof. Let $\mathcal{L}$ denote said set of implications and T be a set which respects $\mathcal{L}$. If $|T| \geq k$, then T must be the closure of a premise of the form $P \cup \{g_{i+1}\}$ whose conclusion is $\{g_1, \ldots, g_i\}$. For the other case, we prove that each $T \subseteq \{g_2, \ldots, g_n\}$ with $|T| = k-1$ does not respect $\mathcal{L}$. Certainly T is non-empty, so we take $g_j \in T$ with the smallest index. Then, $T \setminus \{g_j\}$ belongs to $\binom{\{g_{j+1}, \ldots, g_n\}}{k-2}$, which implies that T is a premise of $\mathcal{L}_{j-1}$, where $\mathcal{L}_{j-1}$ is the implication set with $i = j-1$ in the description of $\mathcal{L}$. Therefore, every respecting set containing T must contain $\{g_1, \ldots, g_j\}$. In particular, T does not respect $\mathcal{L}$.

The following proposition is a well known necessary and sufficient condition for meet-irreducibility in concept lattices. Together with the lemma above, it will be possible to control the number of meet-irreducibles of the lattice associated to the extremal set described in Theorem 3.

Proposition 7. *Let $\mathcal{L}$ be a set of implications over a finite set G. Then, a respecting set T is meet-irreducible if and only if there exists $g \in G$ such that T is maximal amongst the respecting sets which do not have g.*

Proof. If T is meet-irreducible with unique upper cover U, then T is maximal amongst the respecting sets which do not have an arbitrary $g \in U \setminus T$. For the other direction, suppose that T is maximal amongst the respecting sets which do not have $g \in G$. Then, T is not the greatest element and therefore has at least one upper cover. Now, suppose by contradiction, that T has at least two upper covers, say U and V. Because of the property on T, we have that both U and V have the element g. Therefore, $U \cap V$ is a set which respects $\mathcal{L}$, properly contains T and $U \cap V \subsetneq U$, contradicting the fact that U is an upper cover of T.

Now we are able to determine exactly the number of meet-irreducibles of the lattice associated to the extremal set described in Theorem 3:

Theorem 4. *For every pair of parameters n and $k \leq n+1$, there exists at least one (n, k)-extremal lattice with precisely $\binom{n}{k-1} + k - 2$ meet-irreducibles.*

Proof. Let $\mathcal{L}$ denote the (n, k)-extremal set present in Theorem 3 and set $L = \mathfrak{H}(\mathcal{L})$, $G = \{g_1, \ldots, g_n\}$. If $k = 2$, then clearly L is an $n + 1$ element chain with $n = \binom{n}{1} + 2 - 2$ meet-irreducibles. Thus, it is assumed that $k \geq 3$ and the hypothesis forces $n \geq 2$. We establish that no $A \in L$ with $|A| \leq k - 3$ is meet-irreducible. Indeed, since every subset of G with at most $k - 2$ elements respects $\mathcal{L}$, it follows that every subset of G with at most $k - 3$ elements belongs to L and has at least two upper covers. Let $A \in L$ with $k - 2 \leq |A| \leq n - 2$ and set $j = |A| - (k - 2)$. Lemma 2 gives $A = Q \cup \{g_1, \ldots, g_j\}$, with j possibly zero. We show for $j = 0, \ldots, n - k$ that A is meet-irreducible if and only if $g_{j+1} \notin A$. For the converse: suppose that $g_{j+1} \notin A$. Lemma 2 implies that every $B \in L$ with $k - 1 + j$ elements must contain g_{j+1}, so it follows from Proposition 7 that A is meet-irreducible. For the other direction, suppose that $g_{j+1} \in A$. Because $\{g_1, \ldots, g_j\} \subseteq A$ and A has at most $n - 2$ elements, we may take two distinct elements $h, l \in G \setminus A$ with $h, l \notin \{g_1, \ldots, g_j\}$. Thus, both sets $Q_1 := (Q \setminus \{g_{j+1}\}) \cup \{h\}$ and $Q_2 := (Q \setminus \{g_{j+1}\}) \cup \{l\}$ have exactly $k - 2$ elements and empty intersection with $\{g_1, \ldots, g_{j+1}\}$. Therefore, $Q_1 \cup \{g_1, \ldots, g_{j+1}\}$ and $Q_2 \cup \{g_1, \ldots, g_{j+1}\}$ belong to L and are upper covers of A. In particular, A is not meet-irreducible. The number of meet-irreducibles A with $k - 2 \leq |A| \leq n - 2$ is, therefore, $\sum_{i=1}^{n-k+1} \binom{n-i}{k-2} = \binom{n}{k-1} - 1$. Every element $A \in L$ with $|A| = n - 1$ is meet-irreducible (it is a coatom) and, therefore, the total number follows.

Even though Theorem 3 produces extremal sets with arbitrarily given parameters, it is not true that every extremal set is producible by that result. This will be made clear after Theorem 5, to be presented in the next section.

3   A Characteristic Construction of Extremal Sets of Implications

In this section, we completely describe extremal implication sets. First, we show how one may construct an $(n + 1, k + 1)$-extremal set supposing that (n, k), $(n - 1, k), \ldots, (k, k)$-extremal sets are given and that they satisfy some condition of compatibility. Then, we will show that obtaining such (compatible) smaller extremal sets can be done quite easily: it turns out that it is possible to construct an $(n + 1, k + 1)$-extremal set after being given solely one (n, k)-extremal one. Lastly, we proceed to prove that every extremal set must be built through the procedure described in the beginning of the section.

3.1 Construction of a Larger Extremal Set Through Smaller Ones

In order to construct extremal implications sets with increasing parameters, one requires some operation which increases the regularity level of an implication set. This service will be performed by the following operation, which is in fact suggested by Theorem 3. For an implication set $\mathcal{L}$ over G and an element g not in G, we define the *lift of* $\mathcal{L}$ to be $\mathcal{L}^g := \{P \cup \{g\} \to Q \mid P \to Q \in \mathcal{L}\}$. The contrary work is performed by the *drop*, defined as $\mathcal{L}^{-g} := \{P \setminus \{g\} \to Q \mid P \to Q \in \mathcal{L}\}$.

The content of the next lemma is almost predictable. It will, however, be indispensable for further argumentation.

Lemma 3. *Let $\mathcal{L}$ be a set of implications over $G \setminus \{g\}$. Then,*

(i) *$\mathcal{L}$ is straight if and only if its lift $\mathcal{L}^g$ is.*
(ii) *$\mathcal{L}$ is injective if and only if its lift $\mathcal{L}^g$ is.*
(iii) *A set $S \subseteq G \setminus \{g\}$ is a minimal generator with respect to $\mathcal{L}$ if and only if $S \cup \{g\}$ is a minimal generator with respect to $\mathcal{L}^g$.*

Proof. For item (i), let $P \to Q, R \to S \in \mathcal{L}$. Because of $g \notin P \cup Q \cup R \cup S$, we have

$$(P \cup Q \supseteq R \Rightarrow P \cup Q \supseteq S) \Leftrightarrow$$
$$(P \cup \{g\} \cup Q \supseteq R \Rightarrow P \cup \{g\} \cup Q \supseteq S),$$

and the implication on the right-hand side is the condition for straightness in $\mathcal{L}^g$. Regarding item (ii), consider an arbitrary implication $P \to Q \in \mathcal{L}$. We will apply the operators $(\cdot)^{\mathcal{L}}$ to P and $(\cdot)^{\mathcal{L}^g}$ to the premise associated to P in $\mathcal{L}^g$, that is, $P \cup \{g\}$. It is clear that

$$\{B \mid A \to B \in \mathcal{L}, P \supseteq A\} = \{B \mid A \to B \in \mathcal{L}^g, P \cup \{g\} \supseteq A\},$$

which implies $(P \cup \{g\})^{\mathcal{L}^g} = P^{\mathcal{L}} \cup \{g\}$. The same argument may be reapplied:

$$\{B \mid A \to B \in \mathcal{L}, P^{\mathcal{L}} \supseteq A\} = \{B \mid A \to B \in \mathcal{L}^g, P^{\mathcal{L}} \cup \{g\} \supseteq A\},$$

yielding $(P \cup \{g\})^{\mathcal{L}^g \mathcal{L}^g} = P^{\mathcal{L}\mathcal{L}} \cup \{g\}$ and so on. Thus, injectivity of $(\cdot)^{\mathcal{L}^g \mathcal{L}^g \dots \mathcal{L}^g}$ is equivalent to injectivity of $(\cdot)^{\mathcal{L}\mathcal{L}\dots\mathcal{L}}$. For item (iii): a set $S \subseteq G \setminus \{g\}$ is a minimal generator with respect to $\mathcal{L}$ if and only if $S \cap (Q \setminus P) = \emptyset$ for each $P \to Q \in \mathcal{L}$ with $S \supseteq P$, and that holds if and only if $(S \cup \{g\}) \cap [Q \setminus (P \cup \{g\})] = \emptyset$ is true for each $P \cup \{g\} \to Q$ with $S \cup \{g\} \supseteq P \cup \{g\}$.

The union of straight sets is not necessarily straight (an example of this is $\mathcal{L} = \{1 \to 2\}$ and $\mathcal{M} = \{2 \to 3\}$). Of course, straightness is a very desirable property to be maintained during our construction. We therefore define that a family of implication sets $(\mathcal{L}_i)_{i \in I}$ is *compatible* if $\cup_i \mathcal{L}_i$ is straight.

Consider a family of implication sets $(\mathcal{L}_i \mid i \in [a])$. We say that g *separates* $\mathcal{L}_i$ from $\mathcal{L}_{i-1}$ if $\Gamma(\mathcal{L}_{i-1}) \setminus \Gamma(\mathcal{L}_i) = \{g\}$. A family $(\mathcal{L}_i)$ is said to be *cascade* if there exists g_i which separates $\mathcal{L}_i$ from $\mathcal{L}_{i-1}$ for each $2 \leq i \leq a$. In this case, the sequence $g_2, \ldots, g_a$ is called the *separating elements* of the family. If $\mathcal{L} = (\mathcal{L}_i \mid i \in [a])$ is a cascade family and g_1 is any element not in $\Gamma(\mathcal{L}_1)$, then we define the *multi-lift of* $\mathcal{L}$ to be $\hat{\mathcal{L}} := (\mathcal{L}_i^{g_i} \mid i \in [a])$, where $g_2, \ldots, g_a$ are the separating elements of $\mathcal{L}$.

Theorem 5 shows how one constructs an $(n+1, k+1)$-extremal set, provided that extremal sets with parameters $(n, k), (n-1, k), \ldots, (k, k)$ are available and its multi-lift is compatible. A converse to this result will be shown in Corollary 2, where we show that every extremal set has this structure.

Theorem 5. *If $\mathcal{L}_1, \ldots, \mathcal{L}_{n-k+1}$ is a cascade family such that each set $\mathcal{L}_i$ is $(n-i+1, k)$-extremal and $(\hat{\mathcal{L}}_i)$ is compatible, then $\cup_i \hat{\mathcal{L}}_i$ is $(n+1, k+1)$-extremal.*

Proof. Set $\mathcal{M} := \cup_i \hat{\mathcal{L}}_i$. Denote by $g_2, \ldots, g_{n-k+1}$ the separating elements of $(\mathcal{L}_i)$. Let g_1 be the element with $\Gamma(\mathcal{M}) = \Gamma(\mathcal{L}_1) \cup \{g_1\}$. Because $\mathcal{M}$ is straight, one has:

$$\{P^{\mathcal{M}\ldots\mathcal{M}} \mid P \to Q \in \mathcal{M}\} = \{P \cup Q \mid P \to Q \in \mathcal{M}\}$$

$$= \bigcup_{i=1}^{n-k+1} \{P \cup Q \mid P \to Q \in \mathcal{L}_i^{g_i}\}.$$

The union above is disjoint: take $i < j$ and implications $P \to Q$, $R \to S$ belonging, respectively, to $\mathcal{L}^{g_i}$ and $\mathcal{L}^{g_j}$. Then, it is clear that $g_i \in P \cup Q$ and, because the ground set of $\mathcal{L}_j^{g_j}$ does not contain g_i, it holds that $g_i \notin R \cup S$. We develop further:

$$\bigcup_{i=1}^{n-k+1} \{P \cup Q \mid P \to Q \in \mathcal{L}_i^{g_i}\} = \bigcup_{i=1}^{n-k+1} \{P^{\mathcal{L}_i^{g_i} \ldots \mathcal{L}_i^{g_i}} \mid P \to Q \in \mathcal{L}_i^{g_i}\},$$

where the equality above holds because each $\mathcal{L}_i^{g_i}$ is straight, according to Lemma 3. The same lemma gives that every $\mathcal{L}_i^{g_i}$ is injective and, therefore, so is $\mathcal{M}$. To establish that $\mathcal{M}$ is saturated, let T be a subset of $\Gamma(\mathcal{M})$ containing $k+1$ elements. Observe that $|\Gamma(\mathcal{M})| = n+1$. By the pigeonhole principle, T contains some element amongst $\{g_1, \ldots, g_{n-k+1}\}$. Let $g_i \in T$ be such element with the minimum index. Therefore, $T \setminus \{g_i\}$ belongs to the ground set of $\mathcal{L}_i$ and is not a minimal generator, because $\mathcal{L}_i$ is saturated. Lemma 3 assures that T is not a minimal generator with respect to $\mathcal{L}_i^{g_i} \subseteq \mathcal{M}$.

Theorem 5 captures the existence of extremal sets which are not constructible by Theorem 3. For instance, consider $\mathcal{L}_1 = \{3 \to 21, 2 \to 1\}$ and $\mathcal{L}_2 = \{1 \to 2\}$ which are, respectively, $(3, 2)$ and $(2, 2)$-extremal. Then, by calling "4" the new element implicit in the multi-lift, one has that the implication set $\mathcal{M} := \hat{\mathcal{L}}_1 \cup \hat{\mathcal{L}}_2$

equals $\{34 \to 21, 24 \to 1, 13 \to 2\}$ and is straight. By the theorem above, $\mathcal{M}$ is $(4, 3)$-extremal as well. The sets which respect $\mathcal{M}$ form an interordinal scale, depicted in Fig. 3. Note that the sets 14 and 23 respect $\mathcal{M}$ and can not be written in the form described by Lemma 2. Thus, $\mathfrak{H}(\mathcal{M})$ can not be produced by using Theorem 3.

To apply Theorem 5, one needs a family of implication sets whose multi-lift is compatible. The next lemma shows, in particular, that it suffices to find a family which is *itself* compatible, since compatibility is preserved by the multi-lift operation.

Lemma 4. *The multi-lift of a compatible and cascade family is compatible and cascade.*

Proof. Let $(\mathcal{L}_i)$ be a cascade family of implications, g_1 an element not in $\Gamma(\mathcal{L}_1)$ and let $g_2, \ldots, g_a$ be the separating elements. Contraposition: let $i \neq j$ and take two implications $P \cup \{g_i\} \to Q \in \hat{\mathcal{L}}_i$ and $R \cup \{g_j\} \to S \in \hat{\mathcal{L}}_j$ with $P \cup \{g_i\} \cup Q \supseteq R \cup \{g_j\}$ and $P \cup \{g_i\} \cup Q \not\supseteq S$. Notice that the mentioned containment forces $i < j$, because otherwise g_j would not belong to the ground set of $\hat{\mathcal{L}}_i$. On account of $i < j$, we have that $g_i \notin R$. Therefore, we also have $P \cup Q \supseteq R \cup \{g_j\} \supseteq R$ and $P \cup Q \not\supseteq S$, which is precisely the violation of straightness for $\cup_i \mathcal{L}_i$, i.e. $(\mathcal{L}_i)$ is not compatible. Being again cascade follows trivially.

Theorem 5, together with Lemma 4, explains how one comes up with the extremal set shown in Theorem 3. Let $n \geq 1, k \geq 2$ and take the family $(\mathcal{L}_i \mid i = 1, 2, \ldots, n-1)$ with $\mathcal{L}_i = \{j \to [j-1] \mid j = 2, \ldots, n-i+1\}$. It comes with almost no effort that each $\mathcal{L}_i$ is $(i+1, 2)$-extremal. Of course, $(\mathcal{L}_i)$ is cascade. The family $(\mathcal{L}_i)$ is also itself a descending chain, i.e., $\mathcal{L}_1 \supseteq \ldots \supseteq \mathcal{L}_{n-1}$. This implies that the union of all of its members is just $\mathcal{L}_1$ and the family is, therefore, compatible. Lemma 4 delivers that $(\hat{\mathcal{L}}_i)$ is compatible and Theorem 5 may be applied to each subfamily $(\mathcal{L}_i \mid i = 1, \ldots, k)$ for $k = 1, \ldots, n-1$. This yields the family $\hat{\mathcal{L}}_1, \hat{\mathcal{L}}_1 \cup \hat{\mathcal{L}}_2, \ldots, \cup_{i=1}^{n-1} \hat{\mathcal{L}}_i$ which is readily seen as a chain of $(i+2, 3)$-extremal sets. By reapplying Theorem 5, one obtains $(i+3, 4)$-extremal sets and so on. An exercise shows that these extremal sets are obtained by the explicit description present in Theorem 3.

3.2 Being the Union of a Multi-lift is Characteristic

In this subsection, we show that every extremal set is the union of the multi-lift of extremal sets with smaller parameters as described in Theorem 5. We begin with one definition: for an implication set $\mathcal{L}$ over G and $g \in G$, we set $^g\mathcal{L} := \{P \to Q \in \mathcal{L} \mid g \in P\}$.

The next claim lemma appears to be a bit technical, but it is intuitive in case the reader is familiar with the notion of *extremal points* in lattices.

Lemma 5. *Let $\mathcal{L}$ be (n, k)-extremal with ground set G and $P \to Q, R \to S \in \mathcal{L}$ be implications with $R \cup S \subseteq P \cup Q$. Then, for every $g \in P$ it holds that $g \in R$ or $g \notin R \cup S$. In particular, if $P \cup Q = G$, then, for each $g \in P$ it holds that $\mathcal{L} \setminus {}^g\mathcal{L}$ is $(n-1, k)$-extremal with ground set $G \setminus \{g\}$.*

Proof. Suppose, by contradiction, that $g \notin R$ and $g \in R \cup S$. In particular, $R \neq P$ and $k \geq 2$. Set $X = (P \setminus \{g\}) \cup R$. Note that $g \notin X$ and $P \cup Q \supseteq X$. Because $\mathcal{L}$ is straight, we have that $P \cup Q$ respects $\mathcal{L}$ and, therefore, the closure of X is contained in $P \cup Q$. On the other hand, because of $R \to S$, $X \supseteq R$ and $g \in S$, we have that g belongs to $X^{\mathcal{L}}$. Hence, $X^{\mathcal{L}\mathcal{L}} \supseteq P \cup Q$. Combining both, we have that the closure of X is precisely $P \cup Q$, i.e., the closure of P. Now, if X is precisely R, then the contradiction with the injectivity of $\mathcal{L}$ is clear, since R is a premise different than P but with the same closure. Otherwise, X contains R properly and, therefore, has at least k elements. Proposition 4 gives us a premise Y of $\mathcal{L}$, the closure of which is the same as the closure of X and $Y \subseteq X$. Observe that $Y \neq P$ because of $g \notin Y$. The premises P and Y contradict the injectivity of $\mathcal{L}$.

The hard work regarding the converse has been done in Lemma 5. We now collect the reward of nice assertions about the structure of extremal sets (and lattices). The following theorem shows that the construction described in Theorem 5 can be easily bootstrapped: indeed, $\mathcal{L}$ carries inside itself a compatible family, making the hypothesis of that theorem easy to be satisfied. In particular, one is able to construct an $(n + 1, k + 1)$-extremal set by having only one (n, k)-extremal set as initial information.

Theorem 6. *Given any (n, k)-extremal set $\mathcal{L}$ with $k \geq 2$, there exists a chain $\mathcal{L} = \mathcal{L}_0 \supseteq \mathcal{L}_1 \supseteq \ldots \supseteq \mathcal{L}_{n-k}$ such that each $\mathcal{L}_i$ is $(n-i, k)$-extremal. In particular, the families $(\mathcal{L}_i)$ and $(\hat{\mathcal{L}}_i)$ are compatible.*

Proof. Set $G = \Gamma(\mathcal{L})$. If $n \leq k$, the claim holds trivially. Thus, assume $n > k$. Because G respects $\mathcal{L}$ and $\mathcal{L}$ is k-regular and saturated, Proposition 4 gives us an implication $P \to Q \in \mathcal{L}$ with $P \cup Q = G$. By applying the last claim in Lemma 5, one obtains that the subset $\mathcal{L}_1 := \mathcal{L} \setminus {}^g\mathcal{L}$ is an $(n - 1, k)$-extremal set. If $n - 1 = k$, we are done. Otherwise, we make use of Lemma 5 again and obtain $\mathcal{L}_2 := \mathcal{L}_1 \setminus {}^h\mathcal{L}_1$, where h is some element of the premise of the ground set of $\mathcal{L}_1$, i.e. $G \setminus \{g\}$ and so on.

Theorem 7 is the main result of this subsection and its immediate consequence is Corollary 2, which is the converse of Theorem 5. In the result below, there is a mention to the premise of the ground set of $\mathcal{L}$: to make that expression clear, observe that the ground set of $\mathcal{L}$ obviously respects $\mathcal{L}$ and, whenever $\mathcal{L}$ is (n, k)-extremal, there exists precisely one implication $P \to Q \in \mathcal{L}$ with $P \cup Q = \Gamma(\mathcal{L})$. We also employ the term g-*lift* to explicitly refer to the new element required for the lift operation.

Theorem 7. *Let $\mathcal{L}$ be an (n, k)-extremal set and g an element belonging to the premise of its ground set. Then, it holds that ${}^g\mathcal{L}$ is the g-lift of an $(n - 1, k - 1)$-extremal set and $\mathcal{L} \setminus {}^g\mathcal{L}$ is $(n - 1, k)$-extremal.*

Proof. Since $\mathcal{L}$ is straight and straightness is a hereditary property (see Lemma 1), it follows that ${}^g\mathcal{L}$ is straight. Moreover, since $\mathcal{L}$ is injective, it follows with help of straightness of $\mathcal{L}$ and ${}^g\mathcal{L}$ that the latter is injective as well.

Items (i) and (ii) from Lemma 3 assure that the drop $^{g}\mathcal{L}^{-g}$ is straight and injective. It is also clearly $(k-2)$-regular. Lemma 5 guarantees that the ground set of $^{g}\mathcal{L}^{-g}$ is $\Gamma(\mathcal{L}) \setminus \{g\}$ and item (iii) from Lemma 3 shows that it is saturated. The assertion regarding $\mathcal{L} \setminus {^{g}\mathcal{L}}$ follows from the final claim in Lemma 5.

To exemplify Theorem 7, consider again $\mathcal{L} = \{34 \to 21, 24 \to 1, 13 \to 2\}$, which is $(4,3)$-extremal. This set is decomposed in Fig. 3 through the use of Theorem 7 with $g = 4$ (also possible: $g = 3$). The result says $\mathcal{L} = \mathcal{L}_1 \cup \mathcal{L}_2$, where $\mathcal{L}_1 := {^{4}\mathcal{L}}$ and $\mathcal{L}_2 := \mathcal{L} \setminus {^{4}\mathcal{L}}$ is $(3,3)$-extremal. Besides, Theorem 7 says that $\mathcal{L}_1$ has the form $\mathcal{L}_3^4$ with $\mathcal{L}_3$ being the $(3,2)$-extremal set $\{3 \to 21, 2 \to 1\}$. The lattice associated to $\mathcal{L}_2$ is shown inside the dashed region in that figure. The dotted region draws attention to the chain $\uparrow 4$, which is a $(3,2)$-extremal sublattice.

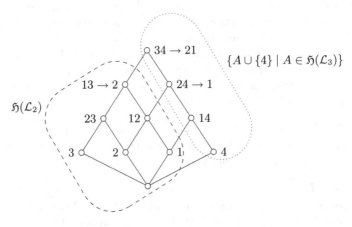

Fig. 3. A $(4,3)$-extremal set as the union of one $(3,3)$-extremal set and the lift of one $(3,2)$-extremal one.

Corollary 2. *For every $(n+1, k+1)$-extremal set $\mathcal{L}$ there exists a cascade family $(\mathcal{L}_i)$ whose multi-lift is compatible and $\mathcal{L} = \cup_{i=1}^{n-k+1} \hat{\mathcal{L}}_i$, where each $\mathcal{L}_i$ is $(n-i+1, k)$-extremal.*

Proof. One applies Theorem 7 repeatedly, by choosing an element h of the premise of the ground set of $\mathcal{L} \setminus {^{g}\mathcal{L}}$.

4   Canonical Bases of Extremal Lattices

Extremal sets were characterized in the last section. Now it is time to establish a strong link between that introduced notion and extremal lattices. Said connection is done through the canonical basis of implications: we show that the canonical basis of every extremal lattice is an extremal set of implications (with the same parameters). Since it was already established by Proposition 6 that the lattice associated to an extremal set is an extremal lattice, the end result is that the notion of extremal implication sets introduced here is consistent with the notion of extremal lattices. Moreover, as it was just shown in Subsects. 3.1 and 3.2, one has a characteristic construction of such lattices.

4.1   Basic Definitions and Results

Consider a lattice L and an element $x \in L$. We use the symbol J_x to denote set of all join-irreducibles which are less or equal x, that is: $J_x = J(L) \cap {\downarrow} x$. Now, suppose that L is finite. Then, every $x \in L$ is the supremum of some subset of $J(L)$: for example, $x = \bigvee J_x$. Such a subset is called a *representation of x through join-irreducible elements* (for brevity: *a representation through irreducibles of x* or a *representation of x*). A representation $S \subseteq J(L)$ of x is called *irredundant* if $\bigvee(S \setminus \{y\}) \neq x$ for every $y \in S$. Of course, every $x \in L$ has an irredundant representation, but it does not need to be unique. Note that irredundant representations are precisely minimal generators of the standard context of L. Hence, any element of a B(k)-free lattice has an irredundant representation of size at most $k - 1$ [AC15].

The scenario here is still the object-sided implication logic of a context and its lattice of extents. Consider a finite context (G, M, I). A set $P \subseteq G$ is called a *pseudo-extent* if $P \neq P''$ and $Q'' \subsetneq P$ for every pseudo-extent $Q \subsetneq P$. In particular, note that if $P \subseteq G$ is not closed but each maximal proper subset of P is closed, then P must be a pseudo-extent. The reader possibly recognizes that, if that is the case, then P is also something called *proper premise* (the definition of which will not be given here but can be looked up in [GW99]). According to [GW99, GD86], a lattice L has a *canonical basis*: $\{S \rightarrow (\bigvee S) \setminus S \mid S \subseteq J(L), \ S$ a pseudo-extent$\}$. In this definition, it is implicit that being a pseudo-extent refers to the standard context of L. This will also be the case whenever we say that a set $S \subseteq J(L)$ is *closed*.

Before we proceed to look into the structure of pseudo-extents in extremal lattices, we need to establish some easy facts regarding straightness. Given an implication set $\mathcal{L}$, its *expansion* is $\mathcal{L}^* := \{P \rightarrow P \cup Q \mid P \rightarrow Q \in \mathcal{L}\}$. Of course, taking the expansion does not change the associated closure system.

Proposition 8. *A set of implications is straight if and only if its expansion is.*

Proof. Let $P \rightarrow Q, R \rightarrow S \in \mathcal{L}$. Define $Q^* = P \cup Q$ and $S^* = R \cup S$ and consider the pair of implications $P \rightarrow Q^*, R \rightarrow S^*$ in $\mathcal{L}^*$. Then, trivially $P \cup Q = P \cup Q^*$, which yields $P \cup Q \supseteq R$ if and only if $P \cup Q^* \supseteq R$ and $P \cup Q \supseteq S$ if and only if $P \cup Q^* \supseteq S$.

The proposition above helps to prove the following:

Lemma 6. *The canonical basis of a finite lattice and its expansion are always straight.*

Proof. Let L be a finite lattice, $\mathcal{L}$ be its canonical basis and set $\mathcal{M} := \mathcal{L}^*$. Suppose, by contradiction, that $\mathcal{M}$ violates the condition for straightness. Then, we take $P \rightarrow Q \in \mathcal{M}$ (note that $Q = \bigvee P$) and $R \rightarrow S \in \mathcal{M}$ with $P \cup Q = Q \supseteq R$ and $P \cup Q = Q \not\supseteq S$. Therefore, Q does not respect $R \rightarrow S$ and, in particular, does not respect $\mathcal{M}$. With symbols, one has that $Q = \bigvee P \notin \mathfrak{H}(\mathcal{M})$. We arrive to a contradiction with the fact that $\mathfrak{H}(\mathcal{M}) = \mathfrak{H}(\mathcal{L}) = \{\bigvee X \mid X \subseteq J(L)\}$. The canonical basis $\mathcal{L}$ is straight as well, on account of Proposition 8.

4.2   The Structure of Canonical Bases of Extremal Lattices

The necessary logical foundations have now been laid and we begin to look more closely to pseudo-extents of extremal lattices. The following fact was in its essence presented in [AC15]:

Proposition 9. *Suppose that L is an (n,k)-extremal lattice. Then, for every $S,T \subseteq J(L)$ with $|S|,|T| \leq k-1$, the implication $\bigvee S = \bigvee T \Rightarrow S = T$ holds. In particular, every $S \subseteq J(L)$ with $|S| = k-2$ is closed.*

Proof. The first part was proved in the paper. For the corollary, take a set S of $k-2$ join-irreducibles. Then, if S were not closed, we would be able to take some $x \in (\bigvee S) \setminus S$ and we would have $\bigvee S = \bigvee(S \cup \{x\})$, contradicting the first part.

As a consequence of the proposition above, we have Proposition 10, which shows the pseudo-extents of an extremal lattice are the sets of exactly $k-1$ irreducibles which are not closed. This means that pseudo-extents of extremal lattices are actually proper premises.

Proposition 10. *Let L be an (n,k)-extremal lattice. Then, every $S \subseteq J(L)$ with precisely $k-1$ elements is either closed or a pseudo-extent. Moreover, every pseudo-extent of L has exactly $k-1$ elements.*

Proof. Suppose that S has $k-1$ elements and is not closed. On account of Proposition 9, it follows that every maximal proper subset of S is closed. Thus, by definition, S is a pseudo-extent. For the second statement, let S be a pseudo-extent of L. Then, S is not closed and Proposition 9 forces $|S| \geq k-1$. Suppose by contradiction that equality does not hold. Because L is $B(k)$-free, we may take $T \subseteq S$ with $|T| = k-1$ and $\bigvee T = \bigvee S$. Combining the just proven statement with the fact that T is not closed, we have that T is a pseudo-extent. This contradicts the fact that S is a pseudo-extent, because of $T \subsetneq S$ and $\bigvee T \supseteq S$.

The next result wraps up the developed theory and introduced notions.

Theorem 8. *A lattice is (n,k)-extremal if and only if its canonical basis is (n,k)-extremal.*

Proof. The "if" direction was already established by Proposition 6. For the other direction, let L be an (n,k)-extremal lattice and $\mathcal{L}$ be its canonical basis. Set $G = J(L)$. Lemma 6 gives that $\mathcal{L}$ is straight. Now, Proposition 10 implies that $\mathcal{L}$ is $(k-1)$-regular. We now prove that $\mathcal{L}$ is saturated. The fact that L is $B(k)$-free guarantees the existence of an irredundant representation of size at most $k-1$ for each element of L. In particular, for every $S \subseteq G$ with $|S| = k$, there exists $T \subseteq S$ with $|T| = k-1$ and $\bigvee T = \bigvee S$. Note that $\emptyset \neq S \setminus T \subseteq (\bigvee T) \setminus T$. Proposition 10 yields that T is a pseudo-extent and therefore $T \to (\bigvee T) \setminus T$ belongs to $\mathcal{L}$. Such implication attests that S is not a minimal generator, since $S \supseteq T$ and $S \cap [(\bigvee T) \setminus T] \neq \emptyset$. According to Proposition 5 and since $\mathcal{L}$ is $(k-1)$-regular and saturated, $|L| = |\mathfrak{H}(\mathcal{L})|$ must be equal to $\sum_{i=0}^{k-2}\binom{|G|}{i} + |\{P^{\mathcal{L}...\mathcal{L}}|$

$P \subseteq G, |P| = k - 1\}|$. This sum must be exactly $\sum_{i=0}^{k-1} \binom{|G|}{i}$, implying that $\mathcal{L}$ is injective.

Extremal lattices are finite and meet-distributive. Thus, they are graded posets, with $x \mapsto |J_x|$ being one valid rank function. The number of elements having rank i is called the i-th *Whitney number* and denoted w_i. The *Whitney numbers* are the sequence $(w_0, \ldots, w_l)$, where l is the length of the lattice. For instance, in Figs. 2 and 3 one can recognize the sequence $(1, 4, 3, 2, 1)$ as being the Whitney numbers of those $(4, 3)$-extremal lattices. The chain present in Fig. 1 has $(1, 1, 1, 1)$ as its Whitney numbers. Both sequences are described by a general formula (see Corollary 3). In particular, extremal lattices are always unimodular.

Corollary 3. *The Whitney numbers of any (n, k)-extremal lattice with $k \geq 2$ are $\left(\binom{n}{0}, \binom{n}{1}, \ldots, \binom{n}{k-2}, \binom{n-1}{k-2}, \ldots, \binom{k-2}{k-2}\right)$.*

Proof. For any fixed n and k, the repeated application of the multi-lift operation yields the same number (and size) of premises and conclusions. Therefore, each (n, k)-extremal lattice must have the same Whitney numbers. In particular, they must be the same as those of the lattice given by Theorem 3. $\qquad\square$

Corollary 4. *Every (n, k)-extremal lattice has precisely $\binom{n-1}{k-1}$ pseudo-extents.*

Proof. Each element of rank $r \geq k$ corresponds to a pseudo-extent. Thus, the total number of pseudo-extents can be calculated using the formula given in Corollary 3, which is $\sum_{i=0}^{n-k} \binom{k-2+i}{k-2} = \binom{n-1}{k-1}$. $\qquad\square$

References

[AC15] Albano, A., Chornomaz, B.: Why concept lattices are large - extremal theory for the number of minimal generators and formal concepts. In: Yahia, S.B., Konecny, J. (eds.) Proceedings of the Twelfth International Conference on Concept Lattices and Their Applications, Clermont-Ferrand, France, 13–16 October 2015, CEUR Workshop Proceedings, vol. 1466, pp. 73–86 (2015). CEUR-WS.org

[GD86] Guigues, J.L., Duquenne, V.: Famille minimale d'implications informatives résultant d'un tableau de données binaires. Math. Sci. Hum. **24**(95), 5–18 (1986)

[GW99] Ganter, B., Wille, R.: Formal Concept Analysis: Mathematical Foundations. Springer, Heidelberg (1999)

[Juk10] Jukna, S.: Extremal Combinatorics: With Applications in Computer Science, 1st edn. Springer Publishing Company, Inc., Heidelberg (2010)

[Sau72] Sauer, N.: On the density of families of sets. J. Comb. Theory Ser. A **13**(1), 145–147 (1972)

[She72] Shelah, S.: A combinatorial problem; stability and order for models and theories in infinitary languages. Pac. J. Math. **41**(1), 247–261 (1972)

[VC71] Vapnik, V.N., Chervonenkis, A.Y.: On the uniform convergence of relative frequencies of events to their probabilities. Theory Probab. Appl. **16**(2), 264–280 (1971)

Making Use of Empty Intersections to Improve the Performance of CbO-Type Algorithms

Simon Andrews[(⊠)] [iD]

Conceptual Structures Research Group,
Communication and Computing Research Centre,
Department of Computing,
Faculty of Arts, Computing, Engineering and Sciences,
Sheffield Hallam University, Sheffield, UK
s.andrews@shu.ac.uk

Abstract. This paper describes how improvements in the performance of Close-by-One type algorithms can be achieved by making use of empty intersections in the computation of formal concepts. During the computation, if the intersection between the current concept extent and the next attribute-extent is empty, this fact can be simply inherited by subsequent children of the current concept. Thus subsequent intersections with the same attribute-extent can be skipped. Because these intersections require the testing of each object in the current extent, significant time savings can be made by avoiding them. The paper also shows how further time savings can be made by forgoing the traditional canonicity test for new extents, if the intersection is empty. Finally, the paper describes how, because of typical optimizations made in the implementation of CbO-type algorithms, even more time can be saved by amalgamating inherited attributes with inherited empty intersections into a single, simple test.

Keywords: Formal Concept Analysis · FCA · FCA algorithms · Computing formal concepts · Canonicity test · Close-by-One · CbO · In-Close

1 Introduction

There have been many advances in Formal Concept Analysis that address complexity, focus the analysis and concentrate on producing manageable and meaningful results. Such approaches include ice-berg lattices [1], nesting lattices [2], creating sub-contexts [3], fault-tolerance [4,5], expandable concept trees [6], rough concepts [7] and approximation [8]. However, there will always be the need for faster performance of the fundamental operations in FCA, such as the computation of formal concepts. Clearly, we need to avoid the production of concept lattices that are indecipherable, 'bird's nests', containing thousands of nodes and FCA results that are over-complex and meaningless. Nevertheless, with FCA being applied ever more to the analysis of large sets of data, performance is an important factor. Even if the set of concepts is small, dynamic applications need

© Springer International Publishing AG 2017
K. Bertet et al. (Eds.): ICFCA 2017, LNAI 10308, pp. 56–71, 2017.
DOI: 10.1007/978-3-319-59271-8_4

fast computational engines to cope with fast changes in the underlying data. Or it may be the case that although the set of computed concepts is large, applications may use this as a data-set of concepts on which to carry out queries and subsequent analysis. Two recent major European projects provide real-world examples involving large and dynamic data bases: The ATHENA project - using social media in crisis management [9,10] and the ePOOLICE project - scanning the electronic environment to detect organised crime [11]. Both projects involved the development of software systems where FCA was implemented as a component. It was found that the most practical means of applying FCA was to compute very large sets of concepts on the fly, and then query these sets for concepts of interest, rather than create pre-processing components to form sub-contexts from the original data set. Both projects involved creating large sets of structured data from the text-processing of Internet-based unstructured data, dynamically and on the fly. Having a fast concept mining engine to convert this data quickly into formal concepts before querying and visualization was essential to the operation of the systems.

In the development of such fast algorithms, the discovery of the so-called 'canonicity test', whereby the attributes in a concept could be examined to determine its newness in the computation [12], has proved to be fundamental. This test has given rise to a number of algorithms based on the original Close-by-One (CbO) algorithm [12] including the CbO algorithm presented in [13], FCbO [14,15] and In-Close2 [16]. FCbO introduced a combined 'breadth and depth' approach to computation that allowed child concepts to fully inherit their parent's attributes. In-Close2 then added a modified, 'partial-closure', canonicity test to reduce the computation required in the test. FCbO also introduced a technique whereby failed canonicity tests could be inherited, thereby avoiding repeated tests.

This paper describes new advances in the performance of CbO-type algorithms by making use of empty intersections. During the computation, if the intersection between the current concept extent and the next attribute-extent is empty, this fact can be simply inherited by subsequent children of the current concept. Because all extents D that are children of a current extent C are sub-sets of C, it is simple to show that intersections of child extents with an attribute-extent $\{j\}^{\downarrow}$ will be empty if the intersection between the same attribute-extent and the parent extent was empty:

$$\textit{if } D \subseteq C \textit{ and } C \cap \{j\}^{\downarrow} = \emptyset \textit{ then } D \cap \{j\}^{\downarrow} = \emptyset$$

Thus subsequent intersections between child extents and the same attribute-extent can be skipped. Because these intersections require the testing of each object in the current extent, significant time savings can be made by avoiding them.

The rest of this paper is structured as follows: The paper will use the algorithm In-Close2 [16] as its starting point, so Sect. 2 is a recap of that algorithm. Section 3 is a recap of the FCbO algorithm. As a fast CbO-type algorithm for computing formal concepts it is a good benchmark for the In-Close variants.

Section 4 describes a new In-Close variant, In-Close4a, that incorporates the inheritance of empty intersections as described above. It should be noted that In-Close1, In-Close2 and In-Close3 are the existing versions of In-Close, as presented in [17]. Thus the new versions presented here are named In-Close4a and In-Close4b. In-Close3 is a version that incorporated FCbO's feature of inheriting failed canonicity tests, but, because of the added complexity and overheads incurred, actually performed less well overall than In-Close2. Thus In-Close2, rather than In-Close3, is used here as the starting point for the new versions. Section 5 describes a refined version of In-Close4a, called In-Close4b, that forgoes the traditional canonicity test for new extents, if the intersection between the current extent and the next attribute-extent is empty. Section 6 describes some key optimizations made in the implementation of CbO-type algorithms, and how further time savings can be made by amalgamating inherited attributes with inherited empty intersections into a single, simple test. Section 7 presents a series of experiments and results, comparing the performance of In-Close2, In-Close4a, In-Close4b and FCbO. Finally, Sect. 8 provides some concluding remarks and ideas for further work.

2   Recap of the In-Close2 Algorithm

Below is a recap of the In-Close2 algorithm, as presented in [17]. In-Close2 was been 'bred' from In-Close and FCbO to combine the efficiencies of a partial-closure canonicity test (as compared to a full closure test) with full inheritance of the parent intent. The full inheritance is achieved by adapting the combined breadth-first and depth-first approach of FCbO [14,15]. The main cycle is completed before passing to the next level, so that all the attributes of a parent intent can be passed down to the next level rather than just some of them. Like In-Close, child intents only have to be 'finished off' by adding attributes when $A = C$, but now additional attributes after j are also inherited and can be skipped. During the main cycle, whilst the current intent is being closed, new extents that pass the canonicity test are stored in a queue, similar to the queue in FCbO, to be processed after the main cycle has completed.

The In-Close2 algorithm is invoked with an initial pair $(A, B) = (X, \emptyset)$, where A is a set of objects (extent) and B is a set of attributes (intent) and X is the set of all objects in the formal context, and initial attribute $y = 0$. Y is the set of all attributes in the formal context and Y_j is the set of all attributes up to (but not including) j.

Line 1 - Iterate across the formal context, from a starting attribute y up to attribute $n - 1$, where n is the number of attributes in the context.
Line 2 - Skip attributes already in B. Because intents now inherit all of their parent's attributes, these can be skipped.
Line 3 - Form an extent, C, by intersecting the current extent, A, with the next column of objects in the context.
Line 4 - If the extent formed, C, equals the extent, A, of the concept whose intent is currently being closed, then...

In-Close2

ComputeConceptsFrom$((A, B), y)$

```
1  for j ← y upto n − 1 do
2  │   if j ∉ B then
3  │   │   C ← A ∩ {j}↓
4  │   │   if C = A then
5  │   │   │   B ← B ∪ {j}
6  │   │   else
7  │   │   │   if B ∩ Yⱼ = C↑ⱼ then
8  │   │   │   │   PutInQueue(C, j)
9  ProcessConcept((A, B))
10 while GetFromQueue(C, j) do
11 │   D ← B ∪ {j}
12 │   ComputeConceptsFrom((C, D), j + 1)
```

Line 5 - ...add the current attribute, j, to the intent being closed, B.

Line 7 - Otherwise, test the canonicity using the partial-closure canonicity test [18]: $\uparrow$ is the standard closure operator in FCA and $\uparrow_j$ is a modification meaning "close up to, but not including attribute j".

Line 8 - If the canonicity test is passed, place the new extent, C, and the location where it was found, j, in a queue for later processing.

Line 9 - Pass concept (A, B) to the notional procedure ProcessConcept to process it in some way (for example, storing it in a set of concepts).

Lines 10 - The queue is processed by obtaining each new extent and associated location from the queue.

Line 11 - Each new partial intent, D, inherits all the attributes from its completed parent intent, B, along with the attribute, j, where its extent was found.

Line 12 - Call ComputeConceptsFrom to compute child concepts from $j + 1$ and to complete the intent D.

3   Recap of the FCbO Algorithm

Below is a recap of the FCbO algorithm [14,15] as presented in [17]. FCbO introduced the feature of inherited canonicity test failures to improve the performance of CbO-type algorithms, along with the combined breadth/depth first approach to enable full inheritance of parent intents. The inheritance of test failure is achieved by recording intents that are not canonical as N^js, where j is the current attribute, thus enabling subsequent levels to test these failed intents against the current one and thus avoid the computation of a redundant extent and intent. FCbO is invoked with the initial concept $(A, B) = (X, X^\uparrow)$, initial attribute $y = 0$ and a set of empty Ns, $\{N^y = \emptyset \mid y \in Y\}$.

FCbO

ComputeConceptsFrom$((A,B), y, \{N^y \mid y \in Y\})$

---

1 ProcessConcept$((A,B))$
2 **for** $j \leftarrow y$ **upto** $n-1$ **do**
3 $\quad M^j \leftarrow N^j$
4 $\quad$ **if** $j \notin B$ **and** $N^j \cap Y_j \subseteq B \cap Y_j$ **then**
5 $\quad\quad C \leftarrow A \cap \{j\}^{\downarrow}$
6 $\quad\quad D \leftarrow C^{\uparrow}$
7 $\quad\quad$ **if** $B \cap Y_j = D \cap Y_j$ **then**
8 $\quad\quad\quad$ PutInQueue $((C,D), j)$
9 $\quad\quad$ **else**
10 $\quad\quad\quad M^j \leftarrow D$

11 **while** GetFromQueue$((C,D), j)$ **do**
12 $\quad$ ComputeConceptsFrom$((C,D), j+1, \{M^y \mid y \in Y\})$

Line 1 - Pass concept (A,B) to the notional procedure ProcessConcept to process it in some way (for example, storing it in a set of concepts).
Line 2 - Iterate across the context, from starting attribute y up to attribute $n-1$.
Line 3 - M^j is set to the latest intent that failed the canonicity test at attribute j, N^j.
Line 4 - Skip attributes in B and those that have an inherited record of failure.
Line 5 - Otherwise, form an extent, C, by intersecting the current extent, A, with the next column of objects in the context.
Line 6 - Close the extent to form an intent, D.
Line 7 - Perform the canonicity test.
Line 8 - If the concept is a new one, store it in a queue along with the attribute it was computed at.
Line 10 - Otherwise set the record of failure for attribute j, M^j, to the intent that failed the canonicity test.
Line 11 - Get each stored concept from the queue...
Line 12 - ...and pass it to the next level, along with the stored starting attribute for the next level and the failed intents from this level.

4 New Algorithm, In-Close4a: Skipping Attributes with Inherited Empty Intersections

In-Close4a adds a new feature to In-Close2 by enabling subsequent levels to skip attributes that have previously resulted in an empty intersection with a parent extent. If a parent extent intersected with an attribute-extent results in an empty set, then any subset (child) of the parent extent intersected with the same attribute-extent will also result in an empty set. Thus, if a record is kept of the parent empty intersections, subsequent children can skip the attributes concerned. Because the intersection between the current extent and the attribute-extent must iterate all of the objects in the current extent, a significant number of operations can potentially be avoided if intersections can be skipped.

The new algorithm, In-Close4a, incorporating this feature is given below. In-Close4a is invoked with an initial pair $(A, B) = (X, \emptyset)$, an initial attribute $y = 0$ and an empty set of attributes, $P = \emptyset$.

In-Close4a

ComputeConceptsFrom$((A, B), y, P)$

```
1   for j ← y upto n − 1 do
2       if j ∉ B and j ∉ P then
3           C ← A ∩ {j}↓
4           if C = A then
5               B ← B ∪ {j}
6           else
7               if C = ∅ then
8                   P ← P ∪ {j}
9               if B ∩ Yⱼ = C↑ʲ then
10                  PutInQueue(C, j)
11  ProcessConcept((A, B))
12  Q ← P
13  while GetFromQueue(C, j) do
14      D ← B ∪ {j}
15      ComputeConceptsFrom((C, D), j + 1, Q)
```

The changes in the algorithm from In-Close2 are as follows:

Line 2 - Skip attributes already in B and also skip attributes in P, as we know they will result in empty intersections.

Line 7 - If the new extent, C, is empty...

Line 8 - ... add the current attribute to P.

Line 12 - Store P in Q ready to pass the attributes resulting in empty intersections to the next level.

Note that the algorithm will compute the concept $(\emptyset, Y)$, if it exists. If $C = \emptyset$, then the canonicity test is still carried out, and, if it is the first occurrence of the empty set of objects, it will pass the canonicity test and the concept will be completed at the next level.

5   New Algorithm, In-Close4b: Forgoing the Canonicity Test After Empty Intersections

The concept $(\emptyset, Y)$ can be viewed as a special case and will exist if there are no objects that have all the attributes in the formal context. If it exists it will be computed by CbO-type algorithms at the first occurrence of $C = \emptyset$, with subsequent empty extents failing the canonicity test (although FCbO will avoid some of these tests via inherited test-failure). However, as an alternative approach, In-Close4b forgoes the canonicity test when $C = \emptyset$, in effect making it a case of automatic failure. Thus, as a refinement of In-Close4b, the new algorithm not only skips inherited empty intersections but also avoids having to carry out a canonicity test whenever an empty intersection occurs. It is quite likely for the empty intersection to occur very frequently in the computation, so there is thus the potential to save a significant number of operations.

Of course, the resulting algorithm is incomplete, in that it will not compute the concept $(\emptyset, Y)$. However, it is a simple task to add it afterwards, if it exists: If $Y^{\downarrow} = \emptyset$ then add $(\emptyset, Y)$ to the set of computed concepts.

The new algorithm, In-Close4b, incorporating this feature is given below. In-Close4b is invoked with an initial pair $(A, B) = (X, \emptyset)$, an initial attribute $y = 0$ and an empty set of attributes, $P = \emptyset$.

In-Close4b

ComputeConceptsFrom$((A, B), y, P)$

```
1  for j ← y upto n − 1 do
2      if j ∉ B and j ∉ P then
3          C ← A ∩ {j}↓
4          if C ≠ ∅ then
5              if C = A then
6                  B ← B ∪ {j}
7              else
8                  if B ∩ Yj = C↑j then
9                      PutInQueue(C, j)
10         else
11             P ← P ∪ {j}
12 ProcessConcept((A, B))
13 Q ← P
14 while GetFromQueue(C, j) do
15     D ← B ∪ {j}
16     ComputeConceptsFrom((C, D), j + 1, Q)
```

It should be noted that forgoing the canonicity test following an empty intersection was actually present in the very first incarnation of the In-Close algorithm [19] (although not in later versions). However, it was not realised as a novel time-saving device at the time and thus was not explicitly described as such. Its effects were not explicitly measured or compared, nor were its implications considered, as far as incompleteness is concerned. Consequently, In-Close4b could be regarded as a new 'best of breed' algorithm, combing time-saving features of In-Close, In-Close2, In-Close4a and FCbO.

An implementation of In-Close4b is available, open-source and free to download from Sourceforge[1].

6   Optimisation

In this section, three key optimisations will be briefly described. The first two are recapping existing major optimisations: the use of bit-arrays and the implementation of the partial-closure canonicity test, and the third explains how the inheritance of attributes and the inheritance of empty intersections can be amalgamated into a single test. The experiments in Sect. 7 use optimised implementations, so it is worth understanding the key optimisations involved.

Recap of the Use of Bit-Arrays. Implementations of CbO-type algorithms, such as In-Close and FCbO, typically use a bit-array to represent the formal context. This allows operations on the formal context, such as closure operations, to be implemented using bit-wise operators in the manner of fine-grained parallel processing. In a typical 64-bit architecture, this means that 64 cells of the formal context can be operated on simultaneously.

Recap of the Implementation of the Partial-Closure Canonicity Test. In practice, it is not necessary to always close the new extent up to the current attribute. It is only necessary to find the *first instance* where $B \cap Y_j$ and $C^{\uparrow_j}$ do not agree. Thus failure is typically detected before j is reached, making even more time savings than the partial-closure would. By comparison, in other algorithms, such as FCbO, a *full-closure* is *always* required because, not only is it the basis of the canonicity test, it also provides the closure of new concepts. In In-Close, with the partial-closure canonicity test, new concepts are closed incrementally whenever $C = A$ by $B \leftarrow B \cup \{j\}$ (lines 4 and 5 of In-Close4a, for example). Furthermore, given that the test $C = A$ is provided 'for free', as a by-product of the intersection in $C \leftarrow A \cap \{j\}^{\downarrow}$, the overheads of this incremental closure are insignificant.

Amalgamation of the Inheritance of Attributes and the Inheritance of Empty Intersections. In typical implementations of fast CbO-type algorithms, the intent, B, of a concept is stored twice: once as a local version, in Boolean (bit) form, with each bit representing the presence or absence of an attribute, and once as integers (the index numbers of the attributes) in a tree data structure. The latter is space-saving and thus used for storage and output, whilst the former is used as a fast method of determining $j \notin B$ to skip inherited

---

[1] https://sourceforge.net/projects/inclose/.

attributes. Thus, because the Boolean form is not required for subsequent storage and output, it can be used to record the inherited attribute *and* the empty intersections - both are tested, recorded and inherited in the same bit-array. The addition of the inheritance of empty intersections does not, therefore, require an additional array for P and incurs very few overheads in the implementation.

7   Evaluation of Performance

In this section, In-Close2, In-Close4a, In-Close4b and FCbO will be evaluated by comparing their performance over a varied range of data sets. The experiments are divided into three groups: (1) real data sets, (2) artificial data sets, and (3) randomised data sets. Following the performance comparison, the algorithms are investigated to compare the number of canonicity tests carried out in each case, as this is the most labour intensive operation involved. Indeed, the intention of features such as skipping attributes and inheriting canonicity failures are specifically designed to reduce the number of canonicity test required.

Optimised implementations of the algorithms were created using C++ and the experiments were conducted on a standard 64-bit Intel architecture, using a PC with an Intel Core i7-2600 3.40 GHz CPU and 8 GB of RAM. To cater for any inconsistency of system performance, each experiment was conducted multiple times and the average time taken for each.

Real Data Set Experiments. Four real data sets were used in the experiments: *Mushroom, Adult* and *Internet Ads*, taken from the UCI Machine Learning Repository [20] and *Student*, an anonymised data set from an internal student experience survey carried out at Sheffield Hallam University, UK. The data sets were selected to represent a broad range of features, in terms of size and density, and the UCI ones, in particular, are well known and used in FCA work.

The results of the experiments are given in Tables 1 (timings) and 2 (canonicity tests). In all timing cases (apart from two ties) In-Close4b was fastest, followed by In-Close4a then In-Close2, and FCbO was the slowest.

The smallest number of canonicity tests was achieved by FCbO for *Mushroom* and *Student* and by In-Close4b for *Adult* and *Internet Ads*. It is clear, from comparing In-Close2 results with those of In-Close4a and 4b that making use of empty intersections has a large effect in reducing the number of canonicity tests required. However, it is also clear that the inherited canonicity failures in FCbO also produce a large reduction. Comparing the timings with the number of canonicity tests carried out (particularly between In-Close2 and FCbO) gives clear evidence of the significance of the partial-closure canonicity test in reducing computation time. Of course, it should be noted that making use of empty intersections prevents canonicity test involving the closure of the empty set of objects. Such closures are obviously less labour intensive than closing non-empty sets of object, but nevertheless a significant number of operations are still necessary, and the timing results clearly show that making use of empty intersections shows a small but significant speed-up for each of the data sets.

Table 1. Real data set results (timings in seconds).

Data set	Mushroom	Adult	Internet Ads	Student
$\lvert G \rvert \times \lvert M \rvert$	8,124 × 125	32,561 × 124	3,279 × 1,565	587 × 145
Density	17.36%	11.29%	0.97%	24.50%
#Concepts	226,921	1,388,469	16,570	22,760,243
FCbO	0.21	1.46	0.31	8.80
In-Close2	0.21	1.05	0.12	5.06
In-Close4a	0.19	0.95	0.08	**4.65**
In-Close4b	**0.17**	**0.88**	**0.07**	**4.65**

Table 2. Real data set results (canonicity tests).

Data set	Mushroom	Adult	Internet Ads	Student
In-Close2	1,164,829	2,614,810	1,867,689	55,038,419
In-Close4a	447,356	1,916,356	682,635	53,859,754
FCbO	**325,986**	2,029,933	363,568	**40,630,663**
In-Close4b	405,584	**1,707,707**	**91,029**	53,162,649

Artificial Data Set Experiments. Artificial data sets were used that, although randomised, the randomisation was constrained by properties of real data sets, such as many-valued attributes having a fixed number of possible values. Three data sets, *M7X10G120K*, *M10X30G120K* and *T10I4D100K*, were used to provide a range of features in terms of size and density.

The timing results of the artificial data set experiments are given in Table 3 and the comparison of the number of canonicity tests carried out is given in Table 4. The timing results reiterate those obtained in the real data set experiments: In-Close4b fastest, then In-Close4a, then In-Close2, then FCbO. In terms of the number of canonicity tests carried out, In-Close4b had the fewest, then FCbO, then In-Close4a, then In-Close2.

Table 3. Artificial data set results (timings in seconds).

Data set	M7X10G120K	M10X30G120K	T10I4D100K
$\lvert G \rvert \times \lvert M \rvert$	120,000 × 70	120,000 × 300	100,000 × 1,000
Density	10.00%	3.33%	1.01%
#Concepts	1,166,326	4,570,493	2,347,376
FCbO	1.35	15.45	23.83
In-Close2	0.98	8.37	9.10
In-Close4a	0.81	7.45	7.79
In-Close4b	**0.77**	**5.60**	**6.56**

Table 4. Artificial data set results (canonicity tests).

Data set	M7X10G120K	M10X30G120K	T10I4D100K
In-Close2	14,621,925	330,546.826	185,296,387
In-Close4a	5,548,392	191,601,668	112,154,256
FCbO	4,640,906	167,814,522	75,281,105
In-Close4b	**2,360,015**	**29,686,007**	**21,262,544**

Random Data Set Experiments. Three series of random data experiments were carried out, testing the effect of changes in the number of attributes, context density, and number of objects:

- Figure 1: *Attributes series* - with 5% density and 5,000 objects, the number of attributes was varied between 300 and 1,000. The number of concepts varied from approximately 1,000,000 to 22,000,000.
- Figure 3: *Objects series* - with 5% density and 200 attributes, the number of objects was varied between 30,000 and 100,000. The number of concepts varied from approximately 4,000,000 to 22,000,000.
- Figure 2: *Density series* - with 200 attributes and 10,000 objects, the density of 1 s in the context was varied between 3 and 10%. The number of concepts varied from approximately 200,000 to 19,000,000.

The results of the random data set timings are given in Figs. 1 (attribute series), 2 (density series) and 3. A comparison of the number of canonicity tests carried out for the random data sets in each algorithm is given in Figs. 4 (attribute series), 5 (density series) and 6.

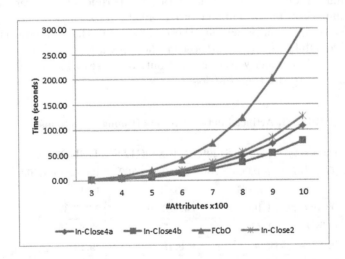

Fig. 1. Comparison of performance with varying number of attributes. 5% density, 5,000 objects. #Concepts range from approx. 1,000,000–22,000,000

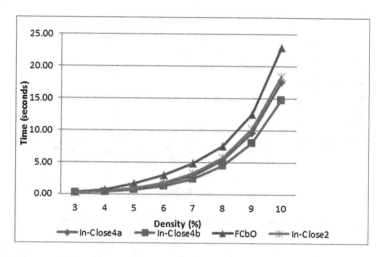

Fig. 2. Comparison of performance with varying context density. 200 attributes, 10,000 objects. #Concepts range from approx. 4,000,000–22,000,000

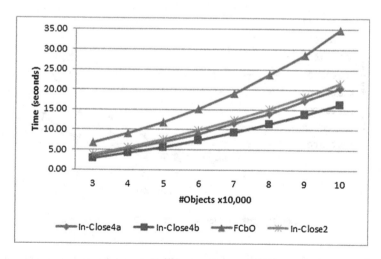

Fig. 3. Comparison of performance with varying number of objects. 5% density, 200 attributes. #Concepts range from approx. 200,000–19,000,000

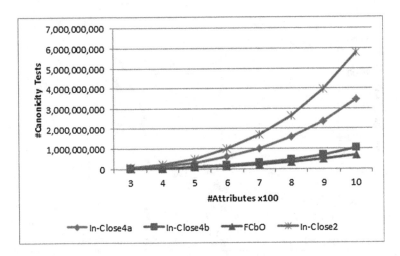

Fig. 4. Comparison of the number of canonicity tests carried out with varying number of attributes. 5% density, 5,000 objects. #Concepts range from approx. 1,000,000–22,000,000

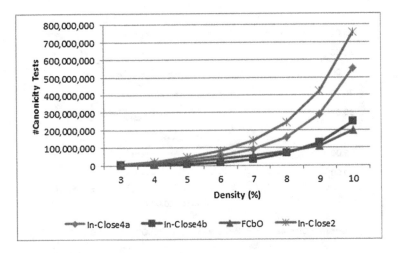

Fig. 5. Comparison of the number of canonicity tests carried out with varying context density. 200 attributes, 10,000 objects. #Concepts range from approx. 4,000,000–22,000,000

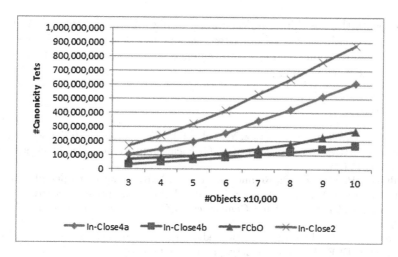

Fig. 6. Comparison of the number of canonicity tests carried out with varying number of objects. 5% density, 200 attributes. #Concepts range from approx. 200,000–19,000,000

Again, the results confirm those from the previous experiments, with In-Close4b showing the best performance, then In-Close4a, then In-Close2, then FCbO. In terms of the number of canonicity tests carried out, overall FCbO showed the greatest reduction, although In-Close4b was a close second and, for lower density data sets in the density series, In-Close4b had greatest reduction.

8 Conclusions

In conclusion, the experimental results clearly show that making use of empty intersections, by skipping inherited empty intersections (In-Close4a) and forgoing the canonicity test following an empty intersection (In-Close4b), provides a small but significant improvement to the performance of CbO-type algorithms. Their implementation is simple, costing almost no overhead and easily optimised to amalgamate inherited empty intersections with inherited attributes.

In terms of further work, the canonicity test results for FCbO are interesting. FCbO shows an equal or better reduction in canonicity tests carried out, compared to the other algorithms, through its inheritance of canonicity test failures, but this is outweighed by the costly 'full-closure' canonicity test it employs (as also evidence in previous papers [16–18]). Although previous work [17] has already produced an algorithm (In-Close3) that incorporates FCbO's inheritance of canonicity test failures, it failed to provide any significant improvements in performance compared to In-Close2, due to the complexity and consequent overheads in computation that it required. Nevertheless, the large reduction in canonicity tests carried out in FCbO makes it tempting to re-visit the inheritance of test failures - either to improve and optimise further its implementation, or to investigate the possibility of designing a simpler method of test failure inheritance. If test failures can be inherited without the penalty of significant overheads in computation, further improvements in the performance of CbO-type algorithms will be possible.

References

1. Stumme, G., Taouil, R., Bastide, Y., Lakhal, L.: Conceptual clustering with iceberg concept lattices. In: Proceedings of GI-Fachgruppentreffen Maschinelles Lernen (FGML01) (2001)
2. Valtchev, P., Grosser, D., Roume, C., Hacene, M.R.: Galicia: an open platform for lattices. In: de Moor, A., Ganter, B. (eds.) Using Conceptual Structures: Contributions to 11th International Conference on Conceptual Structures, pp. 241–254 (2003)
3. Andrews, S., Orphanides, C.: Knowledge discovery through creating formal contexts. In: IEEE Computer Society, pp. 455–460 (2010)
4. Pensa, R.G., Boulicaut, J.-F.: Towards fault-tolerant formal concept analysis. In: Bandini, S., Manzoni, S. (eds.) AI*IA 2005. LNCS, vol. 3673, pp. 212–223. Springer, Heidelberg (2005). doi:10.1007/11558590_22
5. Dau, F.: An implementation for fault tolerance and experimental results. In: CUBIST Workshop, pp. 21–30 (2013)
6. Andrews, S., Hirsch, L.: A tool for creating and visualising formal concept trees. In: CEUR Workshop Proceedings: Proceedings of the Fifth Conceptual Structures Tools and Interoperability Workshop (CSTIW 2016), vol. 1637, pp. 1–9 (2016)
7. Liu, M., Shao, M., Zhang, W., Wu, C.: Reduction method for concept lattices based on rough set theory and its application. Comput. Math. Appl. **53**, 1390–1410 (2007)
8. Gaume, B., Navarro, E., Prade, H.: Clustering bipartite graphs in terms of approximate formal concepts and sub-contexts. Int. J. Comput. Intell. Syst. **6**, 1125–1142 (2013)
9. ATHENA: The European ATHENA Project - use of new smart devices and social media in crisis situations. http://www.projectathena.eu/. Accessed September 2016
10. Andrews, S., Yates, S., Akhgar, B., Fortune, D.: Strategic intelligence management: national security imperatives and information and communication technologies. In: The ATHENA Project: Using Formal Concept Analysis to Facilitate the Actions of Responders in a Crisis Situation, pp. 167–180. Butterworth-Heinemann, Elsevier (2013)
11. Andrews, S., Brewster, B., Day, T.: Organised crime and social media: detecting and corroborating weak signals of human trafficking online. In: Haemmerlé, O., Stapleton, G., Faron Zucker, C. (eds.) ICCS 2016. LNCS, vol. 9717, pp. 137–150. Springer, Cham (2016). doi:10.1007/978-3-319-40985-6_11
12. Kuznetsov, S.O.: Mathematical aspects of concept analysis. Math. Sci. **80**, 1654–1698 (1996)
13. Krajca, P., Outrata, J., Vychodil, V.: Parallel recursive algorithm for FCA. In: Belohavlek, R., Kuznetsov, S. (eds.) Proceedings of Concept Lattices and their Applications (2008)
14. Krajca, P., Vychodil, V., Outrata, J.: Advances in algorithms based on CbO. In: Kryszkiewicz, M., Obiedkov, S. (eds.) CLA 2010. University of Sevilla, pp. 325–337 (2010)

15. Outrata, J., Vychodil, V.: Fast algorithm for computing fixpoints of Galois connections induced by object-attribute relational data. Inf. Sci. **185**, 114–127 (2012)
16. Andrews, S.: In-close2, a high performance formal concept miner. In: Andrews, S., Polovina, S., Hill, R., Akhgar, B. (eds.) ICCS 2011. LNCS (LNAI), vol. 6828, pp. 50–62. Springer, Heidelberg (2011). doi:10.1007/978-3-642-22688-5_4
17. Andrews, S.: A 'Best-of-Breed' approach for designing a fast algorithm for computing fixpoints of Galois connections. Inf. Sci. **295**, 633–649 (2015)
18. Andrews, S.: A partial-closure canonicity test to increase the efficiency of CbO-type algorithms. In: Hernandez, N., Jäschke, R., Croitoru, M. (eds.) ICCS 2014. LNCS, vol. 8577, pp. 37–50. Springer, Cham (2014). doi:10.1007/978-3-319-08389-6_5
19. Andrews, S.: In-close, a fast algorithm for computing formal concepts. In: Rudolph, S., Dau, F., Kuznetsov, S.O. (eds.) ICCS 2009, CEUR WS, vol. 483 (2009). http://sunsite.informatik.rwth-aachen.de/Publications/CEUR-WS/Vol-483/
20. Frank, A., Asuncion, A.: UCI machine learning repository (2010). http://archive.ics.uci.edu/ml

On the Usability of Probably Approximately Correct Implication Bases

Daniel Borchmann[1]($\boxtimes$) (iD), Tom Hanika[2,3] (iD), and Sergei Obiedkov[4] (iD)

[1] Chair of Automata Theory, Technische Universität Dresden,
Dresden, Germany
daniel.borchmann@tu-dresden.de
[2] Knowledge and Data Engineering Group, University of Kassel,
Kassel, Germany
tom.hanika@cs.uni-kassel.de
[3] Interdisciplinary Research Center for Information System Design,
University of Kassel, Kassel, Germany
[4] Faculty of Computer Science,
National Research University Higher School of Economics,
Moscow, Russian Federation
sergei.obj@gmail.com

Abstract. We revisit the notion of *probably approximately correct implication bases* from the literature and present a first formulation in the language of formal concept analysis, with the goal to investigate whether such bases represent a suitable substitute for exact implication bases in practical use cases. To this end, we quantitatively examine the behavior of probably approximately correct implication bases on artificial and real-world data sets and compare their precision and recall with respect to their corresponding exact implication bases. Using a small example, we also provide evidence suggesting that implications from probably approximately correct bases can still represent meaningful knowledge from a given data set.

Keywords: Formal concept analysis · Implications · Query learning · PAC learning

1 Introduction

From a practical point of view, computing implication bases of formal contexts is a challenging task. The reason for this is twofold: on the one hand, bases of formal contexts can be of exponential size [14] (see also an earlier work [12] for the same result presented in different terms), and thus just writing out the result can take a long time. On the other hand, even in cases where implication bases can be small, efficient methods to compute them are unknown in general, and

---

The authors of this work are given in alphabetical order. No priority in authorship is implied.

© Springer International Publishing AG 2017
K. Bertet et al. (Eds.): ICFCA 2017, LNAI 10308, pp. 72–88, 2017.
DOI: 10.1007/978-3-319-59271-8_5

running times may thus be very high. This is particularly true for computing the *canonical basis*, for which only very few algorithms [9, 15] are known, and, in addition to the canonical basis, they compute the entire concept lattice.

Approaches to tackle this problem are to parallelize existing algorithms [13], or restrict attention to implication bases that are more amenable to algorithmic treatment, such as proper premises [16] or D-bases [1]. The latter usually comes with the downside that the number of implications is larger than necessary. A further, rather pragmatic approach is to consider implications as strong association rules and employ highly optimized association rule miners, but then the number of resulting implications increases even more.

In this work, we want to introduce another approach, which is conceptually different from those previously mentioned: instead of computing exact bases that can be astronomically large and hard to compute, we propose to compute *approximately correct* bases that, being easier to obtain, capture essential parts of the implication theory of the given data set. To facilitate algorithmic amenability, it turns out to be a favorable idea to compute bases that are approximately correct *with high probability*. Such bases are called *probably approximately correct bases* (PAC bases), and they can be computed in polynomial time.

PAC bases make it possible to relax the rather strong condition of computing an exact representation of the implicational knowledge of a data set. However, this new freedom comes at the price of the uncertainty that approximation always brings: is the result suitable for my intended application? Of course, the answer to this question depends deeply on the application in mind and cannot be given in general. On the other hand, some general aspects of the usability of PAC bases can be investigated, and it is the purpose of this work to provide first evidence that such bases can indeed be useful. More precisely, we want to show that despite the probabilistic nature of these bases, the results they provide are indeed not significantly different from the actual bases (in a certain sense that we shall make clear later), and that the returned implication sets can contain meaningful implications. To this end, we investigate PAC bases on both artificial and real-world data sets and discuss their relationship with their exact counterparts.

The idea of considering PAC bases is not new [12], but has somehow not received much attention as a different, and maybe tantamount, approach to extract implicational knowledge from formal contexts. Moreover, PAC bases also allow interesting connections between formal concept analysis and *query learning* [3] (as we shall see), a connection that, with respect to attribute exploration, awaits further investigation.

The paper is structured as follows. After a brief review of related work in Sect. 2, we shall introduce probably approximately correct bases in Sect. 3, including a means to compute them based on results from query learning. In Sect. 4, we discuss usability issues, both from a quantitative and a qualitative point of view. We shall close our discussion with summary and outlook in Sect. 5.

2   Related Work

Approximately learning concepts with high probability has first been introduced in the seminal work by Valiant [18]. From this starting point, *probably approximately correct* learning has come a long way and has been applied in a variety of use cases. Work that is particularly relevant for our concerns is by Kautz et al. on Horn approximation of empirical data [12]. In there a first algorithm for computing probably approximately correct implication bases for a given data set has been proposed [12, Theorem 15]. This algorithm has the benefit that all closed sets of the actual implication theory will be among the ones of the computed theory, but the latter may possibly contain more. However, the algorithm requires direct access to the actual data, which therefore must be given explicitly.

Approximately correct bases have also been considered before in the realm of formal concept analysis, although not much. The dissertation by Babin [6] contains results about *approximate bases* and some first experimental evaluations. His notion of approximation is somewhat stronger than the one we employ in this work: Babin defines a set of implications $\mathcal{H}$ to be an approximation of a given set $\mathcal{L}$ if the closure operators of $\mathcal{L}$ and $\mathcal{H}$ coincide on most sets. In our work, $\mathcal{H}$ is an approximation of $\mathcal{L}$ if and only if the number of models in which $\mathcal{H}$ and $\mathcal{L}$ differ is small. Details will follow in the next section. The approach of considering implications with *high confidence* in addition to exact implications can also be seen as a variant of approximate bases [7].

To compute PAC bases, we make use of results from the research field of *query learning* [3]. More precisely, we make use of the work by Angluin et al. on learning Horn theories through query learning [4], where the target Horn theory is accessible only through a *membership* and *equivalence* oracle. Using existing results, this algorithm can easily be adapted to compute probably approximately correct Horn theories, and we give a self-contained explanation of the algorithm in this work. Related to query learning is *attribute exploration* [10], an algorithm from formal concept analysis that allows one to learn Horn theories from domain experts.

3   Probably Approximately Correct Bases via Query Learning

Before introducing approximately correct and probably approximately correct bases in Sect. 3.2, we give a brief (and dense) recall in Sect. 3.1 of the relevant definitions and terminology from formal concept analysis used in this work. We then demonstrate in Sect. 3.3 how probably approximately correct bases can be computed using ideas from query learning.

3.1   Bases of Implications

Recall that a formal context is a triple $\mathbb{K} = (G, M, I)$ where G and M are sets and $I \subseteq G \times M$. We shall denote the derivation operators in $\mathbb{K}$ with the usual $\cdot'$-notation, i.e., $A' = \{\, m \in M \mid \forall g \in A : (g, m) \in I \,\}$ and $B' = \{\, g \in G \mid \forall m \in B : (g, m) \in I \,\}$ for $A \subseteq G$ and $B \subseteq M$. The sets A and B are *closed* in $\mathbb{K}$ if $A = A''$ and $B = B''$, respectively. The set of subsets of M closed in $\mathbb{K}$ is called the set of *intents* of $\mathbb{K}$ and is denoted by $\mathrm{Int}(\mathbb{K})$.

An *implication* over M is an expression $X \to Y$ where $X, Y \subseteq M$. The set of all implications over M is denoted by $\mathrm{Imp}(M)$. A set $A \subseteq M$ is *closed* under

$X \rightarrow Y$ if $X \not\subseteq A$ or $Y \subseteq A$. In this case, A is also called a *model* of $X \rightarrow Y$ and $X \rightarrow Y$ is said to *respect* A. The set A is *closed* under a set of implications $\mathcal{L}$ if A is closed under every implication in $\mathcal{L}$. The set of all sets closed under $\mathcal{L}$, the *models* of $\mathcal{L}$, is denoted by $\mathrm{Mod}(\mathcal{L})$.

An implication $X \rightarrow Y$ is *valid* in $\mathbb{K}$ if $\{g\}'$ is closed under $X \rightarrow Y$ for all $g \in G$ (equivalently: $X' \subseteq Y'$, $Y \subseteq X''$). A set $\mathcal{L}$ of implications is *sound* for $\mathbb{K}$ if every implication in $\mathcal{L}$ is valid in $\mathbb{K}$.

An implication $X \rightarrow Y \in \mathrm{Imp}(M)$ *follows* from $\mathcal{L} \subseteq \mathrm{Imp}(M)$, written $\mathcal{L} \models (X \rightarrow Y)$, if every model of $\mathcal{L}$ is a model of $X \rightarrow Y$. Equivalently, $\mathcal{L} \models (X \rightarrow Y)$ if and only if $Y \subseteq \mathcal{L}(X)$, where $\mathcal{L}(X)$ is the $\subseteq$-smallest superset of X closed under $\mathcal{L}$.

A set $\mathcal{L} \subseteq \mathrm{Imp}(M)$ is *complete* for $\mathbb{K}$ if every implication valid in $\mathbb{K}$ follows from $\mathcal{L}$. A set $\mathcal{L}$ is an *exact implication basis* (or simply *basis*) of $\mathbb{K}$ if $\mathcal{L}$ is sound and complete for $\mathbb{K}$. Alternatively, $\mathcal{L}$ is a basis of $\mathbb{K}$ if the models of $\mathcal{L}$ are the intents of $\mathbb{K}$.

A basis $\mathcal{L}$ of $\mathbb{K}$ is called *irredundant* if no strict subset of $\mathcal{L}$ is a basis of $\mathbb{K}$. It is called *minimal* if there is no basis $\hat{\mathcal{L}}$ of $\mathbb{K}$ with strictly fewer implications. Every minimal basis is clearly irredundant, but the converse is not true in general.

For a context $\mathbb{K} = (G, M, I)$ with finite G and M, one minimal basis can be given explicitly as the *canonical basis* $\mathrm{Can}(\mathbb{K}) = \{ P \rightarrow P'' \mid P$ pseudo-intent of $\mathbb{K} \}$ [11]. Recall that a *pseudo-intent* of $\mathbb{K}$ is a set $P \subseteq M$ such that $P \neq P''$ and each pseudo-intent $Q \subsetneq P$ satisfies $Q'' \subseteq P$.

3.2   Probably Approximately Correct Implication Bases

Exact implication bases provide a convenient way to represent the theory of a formal context in a compact way. However, the computation of such bases is difficult, and currently known algorithms impose an enormous additional overhead on the already high running times. On the other hand, real-world data sets are usually *noisy*, i.e., contain errors and inaccuracies, and computing their exact bases is futile from the very beginning.

Therefore, we consider *approximately correct implication bases* of $\mathbb{K}$, hoping that such approximations still capture essential parts of the theory of $\mathbb{K}$, while being easier to compute. Clearly, the precise notion of approximation determines the usefulness of this approach. In this work, we want to take the stance that a set $\mathcal{H}$ of implications is an *approximately correct basis* of $\mathbb{K}$ if the closed sets of $\mathcal{H}$ are "most often" closed in $\mathbb{K}$ and vice versa. This is formalized in the following definition.

Definition 1. *Let M be a finite set and let $\mathbb{K} = (G, M, I)$ be a formal context. A set $\mathcal{H} \subseteq \mathrm{Imp}(M)$ is called an* approximately correct basis *of $\mathbb{K}$ with inaccuracy $\varepsilon > 0$ if*

$$\mathrm{dist}(\mathcal{H}, \mathbb{K}) := \frac{|\mathrm{Mod}(\mathcal{H}) \bigtriangleup \mathrm{Int}(\mathbb{K})|}{2^{|M|}} < \varepsilon.$$

We call $\mathrm{dist}(\mathcal{H}, \mathbb{K})$ the Horn-distance *between $\mathcal{H}$ and $\mathbb{K}$.*

The notion of Horn-distance can easily be extended to sets of implications: the *Horn-distance* between $\mathcal{L} \subseteq \mathrm{Imp}(M)$ and $\mathcal{H} \subseteq \mathrm{Imp}(M)$ is defined as in the definition above, replacing $\mathrm{Int}(\mathbb{K})$ by $\mathrm{Mod}(\mathcal{L})$. Note that with this definition, $\mathrm{dist}(\mathcal{L}, \mathbb{K}) = \mathrm{dist}(\mathcal{L}, \mathcal{H})$ for every exact implication basis $\mathcal{H}$ of $\mathbb{K}$. On the other hand, every set $\mathcal{L}$ can be represented as a basis of a formal context $\mathbb{K}$, and, in this case, $\mathrm{dist}(\mathcal{H}, \mathcal{L}) = \mathrm{dist}(\mathcal{H}, \mathbb{K})$ for all $\mathcal{H} \subseteq \mathrm{Imp}(M)$.

For practical purposes, it may be enough to be able to compute approximately correct bases with high probability. This eases algorithmic treatment from a theoretical perspective, in the sense that it is possible to find algorithms that run in polynomial time.

Definition 2. *Let M be a finite set and let $\mathbb{K} = (G, M, I)$ be a formal context. Let $\Omega = (W, \mathcal{E}, \mathrm{Pr})$ be a probability space. A random variable $\mathcal{H} : \Omega \to \mathfrak{P}(\mathrm{Imp}(M))$ is called a* probably approximately correct basis *(PAC basis) of $\mathbb{K}$ with* inaccuracy $\varepsilon > 0$ *and* uncertainty $\delta > 0$ *if $\mathrm{Pr}(\mathrm{dist}(\mathcal{H}, \mathbb{K}) > \varepsilon) < \delta$.*

3.3 How to Compute Probably Approximately Correct Bases

We shall make use of query learning to compute PAC bases. The principal goal of query learning is to find explicit representation of *concepts* under the restriction of only having access to certain kinds of *oracles*. The particular case we are interested in is to learn conjunctive normal forms of Horn formulas from *membership* and *equivalence* oracles. Since conjunctive normal forms of Horn formulas correspond to sets of unit implications, this use-case allows learning sets of implications from oracles. Indeed, the restriction to unit implications can be dropped, as we shall see shortly.

Let $\mathcal{L} \subseteq \mathrm{Imp}(M)$ be a set of implications. A *membership oracle* for $\mathcal{L}$ is a function $f : \mathfrak{P}(M) \to \{\top, \bot\}$ such that $f(X) = \top$ for $X \subseteq M$ if and only if X is a model of $\mathcal{L}$. An *equivalence oracle* for $\mathcal{L}$ is a function $g : \mathfrak{P}(\mathrm{Imp}(M)) \to \{\top\} \cup \mathfrak{P}(M)$ such that $g(\mathcal{H}) = \top$ if and only if $\mathcal{H}$ is equivalent to $\mathcal{L}$, i.e., $\mathrm{Mod}(\mathcal{H}) = \mathrm{Mod}(\mathcal{L})$. Otherwise, $X := g(\mathcal{H})$ is a *counterexample* for the equivalence of $\mathcal{H}$ and $\mathcal{L}$, i.e., $X \in \mathrm{Mod}(\mathcal{H}) \triangle \mathrm{Mod}(\mathcal{L})$. We shall call X a *positive counterexample* if $X \in \mathrm{Mod}(\mathcal{L}) \backslash \mathrm{Mod}(\mathcal{H})$, and a *negative counterexample* if $X \in \mathrm{Mod}(\mathcal{H}) \backslash \mathrm{Mod}(\mathcal{L})$.

To learn sets of implications through membership and equivalence oracles, we shall use the well-known HORN1 algorithm [4]. Pseudocode describing this algorithm is given in Fig. 1, where we have adapted the algorithm to use FCA terminology.

The principal way the HORN1 algorithm works is the following: keeping a *working hypothesis* $\mathcal{H}$, the algorithm repeatedly queries the equivalence oracle about whether $\mathcal{H}$ is equivalent to the sought basis $\mathcal{L}$. If this is the case, the algorithm stops. Otherwise, it receives a counterexample C from the oracle, and depending on whether C is a positive or a negative counterexample, it adapts the hypothesis accordingly. In the case C is a positive counterexample, all implications in $\mathcal{H}$ not respecting C are modified by removing attributes not in C from their conclusions. Otherwise, C is a negative counterexample, and $\mathcal{H}$ must be adapted so that C is not a model of $\mathcal{H}$ anymore. This is done by

```
define horn1(M,member?,equivalent?)
   H := ∅
   while C := equivalent?(H) is a counterexample do
      if some A→B∈H does not respect C then
         replace all implications A→B∈H
            not respecting C by A→B∩C
      else
         find first A→B∈H such that
            C∩A≠A and member?(C∩A) returns false
         if A→B exists then
            replace A→B by C∩A→B
         else
            add C→M to H
         end
      end
   end
   return H
end
```

Fig. 1. HORN1, adapted to FCA terminology

searching for an implication $(A \to B) \in \mathcal{H}$ such that $C \cap A \neq A$ is not a model of $\mathcal{L}$, employing the membership query. If such an implication is found, it is replaced by $C \cap A \to B \cup (A \setminus C)$. Otherwise, the implication $C \to M$ is simply added to $\mathcal{H}$.

With this algorithm, it is possible to learn implicational theories from equivalence and membership oracles alone. Indeed, the resulting set $\mathcal{H}$ of implications is always the canonical basis equivalent to $\mathcal{L}$ [5]. Moreover, the algorithm always runs in polynomial time in $|M|$ and the size of the sought implication basis [4, Theorem 2].

We now want to describe an adaption of the HORN1 algorithm that allows to compute PAC bases in polynomial time in size of M, the output $\mathcal{L}$, as well as $1/\varepsilon$ and $1/\delta$. For this we modify the original algorithm of Fig. 1 as follows: given a set $\mathcal{H}$ of implications, instead of checking exactly whether $\mathcal{H}$ is equivalent to the sought implicational theory $\mathcal{L}$, we employ the strategy of *sampling* [3] to simulate the equivalence oracle. More precisely, we sample for a certain number of iterations subsets X of M and check whether X is a model of $\mathcal{H}$ and not of $\mathcal{L}$ or vice versa. In other words, we ask whether X is an element of $\mathrm{Mod}(\mathcal{H}) \triangle \mathrm{Mod}(\mathcal{L})$. Intuitively, given enough iterations, the sampling version of the equivalence oracle should be close to the actual equivalence oracle, and the modified algorithm should return a basis that is close to the sought one.

Pseudocode implementing the previous elaboration is given in Fig. 2, and it requires some further explanation. The algorithm computing a PAC basis of an implication theory given by access to a membership oracle is called *pac-basis*. This function is implemented in terms of *horn1*, which, as explained before, receives as equivalence oracle a sampling algorithm that uses the membership oracle to decide whether a randomly sampled subset is a counterexample. This

```
define  approx-equivalent? (member? , ε, δ)
    i := 0 ;; number of equivalence queries

    return function (H) begin
            i := i + 1
        for ℓᵢ times do
            choose X ⊆ M
            if (member? (X) and X ∉ Mod(H)) or
                (not member? (X) and X ∈ Mod(H)) then
                return X
            end
        end
        return true
    end
end


define  pac-basis (M, member? , ε, δ)
    return  horn1 (M, member? , approx-equivalent? (member? , ε, δ))
end
```

Fig. 2. Computing PAC bases

sampling equivalence oracle is returned by *approx-equivalent?*, and manages an internal counter i keeping track of the number of invocations of the returned equivalence oracle. Every time this oracle is called, the counter is incremented and thus influences the number ℓ_i of samples the oracle draws.

Theorem 3 shows a way to chose ℓ_i so that *pac-basis* computes a PAC basis.

Theorem 3. *Let $0 < \varepsilon \le 1$ and $0 < \delta \le 1$. Set*

$$\ell_i := \left\lceil \frac{1}{\varepsilon} \cdot \left(i + \log_2 \frac{1}{\delta} \right) \right\rceil.$$

Denote with $\mathcal{H}$ the random variable representing the outcome of the call to pac-basis with arguments M and the membership oracle for $\mathcal{L}$. Then $\mathcal{H}$ is a PAC basis for $\mathcal{L}$, i.e., $\Pr\left(\mathrm{dist}(\mathcal{H}, \mathcal{L}) > \varepsilon\right) < \delta$, where $\Pr$ denotes the probability distribution over all possible runs of pac-basis with the given arguments. Moreover, pac-basis finishes in time polynomial in $|M|$, $|\mathcal{L}|$, $1/\varepsilon$, and $1/\delta$.

Proof. We know that the runtime of *pac-basis* is bounded by a polynomial in the given parameters, provided we count the invocations of the oracles as single steps. Moreover, the numbers ℓ_i are polynomial in $|M|$, $|\mathcal{L}|$, $1/\varepsilon$, and $1/\delta$ (since i is polynomial in $|M|$ and $|\mathcal{L}|$), and thus *pac-basis* always runs in polynomial time.

The algorithm *horn1* requires a number of counterexamples polynomial in $|M|$ and $|\mathcal{L}|$. Suppose that this number is at most k. We want to ensure that in ith call to the sampling equivalence oracle, the probability δ_i of failing to find a counterexample (if one exists) is at most $\delta/2^i$. Then the probability of failing to find a counterexample in any of at most k calls to the sampling equivalence oracle is at most

$$\frac{\delta}{2} + \left(1 - \frac{\delta}{2}\right) \cdot \left(\frac{\delta}{4} + \left(1 - \frac{\delta}{4}\right)\left(\frac{\delta}{8} + \left(1 - \frac{\delta}{8}\right) \cdot (\cdots)\right)\right) \leq \frac{\delta}{2} + \frac{\delta}{4} + \cdots + \frac{\delta}{2^k} < \delta.$$

Assume that in some step i of the algorithm, the currently computed hypothesis $\hat{\mathcal{H}}$ satisfies

$$\operatorname{dist}(\hat{\mathcal{H}}, \mathcal{L}) = \frac{|\operatorname{Mod} \hat{\mathcal{H}} \triangle \operatorname{Mod} \mathcal{L}|}{2^{|M|}} > \varepsilon. \tag{1}$$

Then choosing $X \in \operatorname{Mod} \hat{\mathcal{H}} \triangle \operatorname{Mod} \mathcal{L}$ succeeds with probability at least ε, and the probability of failing to find a counterexample in ℓ_i iterations is at most $(1 - \varepsilon)^{\ell_i}$. We want to choose ℓ_i such that $(1 - \varepsilon)^{\ell_i} < \delta_i$. We obtain

$$\log_{1-\varepsilon} \delta_i = \frac{\log_2 \delta_i}{\log_2(1 - \varepsilon)} = \frac{\log_2(1/\delta_i)}{-\log_2(1 - \varepsilon)} \leq \frac{\log_2(1/\delta_i)}{\varepsilon},$$

because $-\log_2(1 - \varepsilon) > \varepsilon$. Thus, choosing any ℓ_i satisfying $\ell_i > \frac{1}{\varepsilon} \log_2 \frac{1}{\delta_i}$ is sufficient for our algorithm to be approximately correct. In particular, we can set

$$\ell_i := \left\lceil \frac{1}{\varepsilon} \log_2 \frac{1}{\delta_i} \right\rceil = \left\lceil \frac{1}{\varepsilon} \log_2 \frac{2^i}{\delta} \right\rceil = \left\lceil \frac{1}{\varepsilon} \left(i + \log_2 \frac{1}{\delta}\right) \right\rceil,$$

as claimed. This finishes the proof.

The preceding argumentation relies on the fact that we choose subsets $X \subseteq M$ uniformly at random. However, it is conceivable that, for certain applications, computing PAC bases for uniformly sampled subsets $X \subseteq M$ might be too much of a restriction, in particular, when certain combinations of attributes are more likely than others. In this case, PAC bases are sought with respect to some *arbitrary distribution* of $X \subseteq M$.

It turns out that such a generalization of Theorem 3 can easily be obtained. For this, we observe that the only place where uniform sampling is needed is in Eq. (1) and the subsequent argument that choosing a counterexample $X \in \operatorname{Mod}(\hat{\mathcal{H}}) \triangle \operatorname{Mod}(\mathcal{L})$ succeeds with probability at least ε.

To generalize this to an arbitrary distribution, let X be a random variable with values in $\mathfrak{P}(M)$, and denote the corresponding probability distribution with Pr_1. Then Eq. (1) can be generalized to

$$\operatorname{Pr}_1(X \in \operatorname{Mod}(\hat{\mathcal{H}}) \triangle \operatorname{Mod}(\mathcal{L})) > \varepsilon.$$

Under this condition, choosing a counterexample in $\operatorname{Mod}(\hat{\mathcal{H}}) \triangle \operatorname{Mod}(\mathcal{L})$ still succeeds with probability at least ε, and the rest of the proof goes through. More precisely, we obtain the following result.

Theorem 4. *Let M be a finite set, $\mathcal{L} \subseteq \operatorname{Imp}(M)$. Denote with X a random variable taking subsets of M as values, and let Pr_1 be the corresponding probability distribution. Further denote with $\mathcal{H}$ the random variable representing the* *results of pac-basis when called with arguments M, $\varepsilon > 0$, $\delta > 0$, a membership oracle for M, and where the sampling equivalence oracle uses the random variable X to draw counterexamples. If Pr_2 denotes the corresponding probability distribution for $\mathcal{H}$, then*

$$\mathrm{Pr}_2\Big(\mathrm{Pr}_1\big(X \in \mathrm{Mod}(\mathcal{H}) \,\triangle\, \mathrm{Mod}(\mathcal{L})\big) > \varepsilon\Big) < \delta.$$

Moreover, the runtime of pac-basis is bounded by a polynomial in the sizes of M, $\mathcal{L}$ and the values $1/\varepsilon$, $1/\delta$.

4   Usability

We have seen that PAC bases can be computed fast, but the question remains whether they are a useful representation of the implicational knowledge embedded in a given data set. To approach this question, we now want to provide a first assessment of the usability in terms of quality and quantity of the approximated implications. To this end, we conduct several experiments on artificial and real-world data sets. In Sect. 4.1, we measure the approximation quality provided by PAC bases. Furthermore, in Sect. 4.2 we examine a particular context and argue that PAC bases also provide a meaningful approximation of the corresponding canonical basis.

4.1   Practical Quality of Approximation

In theory, PAC bases provide good approximation of exact bases with high probability. But how do they behave with respect to practical situations? To give first impressions for the answer to this question, we shall use formal contexts to implement membership oracles: the oracle returns $\top$ for a set X if and only if X is an intent of the used data set.

We have chosen three different collections of data sets for our experiments. First, we shall examine how the *pac-basis* algorithm performs on real-world formal contexts. For this we utilize a data set based on a public data dump of the BibSonomy platform, as described in [8]. Our second experiment is conducted on a subclass of artificial formal contexts. As it was shown in [8], it is so far unknown how to generate formal contexts uniformly at random. Hence, we use the "usual way" of creating artificial formal contexts, with all warnings in place: for a given number of attributes and density, choose randomly a valid number of objects and use a biased coin to draw the crosses. The last experiment is focused on repetition stability: we calculate PAC bases of a fixed formal context multiple times and examine the standard deviation of the results.

The comparison will utilize three different measures. For every context in consideration, we shall compute the Horn-distance between the canonical basis and the approximating bases returned by *pac-basis*. Furthermore, we shall also make use of the usual *precision* and *recall* measures, defined as follows.

Definition 5. *Let M be a finite set and let $\mathbb{K} = (G, M, I)$ be a formal context. Then the* precision *and* recall *of $\mathcal{H}$, respectively, are defined as*

$$\mathrm{prec}(\mathbb{K}, \mathcal{H}) := \frac{|\{(A \to B) \in \mathcal{H} \mid \mathrm{Can}(\mathbb{K}) \models (A \to B)\}|}{|\mathcal{H}|},$$

$$\mathrm{recall}(\mathbb{K}, \mathcal{H}) := \frac{|\{(A \to B) \in \mathrm{Can}(\mathbb{K}) \mid \mathcal{H} \models (A \to B)\}|}{|\mathrm{Can}(\mathbb{K})|}.$$

In other words, precision is measuring the fraction of valid implications in the approximating basis $\mathcal{H}$, and recall is measuring the fraction of valid implications in the canonical basis that follow semantically from the approximating basis $\mathcal{H}$. Since we compute precision and recall for multiple contexts in the experiments, we consider the *macro average* of those measures, i.e., the mean of the values of these measure on the given contexts.

BibSonomy Contexts. This data set consists of a collection of 2835 formal contexts, each having exactly 12 attributes. It was created utilizing a data-dump from the BibSonomy platform, and a detailed description of how this had been done can be found in [8]. Those contexts have a varying number of objects and their canonical bases have sizes between one and 189.

Let us first fix the inaccuracy ε and vary the uncertainty δ in order to investigate the influence of the latter. The mean value and the standard deviation over all 2835 formal contexts of the Horn-distance between the canonical basis and a PAC basis is shown in Fig. 3. A first observation is that for all chosen values of ε, an increase of δ only yields a small change of the mean value, in most cases an increase as well. The standard deviation is, in almost all cases, also increasing. The results for the macro average of precision and recall are shown in Fig. 5. Again, only a small impact on the final outcome when varying δ could be observed. We therefore omitted to show these in favor of the following plots.

Dually, let us now fix the uncertainty δ and vary the inaccuracy ε. The Horn-distances between the canonical basis and a computed PAC basis for this experiment are shown in Fig. 4. From this we can learn numerous things. First, we see that increasing ε always leads to a considerable increase in the Horn-distance, signaling that the PAC basis deviates more and more from the canonical basis.

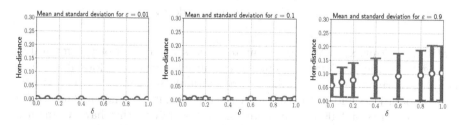

Fig. 3. Horn-distances between the contexts from the BibSonomy data set and corresponding PAC bases for fixed ε and varying δ.

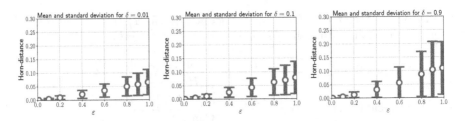

Fig. 4. Horn-distance between the contexts from the BibSonomy data and corresponding PAC bases for fixed δ and varying ε.

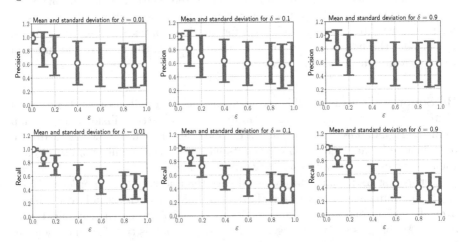

Fig. 5. Measured precision (above) and recall (below) for fixed δ and varying ε for the BibSonomy data set.

However, it is important to note that the mean values are always below ε, most times even significantly. Also, the increase for the Horn-distance while increasing ε is significantly smaller than one. That is to say, the required inaccuracy bound is never realized, and especially for larger values of ε the deviation of the computed PAC basis from the exact implicational theory is less than the algorithm would allow to. We observe a similar behavior for precision and recall. For small values of ε, both precision and recall are very high, i.e., close to one, and subsequently seem to follow an exponential decay.

Artificial Contexts. We now want to discuss the results of a computation analogous to the previous one, but with artificially generated formal contexts. For these formal contexts, the size of the attribute set is fixed at ten, and the number of objects and the density are chosen uniformly at random. The original data set consists of 4500 formal contexts, but we omit all that have a canonical basis with fewer than ten implications, to eliminate the high impact a single false implication in bases of small cardinality would have.

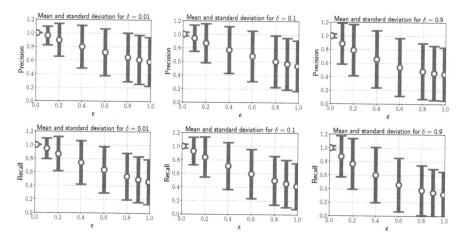

Fig. 6. Measured recall for fixed ε and varying δ (above) and fixed δ and varying ε (below) for 3939 randomly generated formal contexts with ten attributes.

A selection of the experimental results is shown in Fig. 6. We limit the presentation to precision and recall only, since the previous experiments indicate that investigating Horn-distance does not yield any new insights. For $\varepsilon = 0.01$ and $\delta - 1 = 0.01$, the precision as well as the recall is almost exactly one (0.999), with a standard deviation of almost zero (0.003). When increasing ε, the mean values deteriorate analogously to the previous experiment, but the standard deviation increases significantly more.

Stability. In our final experiment, we want to consider the impact of the randomness of the *pac-basis* algorithm when computing bases of fixed formal contexts. To this end, we shall consider particular formal contexts $\mathbb{K}$ and repeatedly compute probably approximately correct implication bases of $\mathbb{K}$. For these bases, we again compute recall and precision as we did in the previous experiments.

We shall consider three different artificial formal contexts with eight, nine, and ten attributes, and canonical bases of size 31, 40, and 70, respectively. In Fig. 7, we show the precision and recall values for these contexts when calculating PAC bases 100 times. In general, the standard-deviation of precision and recall for small values of ε are low. Increasing this parameter leads to an exponential decay of precision and recall, as expected, and the standard-deviation increases as well. We expect that both the decay of the mean value as well as the increase in standard deviation are less distinct for formal contexts with large canonical bases.

Discussion. Altogether the experiments show promising results. However, there are some peculiarities to be discussed. The impact of δ for Horn-distance in the case of the BibSonomy data set was considerably low. At this point, it is not clear whether this is due to the nature of the chosen contexts or to the fact that

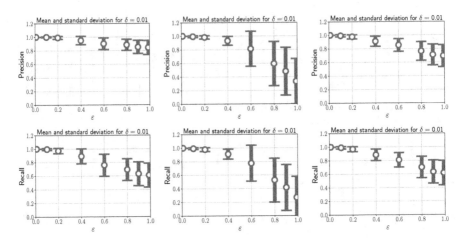

Fig. 7. For fixed δ and varying ε, measured precision (above) and recall (below) stability for 100 runs on the same formal context with eight (left), nine (middle), and ten (right) attributes.

the algorithm is less constrained by δ. The results presented in Fig. 5 show that neither precision nor recall are impacted by varying δ as well. All in all, for the formal contexts of the BibSonomy data set, the algorithm delivered solid results in terms of inaccuracy and uncertainty, in particular when considering precision and recall, see Fig. 5. Both measures indicate that the PAC bases perform astonishingly well, even for high values of ε.

For the experiment of the artificial contexts, the standard deviation increases significantly more than in the BibSonomy experiment. The source for this could not be determined in this work and needs further investigation. The overall inferior results for the artificial contexts, in comparison to the results for the BibSonomy data set, may be credited to the fact that many of the artificial contexts had a small canonical basis between 10 and 30. For those, a small amount of false or missing implications had a great impact on precision and recall. Nevertheless, the promising results for small values of ε back the usability of the PAC basis generating algorithm.

4.2 A Small Case-Study

Let us consider a classical example, namely the *Star-Alliance* context [17], consisting of the members of the Star Alliance airline alliance prior to 2002, together with the regions of the world they fly to. The formal context $\mathbb{K}_{\mathsf{SA}}$ is given in Fig. 8; it consists of 13 airlines and 9 regions, and $\mathrm{Can}(\mathbb{K}_{\mathsf{SA}})$ consists of 13 implications.

In the following, we shall investigate PAC bases of $\mathbb{K}_{\mathsf{SA}}$ and compare them to $\mathrm{Can}(\mathbb{K}_{\mathsf{SA}})$. Note that due to the probabilistic nature of this undertaking, it is hard to give certain results, as the outcomes of *pac-basis* can be different on different invocations, as seen in Sect. 4.1. It is nevertheless illuminating to see

	Latin America	Europe	Canada	Asia Pacific	Middle East	Africa	Mexico	Caribbean	United States
Air Canada	×	×	×	×	×		×	×	×
Air New Zealand		×		×					×
All Nippon Airways		×		×					×
Ansett Australia				×					
The Austrian Airlines Group		×	×	×	×	×			×
British Midlands		×							
Lufthansa	×	×	×	×	×	×	×		×
Mexicana	×	×				×	×	×	
Scandinavian Airlines	×	×		×		×			×
Singapore Airlines		×	×	×	×	×			×
Thai Airways International	×	×		×				×	×
United Airlines	×	×	×	×			×	×	×
VARIG	×	×		×			×	×	×

Fig. 8. Star-Alliance context $\mathbb{K}_{\mathsf{SA}}$

what results are possible for certain values of the parameters ε and δ. In particular, we shall see that implications returned by *pac-basis* are still meaningful, even if they are not valid in $\mathbb{K}_{\mathsf{SA}}$.

As a first case, let us consider comparably small values of inaccuracy and uncertainty, namely $\varepsilon = 0.1$ and a $\delta = 0.1$. For those values we obtained a basis $\mathcal{H}_{0.1,0.1}$ that differs from $\mathrm{Can}(\mathbb{K}_{\mathsf{SA}})$ only in the implication

Africa, Asia Pacific, Europe, United States, Canada $\to$ Middle East

being replaced by

Africa, Latin America, Asia Pacific, Mexico, Europe, United States, Canada $\to \bot$

$$(2)$$

Indeed, for the second implication to be refuted by the algorithm, the only counterexample in $\mathbb{K}_{\mathsf{SA}}$ would have been Lufthansa, which does not fly to the Caribbean. However, in our particular run of *pac-basis* that produced $\mathcal{H}_{0.1,0.1}$, this counterexample had not been considered, resulting in the implication from Eq. (2) to remain in the final basis. Thus, while $\mathcal{H}_{0.1,0.1}$ does not coincide with $\mathrm{Can}(\mathbb{K}_{\mathsf{SA}})$, the only implication in which they differ (2) still only has one counterexample in $\mathbb{K}_{\mathsf{SA}}$. Therefore, the basis $\mathcal{H}_{0.1,0.1}$ can be considered as a good approximation of $\mathrm{Can}(\mathbb{K}_{\mathsf{SA}})$.

As in the previous section, it turns out that increasing the parameter δ to values larger than 0.1 does not change much of resulting basis. This is to be expected, since δ is a bound on the probability that the basis returned by *pac-basis* is not of inaccuracy ε. Indeed, even for as large a value as $\delta = 0.8$, the resulting basis we obtained in our run of *pac-basis* was exactly $\mathrm{Can}(\mathbb{K}_{\mathsf{SA}})$.

Nevertheless, care must be exercised when increasing δ, as this increases the chance that *pac-basis* returns a basis that is far off from the actual canonical basis – if not in this run, then maybe in a latter one.

Conversely to this, and in accordance to the results of the previous section, increasing ε, and thus decreasing the bound on the inaccuracy, does indeed have a notable impact on the resulting basis. For example, for $\varepsilon = 0.5$ and $\delta = 0.1$, our run of *pac-basis* returned the basis

$$(\text{Caribbean} \to \bot), \ (\text{Asia Pacific}, \text{Mexico} \to \bot), \ (\text{Asia Pacific}, \text{Europe} \to \bot),$$
$$(\text{Middle East} \to \bot), \ (\text{Latin America} \to \text{Mexico}, \text{United States}, \text{Canada}).$$

While this basis enjoys a small Horn-distance to $\text{Can}(\mathbb{K}_{\text{SA}})$ of around 0.11, it can hardly be considered usable, as it ignores a great deal of objects in $\mathbb{K}_{\text{SA}}$. Changing the uncertainty parameter δ to smaller or larger values again did not change much of the appearance of the bases.

To summarize, for our example context $\mathbb{K}_{\text{SA}}$, we have seen that low values of ε often yield bases that are very close to the canonical basis of $\mathbb{K}_{\text{SA}}$, both intuitively and in terms of Horn-distance to the canonical basis of $\mathbb{K}_{\text{SA}}$. However, the larger the values of ε get, the less useful bases returned by *pac-basis* appear to be. On the other hand, varying the value for the uncertainty parameter δ within certain reasonable bounds does not seem to influence the results of *pac-basis* very much.

5 Summary and Outlook

The goal of this work is to give first evidence that probably approximately correct implication bases are a practical substitute for their exact counterparts, possessing advantageous algorithmic properties. To this end, we have argued both quantitatively and qualitatively that PAC bases are indeed close approximations of the canonical basis of both artificially generated as well as real-world data sets. Moreover, the fact that PAC bases can be computed in output-polynomial time alleviates the usual long running times of algorithms computing implication bases, and renders the applicability on larger data sets possible.

To push forward the usability of PAC bases, more studies are necessary. Further investigating the quality of those bases on real-world data sets is only one concern. An aspect not considered in this work is the *actual* running time necessary to compute PAC bases, compared to the one for the canonical basis, say. To make such a comparison meaningful, a careful implementation of the *pac-basis* algorithm needs to be devised, taking into account aspects of algorithmic design that are beyond the scope of this work.

We also have not considered relationships between PAC bases and existing ideas for extracting implicational knowledge from data. An interesting research direction could be to investigate whether there exists any connection between implications from PAC bases and implications with *high support* and *high confidence* as used in data mining [2]. One could conjecture that there is a deeper connection between these notions, in the sense that implications in PAC bases with high support could also be likely to enjoy a high confidence in the underlying data set. It is also not too far fetched to imagine a notion of PAC bases that incorporates support and confidence right from the beginning.

The classical algorithm to compute the canonical basis of a formal context can easily be extended to the algorithm of *attribute exploration*. This algorithm, akin to query learning, aims at finding an exact representation of an implication theory that is only accessible through a *domain expert*. As the algorithm for computing the canonical basis can be extended to attribute exploration, we are certain that it is also possible to extend the *pac-basis* algorithm to a form of *probably approximately correct attribute exploration*. Such an algorithm, while not being entirely exact, would be highly sufficient for the inherently erroneous process of learning knowledge from human experts, while possibly being much faster. On top of that, existing work in query learning handling non-omniscient, erroneous, or even malicious oracles could be extended to attribute exploration so that it could deal with erroneous or malicious domain experts. In this way, attribute exploration could be made much more robust for learning tasks in the world wide web.

Acknowledgments. Daniel Borchmann gratefully acknowledges support by the Cluster of Excellence "Center for Advancing Electronics Dresden" (cfAED). Sergei Obiedkov received support within the framework of the Basic Research Program at the National Research University Higher School of Economics (HSE) and within the framework of a subsidy by the Russian Academic Excellence Project '5-100'. The computations presented in this paper were conducted by conexp-clj, a general-purpose software for formal concept analysis (https://github.com/exot/conexp-clj).

References

1. Adaricheva, K., Nation, J.B.: Discovery of the D-basis in binary tables based on hypergraph dualization. Theoret. Comput. Sci. **658**, 307–315 (2017)
2. Agrawal, R., Imielinski, T., Swami, A.N.: Mining association rules between sets of items in large databases. In: Proceedings of ACM SIGMOD International Conference on Management of Data, pp. 207–216 (1993)
3. Angluin, D.: Queries and concept learning. Mach. Learn. **2**(4), 319–342 (1988)
4. Angluin, D., Frazier, M., Pitt, L.: Learning conjunctions of Horn clauses. Mach. Learn. **9**(2–3), 147–164 (1992)
5. Arias, M., Balcázar, J.L.: Construction and learnability of canonical Horn formulas. Mach. Learn. **85**(3), 273–297 (2011)
6. Babin, M.A.: Models, methods, and programs for generating relationships from a lattice of closed sets. Ph.D. thesis. Higher School of Economics, Moscow (2012)
7. Borchmann, D.: Learning terminological knowledge with high confidence from erroneous data. Ph.D. thesis, Technische Universität Dresden, Dresden (2014)
8. Borchmann, D., Hanika, T.: Some experimental results on randomly generating formal contexts. In: Huchard, M., Kuznetsov, S. (eds.) Proceedings of 13th International Conference on Concept Lattices and Their Applications (CLA 2016), CEUR Workshop Proceedings, vol. 1624, pp. 57–69. CEUR-WS.org (2016)

9. Ganter, B.: Two basic algorithms in concept analysis. In: Kwuida, L., Sertkaya, B. (eds.) ICFCA 2010. LNCS, vol. 5986, pp. 312–340. Springer, Heidelberg (2010). doi:10.1007/978-3-642-11928-6_22

10. Ganter, B., Obiedkov, S.: Conceptual Exploration. Springer, Heidelberg (2016)

11. Guigues, J.-L., Duquenne, V.: Famille minimale d'implications informatives résultant d'un tableau de données binaires. Mathématiques et Sciences Humaines 24(95), 5–18 (1986)

12. Kautz, H., Kearns, M., Selman, B.: Horn approximations of empirical data. Artif. Intell. 74(1), 129–145 (1995)

13. Kriegel, F., Borchmann, D.: NextClosures: parallel computation of the canonical base. In: Yahia, S.B., Konecny, J. (eds.) Proceedings of 12th International Conference on Concept Lattices and Their Applications (CLA 2015), CEUR Workshop Proceedings, vol. 1466, pp. 182–192. CEUR-WS.org, Clermont-Ferrand (2015)

14. Kuznetsov, S.O.: On the intractability of computing the Duquenne-Guigues base. J. Univers. Comput. Sci. 10(8), 927–933 (2004)

15. Obiedkov, S., Duquenne, V.: Attribute-incremental construction of the canonical implication basis. Ann. Math. Artif. Intell. 49(1–4), 77–99 (2007)

16. Ryssel, U., Distel, F., Borchmann, D.: Fast algorithms for implication bases and attribute exploration using proper premises. Ann. Math. Artif. Intell. Special Issue 65, 1–29 (2013)

17. Stumme, G.: Off to new shores - conceptual knowledge discovery and processing. Int. J. Hum.-Comput. Stud. (IJHCS) 59(3), 287–325 (2003)

18. Valiant, L.G.: A theory of the learnable. Commun. ACM 27(11), 1134–1142 (1984)

FCA in a Logical Programming Setting for Visualization-Oriented Graph Compression

Lucas Bourneuf[(⊠)] and Jacques Nicolas[(⊠)]

Université de Rennes 1 and INRIA centre de Rennes, Campus de Beaulieu,
35042 Rennes cedex, France
{lucas.bourneuf,jacques.nicolas}@inria.fr

Abstract. Molecular biology produces and accumulates huge amounts of data that are generally integrated within graphs of molecules linked by various interactions. Exploring potentially interesting substructures (clusters, motifs) within such graphs requires proper abstraction and visualization methods. Most layout techniques (edge and nodes spatial organization) prove insufficient in this case. Royer et al. introduced in 2008 *Power graph* analysis, a dedicated program using classes of nodes with similar properties and classes of edges linking node classes to achieve a lossless graph compression. The contributions of this paper are twofold. First, we formulate and study this issue in the framework of Formal Concept Analysis. This leads to a generalized view of the initial problem offering new variants and solving approaches. Second, we state the FCA modeling problem in a logical setting, Answer Set programming, which provides a great flexibility for the specification of concept search spaces.

Keywords: Graph compression · Graph visualization · Bioinformatics · ASP

1 Introduction: Graph Compression for Graph Visualization

Large graphs are a common entry of many application domains including information systems, program dependency graphs, social networks, and experimental data. We are particularly interested in molecular biology that accumulates huge amount of data, generally integrated within graphs of molecule interactions. Understanding the main structures present in such graphs is a source of knowledge that goes far beyond general statistics on the graph topology and can lead to the discovery of key organization schema reflecting disease determinants, regulation mechanisms or active domains. A common way of studying them uses visualization methods that focus on smart displays organizing spatially the edges and the nodes or using virtual nodes aggregating structural information [12,18]. A more powerful approach consists in first summarizing the graph and then using this compressed representation for visualization or graph mining.

© Springer International Publishing AG 2017
K. Bertet et al. (Eds.): ICFCA 2017, LNAI 10308, pp. 89–105, 2017.
DOI: 10.1007/978-3-319-59271-8_6

1.1 Graph Compression

Graph compression looks for possible node or edge aggregations, i.e. connections between clusters of nodes instead of connections between individual nodes. Not surprisingly, it has been shown that graph readability increases with edge compression [9]. Among early works in this direction, Agarwal *et al.* [1] have shown that visibility graphs, a type of graph commonly used in computational geometry, can be represented compactly as a union of cliques and bicliques. The decomposition (partition of the edges) or the covering (multiple use of edges) of graphs into subgraphs belonging to a particular family have been the subject of many studies. Bounds on the size complexity of such coverings have been early established for the important particular case of complete bipartite subgraphs [7] but many interesting open combinatorial problems remain in this area [17]. From an algorithmic perspective, the generation of all maximal bicliques of a graph is related to and may be considered as an important subtask of the covering problem. Apart from algorithms developed in Formal Concept Analysis, applied mathematics have also worked on classes of graphs for which it is possible to find the set of all maximal bicliques in time polynomial in the size of the graph. For instance, it is possible to find linear time algorithms for the case of graphs of bounded arboricity [11] or for domino-free graphs [4]. For general graphs, the best one can hope is to get total polynomial algorithms, i.e., polynomial with respect to the size of the union (input + output). This has been proposed in [3], with an algorithm derived from the consensus method.

On the practical side, a nice visualization of compressed graphs introduces additional constraints on the choices of subgraphs that add a complexity level in the covering or decomposition problem. We have already mentioned that it is useful to allow both cliques and bicliques for more compact representations. Introduced by Royer *et al.*, the *Power graph* analysis is a clustering and visualization method [23] that starts from this requirement and has been specifically designed to show these subgraphs typically arising in bioinformatics. Indeed, they have been associated to important structures in biological networks, particularly for protein interactions [2,12,21]. Bicliques also show interactions induced by specific protein domains in case of protein-protein interactions, or for multi-target drugs in case of drug/target/disease networks [8]. Furthermore, this approach has been used as an alternative approach to compare two biological networks by measuring the compression ratio of the compressed union of the graphs [22]. Due to the genericity of the subgraph motifs, *Power graph* analysis has been used for applications in other research fields like reliable community detection in social networks [26].

1.2 Power Graph

Given a graph $G = (V, E)$, a power graph is defined as a special graph $PG = (PV, PE)$ where the nodes PV are subsets of V and the edges PE are subsets of E. A power graph must fulfill the three following conditions:

subgraph condition: Any pair of power nodes connected by a power edge represents a biclique in graph G. As a special case, a clique in G is represented by a single power node and a power edge looping on this node.

power node hierarchy condition: Any two power nodes are either disjoint, or one is included in the other. From the point of view of classification, the sets of vertices clustered in power nodes form a hierarchy.

power edge decomposition condition: Power edges form a partition of the set of edges.

An example of graph compression is shown in Fig. 1.

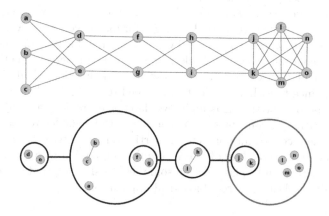

Fig. 1. A graph with 15 nodes labelled from a to o and 35 edges. A smart layout of this graph allows one to understand its underlying structure, something that becomes hard with a growing number of edges. The bottom compressed version of this graph has been produced by *Power graph* analysis and printed with Cytoscape [25], through a plug-in developed by Royer *et al.*. Power edges are shown as thick and black lines linking power nodes (thick circles, black for bicliques and green/grey for cliques). Some edges like (h, i) remain uncompressed. Concept $(\{h, i\}, \{f, g, j, k\})$, despite being a maximal biclique, can't be associated to a single power edge without breaking the power node hierarchy condition. Instead, two power edges are generated, corresponding to bicliques $(\{h, i\}, \{f, g\})$ and $(\{h, i\}, \{j, k\})$. The same way, the subgraph on the subset of vertices $\{d, e, f, g, h, i, j, k\}$ can't be covered by power edges $(\{f, g\}, \{d, e, h, i\})$ and $(\{h, i\}, \{f, g, j, k\})$ since it would break the power edge decomposition condition. (Color figure online)

The issue is to exhibit a power graph with a minimal number of power edges. It is not necessarily unique. It has been shown to be a NP-complete problem [10]. An algorithm and a software are described in [23]. It implements a two-phase approach, first processing the possible power nodes by a hierarchical clustering of the nodes using the Jaccard index on the sets of neighbours, and then building the power edges than can be drawn between any pair of power nodes following a greedy incremental approach, choosing at each step a maximal subgraph in the number of covered edges. The algorithm is very fast but remains heuristic and only computes an approximation of the minimal powergraphs. It also has been implemented with slight variations, for instance to work on directed graphs, or to enable overlapping power nodes for visualization of relations between non-disjoints sets [2], or to enable an edge to be used multiple times for a faster search for near-to-optimal compression [10].

The paper contributions are twofold. First, we formulate and study this issue in the framework of FCA. The goal is to be able to benefit from the advances in this domain to get a better view of the structure of the search space, suggesting variants and solving approaches and conversely to offer a playground for new studies in FCA. Second, we have stated the FCA modeling problem in a logical setting, Answer Set programming (ASP). The first proposition of ASP program for FCA seems due to CV Damásio and published only in [16], where the focus is on the developing expressive query languages for formal contexts. A more recent study extends the search for n-adic FCA [24] and is focused on the issue of filtering large concept spaces by checking additional membership queries. In our case, the particular task of graph compression needs heavier calculations and a specific code since the whole concept space need to be explored.

Our goal is to show that this high level language provides a great flexibility in the specification of concept search spaces and thus enables to easily explore the effect of new constraints or new properties on the lattice of concepts. We have implemented a modeling of *Power graph* compression as a formal concept search. We illustrate in the last section some results showing that the developed tool is already applicable to real biological applications.

2   Power Graph as a Formal Concept Search

This section first defines how the search space for *Power graph* compression can be formalized with FCA. The section ends on a proposition working on formal contexts, allowing one to split the search space of power graphs.

2.1   Motifs as Formal Concepts

Power graph compression leads to edge reduction through motif recognition: bicliques and cliques are abstracted in power nodes and power edges. A power node is a node that represents a non empty set of (power) nodes. Power edges link (power) nodes. A biclique is compressed in two power nodes, one for each set, linked by a power edge. Cliques are represented by a single power node with a reflexive power edge. Stars are particular case of bicliques where a power edge links a single node with a power node. These motifs can be ordered by edge coverage: a motif is greater than another if it covers more edges. For instance, a biclique of 2×4 nodes has an edge coverage of 8, which is larger than a clique of 4 nodes that covers 6 edges. This is the spirit of the Royer *et al.* heuristic.

A *graph context* is a formal context where objects and attributes are nodes in the graph and the binary relation represents edges. For a bipartite graph, the relation may be oriented and objects and attributes are distinct. In the general case, objects and attributes represent the same elements and the graph context is symmetric, comparable to an adjacency matrix. We use lower case for objects and upper case for attributes to distinguish them (see Table 1).

Building a power graph consists in iteratively looking for the largest motifs in the graph (cliques or bicliques) until all edges are covered. Largest bicliques correspond to formal concepts in the graph context where objects and attributes are disjoint (e.g. $\{a, b\} \times \{g, h, i\}$ in Table 1). Largest cliques correspond to formal concepts where objects and attributes are equal. The other cases where objects and attributes partially overlap are not allowed.

Yet the motifs compressed by *Power graph* compression are not always formal concepts themselves, because, in order to respect the power node hierarchy and power edge decomposition conditions, maximal (bi)cliques are not always fully compressed, as indicated in the introduction and illustrated in Fig. 1.

Let (A, B) and (C, D) be two formal concepts, the power node hierarchy condition is not fulfilled if two sets are partially overlapping, that is $X \cap Y \notin \{\emptyset, X, Y\}$ for $X \in \{A, B\}$ and $Y \in \{C, D\}$. This is illustrated in Table 1 for $X = B$ and $Y = C$, with $A = \{a, b\}$, $B = \{g, h, i\}$, $C = \{e, f, i\}$, and $D = \{c, d\}$. In such a case, one of the concept has to be covered by two power edges, using a simple set algebra: (C, D) can be covered by $D \times (C \setminus B)$ and $D \times (C \cap B)$).

Similarly, let (A, B) and (C, D) be two formal concepts, the power edge decomposition condition is not fulfilled if two pairs of sets have a common intersection. Such a situation commonly occurs when the concepts have crossed dependencies: $A \subset D$ and $C \subset B$. In Fig. 1 for instance, with $A = \{f, g\}$, $B = \{d, e, h, i\}$, $C = \{h, i\}$, $D = \{f, g, j, k\}$ the same edges $(f, h), (g, h), (f, i), (g, i)$ are represented multiple times. In such a case, one of the concept has to be restricted: (C, D) can be covered by $(C, D \setminus A)$.

In summary, power graph compression may be seen as a contextual search of concepts in the graph. The effect of the two admissibility constraints required by power graphs are different on the concepts associated to power edges: the power

Table 1. Illustration of the power node hierarchy condition on a graph context: assume the grey cells mark an already selected motif (here, biclique $\{a, b\} \times \{g, h, i\}$). Although the concept $\{c, d\} \times \{e, f, i\}$ is valid as a standalone motif, it overlaps the previous one and these two motifs cannot occur together.

	A	B	C	D	E	F	G	H	I
a							x	x	x
b							x	x	x
c				x	x				x
d				x	x				x
e			x	x					
f			x	x					
g	x	x							
h	x	x							
i	x	x	x	x					

edge decomposition condition does not increase the number of bicliques in the decomposition while the power node hierarchy condition has a more drastic effect by splitting concepts. Finally, it has to be noticed that this problem introduces an interesting covering problem in FCA: how to extract a cover of all edges by a minimum number of concepts.

2.2   Heuristics Modeling

As already discussed, the Royer *et al.* heuristics consists of an iterative search of the largest motifs to compress. From the FCA point of view, this can be modeled by an iterative search of the largest concept and the suppression of covered edges at each step from the graph context until an empty context is reached. Note that in this way, we can possibly obtain a better decomposition solution than Power graph Analysis since the choice of the best concept is evaluated at each step in our case whereas all bicliques are ranked only once during the initial step in the other case. We give in Sect. 3 the programs that implement this search.

The Dwyer *et al.* heuristics is much more complex and can reach an optimal compression. The idea is basically to first create all stars associated with each node, then iteratively merge two (power) nodes involving a maximal number of common neighbors. The approach seems to be compatible with Chein algorithm [6] but it is not clear that it fulfill all conditions of powergraphs and we have not further explored this track of research. Instead, we have designed a concept generation heuristic, the *HDF* method (for High Degree First). At each step, the HDF heuristics selects first nodes of highest degree, then the largest concept using one of these nodes. The degree is just one measure on the graph topology and could be replaced by any notion of *node degree of interest*. We have tried a variant, *K2HDF*, directly inspired from Dwyer *et al.* heuristics. In this variant, the selected nodes include nodes of highest degree and nodes with the highest number of neighbours in common with another node. HDF is quicker than the *greater-first* heuristics, to the expense of a slightly smaller edge reduction (see Sect. 4). We have tested another variant, called Fuzzy HDF (FHDF). The idea is to directly use the degree of interest for scoring. FHDF tries to maximize the total degree of interest of nodes involved in the concept. It provides an upper bound on the real edge coverage.

For small graphs, it is possible to look for a global optimum with our approach by exploring non deterministically all admissible decompositions. In practice, we have limited our study to incremental, locally optimal searches in order to manage the large graphs that occur in bioinformatics. An intermediate approach is possible. At each step, the local optimum is not necessarily unique. If two optimal concepts are compatible, they can be chosen in a same step. If they are not compatible, it introduces a choice point that can be subject to backtracking. In fact, even in this case we need to better manage the dependencies in the graph contexts in relation with the constraints to be checked in order to reduce the complexity of the search. The next proposition is a first step in this direction.

2.3 Exploitation of the Graph Context to Reduce the Search Space

If the number of nodes of a graph is reduced by a factor k, its graph context is reduced by a factor k^2 and the search space of concepts is reduced by a factor 2^{k^2}. This emphasizes the well known importance of reduction strategies as a preprocessing step of a concept search. Simple techniques include the application of standard clarification and reduction procedures and the decomposition of the graph into connected components. Concerning the FCA reduction procedures and since we are interested in largest concepts, the computation of the edge cover size requires all reduced nodes to have an associated weight counting the number of nodes they represent. Concerning the decomposition of the graph, we propose a generalization of the connected component property aiming at further splitting the graph context. This "bipartite split" property can be applied recursively, in order to work on increasingly smaller contexts, but the gain is generally weak.

Dot Operator: The dot operator on sets of objects or attributes is introduced in order to ease the bipartite split expression. It is a relaxed variant of the derivation operator of FCA where the universal quantification is replaced by an existential one. Given a set of objects X (resp. attributes Y), the set $\dot{X}$ (resp. $\dot{Y}$) is made of all attributes (resp. objects) related to *at least one* attribute in X (resp. Y):

$$\dot{X} = \{y \in Y \,|\, \exists x \in X,\ r(x,y)\} \qquad \dot{Y} = \{x \in X \,|\, \exists y \in Y,\ r(x,y)\} \quad (1)$$

As for derivation, the dot operator can be combined multiple times:

$$\ddot{X} = \{x \in X \,|\, \exists y \in \dot{X},\ r(x,y)\} \qquad \ddot{Y} = \{y \in Y \,|\, \exists x \in \dot{Y},\ r(x,y)\} \quad (2)$$

Proposition 1: Given a set of objects O and a set of attributes A, let $P = \{O_1, O_2\}$ be a partition of O and $Q = \{A_1, A_2\}$ be a partition of A. Let $LC(O, A)$ denotes a largest concept of the formal context $\mathcal{C}(O, A)$, i.e. a concept corresponding to a submatrix of largest size. Then, the following property holds:

$$LC(O, A) = \max\left(LC(O_1, A_1), LC(\dot{A}_2 \cup \ddot{O}_2, \dot{O}_2 \cup \ddot{A}_2)\right) \quad (3)$$

Moreover, this equation may be refined if no relation holds over $O_1 \times A_1$:

$$LC(O, A) = \max\left(LC(\dot{A}_2, A_2), LC(O_2, \dot{O}_2)\right) \quad (4)$$

Proof: Let $P = \{O_1, O_2\}$ be a partition of a set of objects O and $Q = \{A_1, A_2\}$ be a partition of a set of attributes A. If the largest concept $LC(O, A)$ is in the context $\mathcal{C}(O_1, A_1)$, then by definition, it will take the right value $LC(O_1, A_1)$. Else, there exists an element of the largest concept either in O_2 or in A_2. If the element is in O_2, then the attributes of $LC(O, A)$ have to be included in $\dot{O}_2$. The same way, the objects of $LC(O, A)$ have to be included in the objects sharing at least one relation with attributes of $\dot{O}_2$, that is, $\ddot{O}_2$. With a symmetric argument, if there is an element of A_2 in the largest concept, then attributes of $LC(O, A)$

Table 2. A partitioned context with no relation over $\mathcal{C}(O_1, A_1)$. The five possible positions of the largest concept $\mathcal{L}$ are shown. $\mathcal{L}$ could be in $\mathcal{C}(O_2, A_1)$ (e.g. $(\{e, f\}, \{H\})$), $\mathcal{C}(O_2, A_2)$ (e.g. $(\{f, g\}, \{M, N\})$), $\mathcal{C}(O_1, A_2)$ (e.g. $(\{a\}, \{L, M\})$), $\mathcal{C}(O_2, A_1 \cup A_2)$ (with an element in A_1 and an element in A_2, e.g. $(\{e, f, g\}, \{J, K\})$) or $\mathcal{C}(O_1 \cup O_2, A_2)$ (e.g. $(\{c, d\}, \{L, M, N\})$).

		A₁			A₂			
		H	I	J	K	L	M	N
O_1	a					x	x	
	b							
	c				x	x	x	x
O_2	d			x	x	x	x	
	e	x	x	x				
	f	x	x	x		x	x	
	g		x	x		x	x	

have to be included in $\dot{A}_2$ and objects in $\ddot{A}_2$. Altogether $LC(O, A)$ must be a concept of the context $\mathcal{C}(\dot{A}_2 \cup \ddot{O}_2, \dot{O}_2 \cup \ddot{A}_2)$.

If no relation holds over $O_1 \times A_1$, then every concept has either all its elements in O_2 or all its elements in A_2. In the first case they are in the formal context $\mathcal{C}(O_2, \dot{O}_2)$, and in the second case they are in the formal context $\mathcal{C}(\dot{A}_2, A_2)$. The largest concept is the largest of the largest concepts of the two contexts. Table 2 gives details on the way the search can be split in this case. $\qquad\square$

From the point of view of graph modeling, the fact that no relation holds over $O_1 \times A_1$ means that it is stable set of the graph (note however that O_1 and A_1 may overlap). If moreover no relation holds over $O_2 \times A_2$, it corresponds to the existence of at least two connected components in the graph. If $O_1 = A_1$, O_1 is a stable set and it can be searched by looking for cliques in the graph's complement. Moreover, we get $O_2 = A_2$, so $LC(\dot{A}_2, A_2) = LC(O_2, \dot{O}_2)$ and it is sufficient to consider the graph context $\mathcal{C}(O, \dot{O}_2)$. Since the aim is to speedup the search for concepts, the search in bounded time of a "best" clique (not necessarily maximal) has to be achieved, a feature allowed by ASP.

3   *PowerGrASP*, Graph Compression Based on FCA

We propose an implementation of *Power graph* compression[1] as a formal concept search using a non-monotonic logical formalism, ASP, to model the constraints on the power nodes and power edges introduced in Sect. 1.2.

ASP (Answer Set Programming) is a form of purely declarative programming oriented towards the resolution of combinatorial problems [20]. It has been successfully used for knowledge representation, problem solving, automated reasoning, and search and optimization. In the sequel, we rely on the input language

---

[1] Repository at http://powergrasp.bourneuf.net.

of the ASP system Potassco (*Potsdam Answer Set Solving Collection* [13] developed in Potsdam University. An ASP program consists of Prolog-like rules h:-$b_1, \ldots, b_m, not\ b_{m+1}, \ldots, not\ b_n$, where each b_i and h are literals and *not* stands for *default negation*. Mainly, each literal is a predicate whose arguments can be constant atoms or variables over a finite domain. Constants start with a lowercase letter, variables start with an uppercase letter or an underscore (don't care variables). The rule states that the head h is proved to be true (h is in an answer set) if the body of the rule is satisfied, i.e. $b_1, \ldots, b_m$ are true and one can not prove that $b_{m+1}, \ldots, b_n$ are true. Note that the result is independent on the ordering of rules or of the ordering of literals in their body. An ASP solver can compute one, several, or all the answer sets that are solutions of the encoded problem. If the body is empty, h is a fact while an empty head specifies an integrity constraint. Together with model minimality, interpreting the program rules this way provides the stable model semantics (see [15] for details). In practice, several syntactical extensions to the language are available and we will use two of them in this paper: the choice rules and optimization statements.

A *choice rule* of the form $i\{h : b_1, \ldots, b_n\}j$, where i and j are integers, states that at least i and at most j grounded h are true among those such that the body is satisfied. In the body part, $N = \{h\}$ evaluates N to the cardinal of the set of h. An optimization statement is of the form $\#operator\{K@P, X : b_1, \ldots, b_n\}$, where *operator* is the type of optimization (either maximize or minimize), K refers to integer weights whose sum as to be optimized, P is an optional priority level in case of multiple optimization (the highest is optimized first) and X is a tuple of variables such that one weight K is associated to each tuple of values of these variables. The Potassco system [14] proposes an efficient implementation of a rich yet simple declarative modeling language relying on principles that led to fast SAT solvers. ASP processing implies two steps, grounding and solving. The grounder generates a propositional program replacing variables by their possible values. The solver is in charge of producing the stable models (answer sets) of the propositional program.

The main objective of this section is to show, through its application to the stated graph compression problem, the effectiveness and flexibility of ASP for modeling various searches in the concept lattice. Of course, it remains always preferable from the point of view of efficiency to design a specific algorithm for a particular FCA problem. However, ASP systems are not "toy" environments and are useful for the design of efficient prototypes or to take into account a knowledge-rich environment while keeping a simple code. ASP has shown to be an attractive alternative to standard imperative languages that enable fast developments and is most of the time sufficient in real applications where many constraints have to be managed. In fact, Potassco proposes an integration of ASP and Python that allows the development of hybrid codes. The reminder of this section provides the relevant part of the encoding for solving variants of *Power graph* compression. All lines starting with % are comments.

The initial graph is assumed to be coded with facts $edge(X, Y)$, where X and Y are vertices (e.g. $edge(a, d)$. $edge(a, e)$.) If the graph is non oriented

```
 1 % Choice of elements in set 1 and 2 of a concept
 2 1 { concept(1,X): edge(X,_)}.
 3 1 { concept(2,Y): edge(_,Y)}.
 4 % Bipartite subgraph: The two sets are disjoint .
 5 :− concept(1,X), concept(2,X).
 6 % A node is impossible in a set if not linked to some node in the other set
 7 imp(1,X):− edge(X,_), concept(2,Y), not edge(X,Y), X!=Y.
 8 imp(2,Y):− edge(_,Y), concept(1,X), not edge(X,Y), X!=Y.
 9 % Consistency ; no impossible element can be added to the concept
10 :− imp(T,X), concept(T,X).
11 % Maximality ; all possible elements have to be added to the concept
12 :− not imp(1,X), not concept(1,X), edge(X,_).
13 :− not imp(2,Y), not concept(2,Y), edge(_,Y).
```

Listing 1.1. Search of concepts: maximal bicliques

and not bipartite, edges are made symmetric by adding a clause $edge(X, Y)$ $:-edge(Y, X)$. Many input graph examples are available in the *PowerGrASP* source code[2].

3.1 Looking for a Formal Concept and for Graph Motifs

Listing 1.1 provides a code specifying a formal concept. They are represented by a predicate of arity 2: $concept(S, X)$ holds if the node X is in the set number S. For instance, the concept $(\{a\}, \{b, c\})$ is represented by $concept(1, a)$, $concept(2, b)$ and $concept(2, c)$. The elements of a concept (at least 1 in each set) are chosen among those linked by an edge (lines 2–3). The set of admissible concepts is restricted by three constrains: the necessity to have two disjoint sets, the necessity to have a biclique between these sets (line 10) and the necessity to have a maximal biclique (lines 12–13). The two last constraints make use of a common symmetric predicate $imp()$ that lists elements that cannot be in a set if another element is chosen in the other set (lines 7–8).

Powergraphs contain two types of motifs, bicliques and cliques. The first one is searched by looking for all solutions of the previous code. Cliques can be considered as a simplification where the two sets are equal and deserves a special code (Listing 1.2). A clique must contain at least 3 elements. The loops on each elements are not required since in most applications they are not present.

The code that ensures generated concepts do not break the power node hierarchy is not presented here for the sake of brevity (see the *PowerGrASP* source code[3]). It mainly consists in the maintenance of an inclusion tree of power nodes. The power edge decomposition condition is met by design: once a concept is chosen, its edges are removed from the compressed graph and cannot therefore be involved in the remaining concepts.

---

[2] `powergrasp/tests/`.
[3] `powergrasp/ASPSources/postprocessing.lp`.

```
1 % Choice of elements in the clique
2 3 { clique(X): edge(X,_)}.
3 % A node is impossible in a set if not linked to some node in the other set
4 imp(X):− edge(X,_), clique(Y), not edge(X,Y), X!=Y.
5 imp(Y):− edge(_,Y), clique(X), not edge(X,Y), X!=Y.
6 % Consistency ; no impossible element can be added to the clique
7 :− imp(X), clique(X).
8 % Maximality ; all possible elements have to be added to the clique
9 :− not imp(X), not clique(X), edge(X,_).
10 :− not imp(Y), not clique(Y), edge(_,Y).
```

Listing 1.2. Search of concepts: maximal cliques

```
1 % The concept score is its edge cover.
2 score(N1*N2):− N1={concept(1,X)}, N2={concept(2,X)}.
3 % Exclusion of concepts with unbounded score.
4 :− score(S) ; S>upperbound.
5 :− score(S) ; S<lowerbound.
6 % Maximize the score.
7 #maximize{S@1,S:score(S)}.
```

Listing 1.3. Concept ordering

3.2 Implementing Concept Scoring

Listing 1.3 implements the scoring of concepts. It simply maximizes the edge cover of found concepts (line 7), which is computed as the product of the number of nodes in each set (line 2) and contained between two constant parameters of the program, *lowerbound* and *upperbound*. By default the lowerbound equals 2 (edge coverage of a star involving three nodes) and the upperbound is the minimum between the size of the previous chosen concept and the number of remaining edges in the connected component.

Note that in the real code, maximization is replaced by a minimization of uppperbound - score. This leads to a significant speedup because the solver converts maximization to minimization and uses higher upperbound values. Likewise, the code in Listing 1.3 is able to handle directed graph but a dedicated slightly more efficient version exists for this type of graph[4].

3.3 Implementing the Stable Search and Heuristics

Stables are used to split the search space as exposed in *Proposition 1* (see Sect. 2.3). They are searched in the complement context, i.e. with 0 and 1 inverted, using the code defined in Listing 1.1 without the bipartite constraint (line 5).

---

[4] powergrasp/ASPSources/findbestorientedbiclique.lp.

```
1 % At least one node of maximal priority is involved in the concept.
2 interest_ok :- concept(_,X) ; max_interest(X).
3 :- not interest_ok.
```

Listing 1.4. Addon to the (bi)clique search program: HDF and K2HDF heuristics.

```
1 % Maximize the concept priority as a sum of node degrees of set 1.
2 #maximize{P@2,X: concept(_,X), interest_level(X,P)}.
```

Listing 1.5. Addon to the (bi)clique search program: FHDF heuristics.

The implementation of heuristics takes only a few ASP lines. The code of HDF is provided in Listing 1.4 (*PowerGrASP*[5]) predicate *interest_ok* states that one of the concept nodes must be chosen among top interesting nodes. It uses *max_interest*, a predicate true only for nodes with highest interest in the current graph. In the *PowerGrASP* source, the computation of this atom is performed on the Python side. The code of FHDF is provided in Listing 1.5 (*PowerGrASP*[6]). It needs each node to get a degree of interest computed in Python and coded in predicate *interest_level*. Line 2 maximizes the sum of node degrees belonging to the concept. Note that the priority of the interest maximization is greater than the priority of the score maximization, so that the heuristics is used as a first selection criterion. For the K2HDF heuristics, the code remains the same than in Listing 1.4 but the Python part is relaxing the constraint by generating *max_interest* for the nodes of maximal degree plus the nodes sharing a maximal number of common neighbors with another node.

4   Biological Benchmarks

The powergraph compression code is benchmarked here using three networks coming from biological data. Our aim is just to show that our ASP code is already useful in managing real graphs. A complete evaluation on a larger benchmark is out of the scope of this paper, which mainly tries to take the first steps in FCA applied to the powergraph issue.

The network named **rna** comes from a study on the pea aphid (*A. pisum*) [27]. It is a bipartite interaction graph linking two disjoint populations of molecules (15 × 1810). The second network, **mdb**, comes from the database MatrixDB [19] describing interactions between extracellular proteins. It contains 5 connected components, the largest one involving 273 nodes and 642 edges. The third graph, **sbind** comes from biological data of Royer *et al.*, and contains one large connected component (205 nodes, 335 edges) and 188 with less than 20 nodes. These three graphs are described in Table 3, and the compression results in

---

[5] powergrasp/ASPSources/by_priority.lp.
[6] powergrasp/ASPSources/by_fuzzy_priority.lp.

Table 4. A graphical representation of the rna network is shown in Fig. 2. All benchmarks have been run on one core of an *Intel i7-6600U* (2.60 GHz).

Bipartite Stable Search. As shown in Sect. 2.3, the bipartite stable is the object needed by *Proposition* 1 in order to restrict the search space. It is computed with a time limit of 5 s; its optimality is thus not guaranteed, but not required either. Results show that stables of consequent size are found (see Table 3). In case of the rna bipartite network, 9 out of 16 nodes of the first set and one third of the second are involved in the stable. This result suggests that finding a large stable is an interesting application of *Proposition* 1.

Table 3. Network statistics. #*node*: number of nodes in the graph; #*edge*: number of edges; *density*: global density of the graph; #*cc*: number of connected components in the graph; *stable*: size of the stable found in the largest connected component.

	#node	#edge	Density	#cc	Bipartite	Stable
rna	1825	2250	0.0013	1	Yes	9×765
mdb	286	652	0.016	5	No	171×213
sbind	864	1241	0.003	189	No	102×122

Table 4. Network compression results. *Compression time*: time needed to fully compress the graph (iterative search of all concepts); *edge reduction*: given by $\frac{\#initial\ edges - \#poweredges}{\#initial\ edges}$; #*powernode*: number of power nodes in the compressed graph; #*poweredge*: number of power edges; *constraint ratio*: the ratio between constraint numbers for the first biclique search, and edge cardinality; *conflict ratio*: the ratio between conflict numbers for the first biclique search, and edge cardinality.

	Regular			HDF			FHDF			K2HDF		
	rna	mdb	sbind	rna	mdb	sbind	rna	mdb	sbind	rna	mdb	sbind
Time (s)	380	80	50	177	25	25	5460	150	40	320	40	25
Edge reduction	96%	64%	61%	94%	47%	50%	77%	50%	53%	95%	54%	52%
#powernode	48	83	206	55	41	231	409	103	244	64	76	233
#poweredge	49	96	182	55	45	200	430	144	229	64	66	193
#remain	50	133	301	81	300	414	91	182	356	57	226	406
Constraint ratio	1293.3	124	49.2	32.0	124	37.5	1293.3	124	37.5	1293.3	124	37.5
Conflict ratio	1.49	2.32	0.2	0.12	2.18	0.11	0.13	1.1	0.27	0.1	2.81	0.16

Comparison with the Royer et al. Implementation Oog: As shown in Table 5, *PowerGrASP* reaches a score equivalent to *Oog*. In fact, since both implement the same strategy, edge reduction is very similar up to slight variations due to non-determinism of motif choices with equal scores. *Oog* uses a dedicated algorithm, making it much more scalable. Our goal at this step was to produce a much more flexible framework however so that it could lead to qualitatively better results with different strategies and the inclusion of background knowledge.

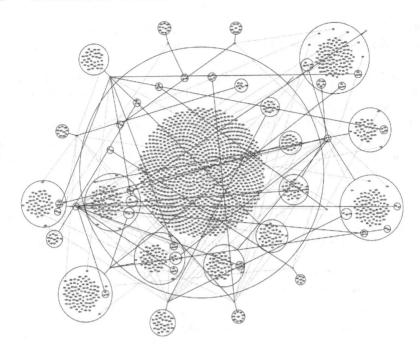

Fig. 2. Cytoscape view of a bipartite RNA network compressed with *PowerGrASP* where two types of molecules interact. The 15 nodes with a triangle shape belongs to the so-called *microRNA* type and appear around the central power node that contains most of the molecules of the *mRNA* type.

Table 5. Comparison of *PowerGrASP* and Royer *et al.* implementation.

	Time (s)		Edge reduction	
	Royer *et al.*	*PowerGrASP*	Royer *et al.*	*PowerGrASP*
rna	4	380	96%	96%
mdb	2	80	64%	64%
sbind	1	50	61%	61%

Benchmarks of the Four Methods. Benchmarks show that *regular* compression (Royer *et al.* heuristic) yields best edge reduction in all cases, but is generally slower. The other heuristics are faster, but reach a smaller edge reduction. *HDF* is the fastest, which is expected because of (1) the easy computation of a node degree of interest and (2) the important restriction of search space it induces, as shown by the number of constraints and conflicts generated (the number of constraints is related to the size of the problem to be solved, conflicts express the practical combinatorial complexity of the search). *HDF* yields the smaller edge reduction. *FHDF* and *K2HDF* bring slight improvements of edge reduction over *HDF*, at the cost of more compression time. The time needed to compress rna graph with *FHDF* is explained by the computation cost that increases with the number of nodes (1825 for **rna**). *K2HDF* reaches a slightly higher edge reduction mostly because the constraint is relaxed, allowing the exploration of a slightly larger subset of concepts.

5   Discussion and Conclusion

The main goal of this paper was to introduce the *Power graph* compression issue, a knowledge-discovery oriented kind of graph compression, in terms of a decomposition problem in the formal concept analysis framework. The Galois concept lattice is the fundamental structure underlying the search space of this problem. If Power Graph compression is reducible to a choice (and an ordering) of concepts, one can thus use FCA to formalize (1) algorithms of motifs enumeration (HDF for instance), (2) reduction methods of the search space (prop. 1 for instance), and (3) other approaches using FCA extensions like TCA and pattern structure to obtains different visualizations.

For real applications, it is reasonable to look for an approximation of optimal decompositions. We have proposed both concept generation heuristics and a splitting strategy to reduce this search space. From the point of view of implementation, we have shown that a high level purely declarative approach using Answer Set Programming allows a compact and flexible encoding of combinatorial searches in the space of concepts. Moreover, the resulting program is efficient enough to handle formal contexts of reasonable size. Large biological networks can be compressed in few minutes as shown in Sect. 4, allowing a hierarchical clustering of many biological entities based on network topology.

Optimal compression of graphs including thousands of nodes is the ultimate goal of this research, a goal that is not reached by the Royer *et al.* heuristic, nor by the variants described in Sect. 3. Our current implementation does not works efficiently on large dense graphs. This is partly due to ASP memory consumption since grounding needs all data to be encoded and loaded in memory in order to explore the search space. Thus reduction strategies are of uttermost importance to split the search space.

A number of improvements can be studied in future work along two main research tracks. From a theoretical perspective, the links between an optimal covering and an optimal decomposition in bicliques and cliques should be considered. It could be interesting to start from a solution for the covering problem, allowing some edges to be covered several times and then refine this solution to comply with all the constraints. This approach would be more compatible with the FCA framework and a search for a global optimum. Moreover, some overlapping configurations may be easily understandable in terms of application and it may be interesting to relax some of the powergraph conditions. From an application perspective, one could search for quasi-motifs, that is, motifs with a few missing edges, which are known to groups proteins of the same family in proteomes [5], and improve the resistance against noisy an incomplete data. This method was used to treat various real-life graphs to reveals important patterns [2]. Finally, knowledge discovery requires to integrate domain specific knowledge like annotations, and metadata on the quality or significance of data.

Acknowledgments. We wish to thank D. Tagu (INRA Le Rheu) and N. Théret (Inserm) for providing us the networks used in the results section. We would also like to express our gratitude to the reviewers for their feedbacks.

References

1. Agarwal, P.K., Alon, N., Aronov, B., Suri, S.: Can visibility graphs be represented compactly? Discret. Comput. Geom. **12**(3), 347–365 (1994)
2. Ahnert, S.E.: Generalised power graph compression reveals dominant relationship patterns in complex networks. Sci. Rep. **4**, Article no. 4385 (2014). https://www.nature.com/articles/srep04385
3. Alexe, G., Alexe, S., Crama, Y., Foldes, S., Hammer, P.L., Simeone, B.: Consensus algorithms for the generation of all maximal bicliques. Discret. Appl. Math. **145**(1), 11–21 (2004). Graph Optimization IV
4. Amilhastre, J., Vilarem, M., Janssen, P.: Complexity of minimum biclique cover and minimum biclique decomposition for bipartite domino-free graphs. Discret. Appl. Math. **86**(2–3), 125–144 (1998)
5. Bu, D., Zhao, Y., Cai, L., Xue, H., Zhu, X., Lu, H., Zhang, J., Sun, S., Ling, L., Zhang, N., Li, G., Chen, R.: Topological structure analysis of the protein–protein interaction network in budding yeast. Nucleic Acids Res. **31**(9), 2443–2450 (2003)
6. Chein, M.: Algorithme de recherche des sous-matrices premiÈres d'une matrice. Bulletin mathématique de la Société des Sciences Mathématiques de la République Socialiste de Roumanie **13**(61)(1), 21–25 (1969)
7. Chung, F.: On the coverings of graphs. Discret. Math. **30**(2), 89–93 (1980)
8. Daminelli, S., Haupt, V.J., Reimann, M., Schroeder, M.: Drug repositioning through incomplete bi-cliques in an integrated drug–target–disease network. Integr. Biol. **4**(7), 778–788 (2012)
9. Dwyer, T., Henry Riche, N., Marriott, K., Mears, C.: Edge compression techniques for visualization of dense directed graphs. IEEE Trans. Vis. Comput. Graph. **19**(12), 2596–2605 (2013)
10. Dwyer, T., Mears, C., Morgan, K., Niven, T., Marriott, K., Wallace, M.: Improved optimal and approximate power graph compression for clearer visualisation of dense graphs. CoRR, abs/1311.6996 (2013)
11. Eppstein, D.: Arboricity and bipartite subgraph listing algorithms. Inf. Process. Lett. **51**(4), 207–211 (1994)
12. Gagneur, J., Krause, R., Bouwmeester, T., Casari, G.: Modular decomposition of protein-protein interaction networks. Genome Biol. **5**(8), R57 (2004)
13. Gebser, M., Kaminski, R., Kaufmann, B., Lindauer, M., Ostrowski, M., Romero, J., Schaub, T., Thiele, S.: Potassco User Guide (2015)
14. Gebser, M., Kaminski, R., Kaufmann, B., Ostrowski, M., Schaub, T., Schneider, M.: Potassco: the Potsdam answer set solving collection. AI Commun. **24**(2), 107–124 (2011)
15. Gelfond, M., Lifschitz, V.: Logic programs with classical negation. In: Proceedings of 7th International Conference on Logic Programming (ICLP), pp. 579–97 (1990)
16. Hitzler, P., Krötzsch, M.: Querying formal contexts with answer set programs. In: Schärfe, H., Hitzler, P., Øhrstrøm, P. (eds.) ICCS-ConceptStruct 2006. LNCS, vol. 4068, pp. 260–273. Springer, Heidelberg (2006). doi:10.1007/11787181_19

17. Jukna, S., Kulikov, A.: On covering graphs by complete bipartite subgraphs. Discret. Math. **309**(10), 3399–3403 (2009)
18. King, A.D., Pržulj, N., Jurisica, I.: Protein complex prediction via cost-based clustering. Bioinformatics **20**(17), 3013–3020 (2004)
19. Launay, G., Salza, R., Multedo, D., Thierry-Mieg, N., Ricard-Blum, S.: Matrixdb, the extracellular matrix interaction database: updated content, a new navigator and expanded functionalities. Nucleic Acids Res. **43**(D1), D321–D327 (2015)
20. Lifschitz, V.: What is answer set programming? In: Proceedings of 23rd National Conference on Artificial Intelligence, AAAI 2008, vol. 3, pp. 1594–1597. AAAI Press (2008)
21. Navlakha, S., Schatz, M.C., Kingsford, C.: Revealing biological modules via graph summarization. J. Comput. Biol. **16**(2), 253–264 (2009)
22. Ogata, H., Fujibuchi, W., Goto, S., Kanehisa, M.: A heuristic graph comparison algorithm and its application to detect functionally related enzyme clusters. Nucleic Acids Res. **28**(20), 4021–4028 (2000)
23. Royer, L., Reimann, M., Andreopoulos, B., Schroeder, M.: Unraveling protein networks with power graph analysis. PLoS Comput. Biol. **4**(7), e1000108 (2008)
24. Rudolph, S., Săcărea, C., Troancă, D.: Membership constraints in formal concept analysis. In: Proceedings of 24th International Conference on Artificial Intelligence, IJCAI 2015, pp. 3186–3192. AAAI Press (2015)
25. Shannon, P., Markiel, A., Ozier, O., Baliga, N.S., Wang, J.T., Ramage, D.E.A.: Cytoscape: a software environment for integrated models of biomolecular interaction networks. Genome Res. **13**(11), 2498–2504 (2003)
26. Tsatsaronis, G., Reimann, M., Varlamis, I., Gkorgkas, O., Nørvåg, K.: Efficient community detection using power graph analysis. In: Proceedings of 9th Workshop on Large-Scale and Distributed Informational Retrieval, pp. 21–26. ACM (2011)
27. Wucher, V.: Modeling of a gene network between mRNAs and miRNAs to predict gene functions involved in phenotypic plasticity in the pea aphid. Thesis, Université Rennes 1, November 2014

A Proposition for Sequence Mining Using Pattern Structures

Victor Codocedo[1,3(✉)], Guillaume Bosc[2], Mehdi Kaytoue[2],
Jean-François Boulicaut[2], and Amedeo Napoli[3]

[1] Inria Chile, Las Condes, Chile
victor.codocedo@inria.cl
[2] Université de Lyon, CNRS, INSA-Lyon, LIRIS, Lyon, France
{guillaume.bosc,mehdi.kaytoue,jean-francois.boulicaut}@insa-lyon.fr
[3] LORIA (CNRS – INRIA Nancy Grand-Est – Université de Lorraine),
Nancy, France
amedeo.napoli@loria.fr

Abstract. In this article we present a novel approach to rare sequence mining using pattern structures. Particularly, we are interested in mining closed sequences, a type of maximal sub-element which allows providing a succinct description of the patterns in a sequence database. We present and describe a sequence pattern structure model in which rare closed subsequences can be easily encoded. We also propose a discussion and characterization of the search space of closed sequences and, through the notion of sequence alignments, provide an intuitive implementation of a similarity operator for the sequence pattern structure based on directed acyclic graphs. Finally, we provide an experimental evaluation of our approach in comparison with state-of-the-art closed sequence mining algorithms showing that our approach can largely outperform them when dealing with large regions of the search space.

1 Introduction

Sequence mining is an interesting application of data analysis through which we aim at finding patterns in strings of symbols, sets or events [1]. One of the simplest incarnations of this problem is finding, within a database of words, a set of substrings that appear more frequently among them, e.g. prefixes or suffixes are usually "frequent substrings".

Traditional algorithms for sequence mining rely on prefix enumeration [11–13] exploiting the fact that the longest a prefix is, the fewer words contain it (e.g. prefix pr- is contained in prehistoric, prehispanic and primitive, while the prefix pri- is only contained in primitive).

While these techniques are usually very efficient at finding frequent subsequences, there are some issues they do not address. Firstly, frequent sequences are usually short and very simple in structure. Because of this, current sequence miners are not able to find patterns with certain complexity restrictions such as minimal length. Secondly, prefix enumeration techniques usually provide numerous sequences as a result of the mining process. In this regard, we may be interested in a subset of the results using notions of maximality (what are known as

© Springer International Publishing AG 2017
K. Bertet et al. (Eds.): ICFCA 2017, LNAI 10308, pp. 106–121, 2017.
DOI: 10.1007/978-3-319-59271-8_7

closed subsequences) which provide a succinct representation of the patterns in a database (CloSpan, ClaSP or BIDE [7,11,12]).

In this article we present a novel approach for rare subsequence mining using the FCA framework. We adapt the pattern structures extension to deal with complex object descriptions [6] to build a mining tool that naturally models "closed sequences". By defining a similarity operator between sequence sets we are able to derive a *rare* sequence mining tool that is able to retrieve large subsequences within a dataset.

The remainder of this article is organized as follows. Section 2 introduces the main notions behind the problem of sequence mining. Section 3 models the problem in the framework of pattern structures. Section 4 formalizes the implementation of our mining technique. Section 5 presents a discussion on the state-of-the-art algorithms for sequence mining. Section 6 presents the experimental evidence to support our findings and a discussion about their implications. Finally, Sect. 7 concludes the article with a summary of our work.

2   Formalization

Before introducing our contributions, let us provide some basic definitions of the sequence mining problem. Let M be a set of items or symbols, a sequence of itemsets is an ordered set denoted as $< \alpha_1, \alpha_2, \alpha_3 ..., \alpha_n >$ where $\alpha_i \subseteq M$, $i \in [1, n]$. Alternatively, we will use a *bar notation* $\alpha_1|\alpha_2|\alpha_3|...|\alpha_n$. A sequence with ID (SID) α is denoted as $\alpha := \alpha_1|\alpha_2|\alpha_3|...|\alpha_n$ and is referred as well as sequence α. The *size* of sequence α is denoted as $size(\alpha) = n$ while its length is denoted as $len(\alpha) = \sum |\alpha_i|$ with $i \in [1, n]$ where $| \cdot |$ indicates set cardinality. A set of sequences $\mathcal{A}$ is called a *sequence database*. For the sake of simplicity we will drop the set parentheses for the itemsets within a sequence using the bar notation, e.g. $\{a, b\}|\{c, d\}$ is denoted as $ab|cd$.

Definition 1. SUBSEQUENCE. *Sequence $\beta := \beta_1|\beta_2|....|\beta_m$ is a subsequence of sequence $\alpha := \alpha_1|\alpha_2|...|\alpha_n$ (denoted as $\beta \preceq \alpha$) iff $m \leq n$ and if there exists a a sequence of natural numbers $i_1 < i_2 < i_3 < ... < i_m$ such that $\beta_j \subseteq \alpha_{i_j}$ with $j \in [1, m]$ and $i_j \in [1, n]$. We denote as $\beta \prec \alpha$ when $\beta \preceq \alpha$ and $\beta \neq \alpha$.*

For example, consider sequence $\beta := a|b|a$ and sequence with SID α^2 in Table 1. In this particular case, there exists a sequence of numbers $\langle 1, 3, 4 \rangle$ where we have: The first itemset of β is a subset of the *first* itemset of α^2 ($\{a\} \subseteq \{a, d\}$). The second itemset of β is a subset of the *third* itemset of α^2 ($\{b\} \subseteq \{b, c\}$). The third itemset of β is a subset of the *fourth* itemset of α^2 ($\{a\} \subseteq \{a, e\}$). We conclude that $\beta \prec \alpha^2$ (since $\beta \neq \alpha^2$). Given a sequence database $\mathcal{A}$ we will consider the set of all sub-sequences of elements in $\mathcal{A}$ denoted as $\mathcal{S}$ where $\beta \in \mathcal{S} \iff \exists \alpha \in \mathcal{A}$ s.t. $\beta \preceq \alpha$ ($\mathcal{A} \subseteq \mathcal{S}$).

Definition 2. SUPPORT OF A SEQUENCE. *Let $\beta \in \mathcal{S}$ be a sequence. The support of β w.r.t. $\mathcal{A}$ is defined as: $\sigma(\beta) = |\{\alpha \in \mathcal{A} \mid \beta \preceq \alpha\}|$*

With $\beta := a|b|a$ we have that $\beta \prec \alpha^1, \alpha^2$ and thus $\sigma(\beta) = 2$.

Definition 3. CLOSED SEQUENCE. $\beta^1 \in \mathcal{S}$ *is a closed sequence w.r.t* $\mathcal{A}$ *iff* $\nexists \beta^2 \in \mathcal{S}$ *such that* $\beta^1 \prec \beta^2$ *and* $\sigma(\beta^1) = \sigma(\beta^2)$.

Consider sequences $\beta^1 := a|b|a$ and $\beta^2 := a|bc|a$ as subsequences of elements in $\mathcal{A}$ of Table 1. It is easy to show that $\sigma(\beta^2) = 2$. Given that $\beta^1 \prec \beta^2$ and that they have the same support, we conclude that β^2 is not a closed sequence w.r.t. $\mathcal{A} = \{\alpha^1, \alpha^2, \alpha^3, \alpha^4\}$.

When mining subsequences, those that are closed in the sense of Definition 3 provide a succinct result, critically compacter than the set of all possible subsequences [11,12]. Consequently, in what follows we will be interested only in sets of closed sequences.

3   The Pattern Structure of Sequences

The pair $(\mathcal{S}, \preceq)$ composed of a sequence set and a subsequence order constitutes a partially ordered set. However, it does not conform a proper lattice structure, since in general, two sequences have more than one common subsequence (the infimum of two sequences is not unique).

In the context of sequence mining, this fact has been already noticed in [13], where the space of sequences and their order is denominated a *hyper-lattice*. In the context of FCA, this problem is analogous to the one introduced in [6] for graph pattern mining. Indeed, we can properly embed the *hyper-lattice* of partially ordered sequences into a lattice of sequence sets using the framework of pattern structures.

Definition 4. SET OF CLOSED SEQUENCES. *Let* $\mathbf{d} \subseteq \mathcal{S}$ *be a set of sequences, we call* $\mathbf{d}^+ \subseteq \mathbf{d}$ *a set of closed sequences iff* $\mathbf{d}^+ = \{\beta^i \in \mathbf{d} \mid \nexists \beta^j \in \mathbf{d}\ s.t.\ \beta^i \prec \beta^j\}$.

The intuition behind Definition 4 is that given a set of sequences $\mathbf{d}$, we can consider every sequence inside it as having the same support. Thus, $\mathbf{d}^+$ contains *closed sequences* in the sense of Definition 3.

Definition 5. SET OF COMMON CLOSED SUBSEQUENCES (SCCS). *Consider two sequences* $\beta^i, \beta^j \in \mathcal{S}$, *we define their meet* $\wedge$ *as follows:*

$$\beta^i \wedge \beta^j = \{\beta \in \mathcal{S} \mid \beta \preceq \beta^i\ and\ \beta \preceq \beta^j\}^+$$

Intuitively, $\beta^i \wedge \beta^j$ corresponds to a set of common subsequences to β^i and β^j. As indicated by the notation, we will require this set to contain only closed sequences (Definition 4) and thus we will denote it as the set of common closed subsequences (SCCS) of β^i and β^j.

Definition 6. SCCS FOR SEQUENCE SETS. *Given two sets of closed sequences* $\mathbf{d}^1, \mathbf{d}^2 \subseteq \mathcal{S}$, *the similarity operator between them is defined as follows:*

$$\mathbf{d}^1 \sqcap \mathbf{d}^2 = \left\{ \bigcup_{\substack{\beta^i \in \mathbf{d}^1 \\ \beta^j \in \mathbf{d}^2}} \beta^i \wedge \beta^j \right\}^+$$

In a nutshell, $\sqcap$ is the SCCS between the elements within two sets of closed sequences. Let $D = \{d \subseteq \mathcal{S} \mid d = d^+\}$ denominate the *set of all sets of closed sequences in $\mathcal{S}$*, then we have $\wedge : \mathcal{S}^2 \rightarrow D$ and $\sqcap : D^2 \rightarrow D$.

For a sequence $\beta \in \mathcal{S}$, $\delta(\beta) = \{\beta^i \in \mathcal{S} \mid \beta^i \preceq \beta\}^+$ denotes its set of closed subsequences. Trivially, the set of closed subsequences of a single sequence β contains a single element which is itself, i.e. $\delta(\beta) = \{\beta\}$. From this, it follows that given any three sequences $\beta^i, \beta^j, \beta^k \in \mathcal{S}$ we have that $\beta^i \wedge \beta^j = \delta(\beta^i) \sqcap \delta(\beta^j)$ and $(\delta(\beta^i) \sqcap \delta(\beta^j)) \sqcap \delta(\beta^k)$ properly represents the SCCS of β^i, β^j and β^k. For example, consider sequences with α^1 and α^3 in Table 1.

$$\delta(\alpha^1) \sqcap \delta(\alpha^3) = \{a|abc|ac|d|cf\} \sqcap \{ef|ab|df|c|b\} = \{ab|d|c, ab|f, a|b\}$$

Clearly, $\sqcap$ is idempotent and commutative as it inherits these properties from $\wedge$. Furthermore, it can be easily shown that $\sqcap$ is also associative. Using these properties, we can provide a more complex example. Consider $\delta(\alpha^1) \sqcap \delta(\alpha^2) \sqcap \delta(\alpha^3) \sqcap \delta(\alpha^4)$ from Table 1. Since we already have the result for $\delta(\alpha^1) \sqcap \delta(\alpha^3)$, we re-arrange and proceed as follows:

$$\delta(\alpha^1) \sqcap \delta(\alpha^2) \sqcap \delta(\alpha^3) \sqcap \delta(\alpha^4) = (\delta(\alpha^1) \sqcap \delta(\alpha^3)) \sqcap (\delta(\alpha^2) \sqcap \delta(\alpha^4))$$
$$\delta(\alpha^1) \sqcap \delta(\alpha^3) = \{ab|d|c, ab|f, a|b\}$$
$$\delta(\alpha^2) \sqcap \delta(\alpha^4) = \{a|c|b, a|c|c, e\}$$

Now, we need to calculate $\{ab|d|c, ab|f, a|b\} \sqcap \{a|c|b, a|c|c, e\}$ which requires 9 SCCS calculations. We show those with non-empty results.

$$ab|d|c \wedge a|c|b = \{a|c, b\} \qquad ab|d|c \wedge a|c|c = \{a|c\} \qquad ab|f \wedge a|c|b = \{a, b\}$$
$$ab|f \wedge a|c|c = \{a\} \qquad a|b \wedge a|c|b = \{a|b\} \qquad a|b \wedge a|c|c = \{a\}$$

The union of these results gives us $\{a, b, a|b, a|c\}$ from which we remove a and b which are not closed subsequences in the set. Finally, we have:

$$\delta(\alpha^1) \sqcap \delta(\alpha^2) \sqcap \delta(\alpha^3) \sqcap \delta(\alpha^4) = \{a|b, a|c\}$$

Indicating that the set of closed subsequences common to all sequences in Table 1 only contains $a|b$ and $a|c$.

Definition 7. ORDER IN D. *Let $d^1, d^2 \in D$, we say that d_1 is subsumed by d_2 (denoted as $d_1 \sqsubseteq d_2$) iff:*

$$d_1 \sqsubseteq d_2 \iff \forall \beta^i \in d_1 \; \exists \beta^j \in d_2 \; s.t. \; \beta^i \preceq \beta^j$$
$$d_1 \sqsubset d_2 \iff d_1 \sqsubseteq d_2 \; and \; d_1 \neq d_2$$

Proposition 1. SEMI-LATTICE OF SEQUENCE DESCRIPTIONS. *Let D be the set of all sets of closed sequences (hereafter indistinctly referred to as "the set of sequence descriptions") built from itemset M. The pair $(D, \sqcap)$ is a semi-lattice of sequence descriptions where $d_1 \sqsubseteq d_2 \iff d_1 \sqcap d_2 = d_1$ for $d_1, d_2 \in D$.*

The proof of $d_1 \sqsubseteq d_2 \iff d_1 \sqcap d_2 = d_1$ in Proposition 1 is straightforward and follows from Definitions 6 and 7.

Definition 8. THE PATTERN STRUCTURE OF SEQUENCES. *Let $\mathcal{A}$ be a sequence database, $(D, \sqcap)$ the semi-lattice of sequence descriptions and $\delta : \mathcal{S} \to D$ a mapping assigning to a given sequence its corresponding set of closed subsequences (recall that $\mathcal{A} \subseteq \mathcal{S}$). The pattern structure of sequences is defined as:*

$$\mathcal{K} = (\mathcal{A}, (D, \sqcap), \delta)$$

For the sake of brevity we do not provide the development of the pattern structure framework which can be found in [6]. However, let us introduce some concepts and notation that will result important in the remainder of the article. The pair (A, d) with $A \subseteq \mathcal{A}$ and $d \in D$ is a *sequence pattern concept* where $A^{\square} = d$ and $d^{\square} = A$. The derivation operator $(\cdot)^{\square}$ is defined dually for extents and pattern descriptions. The order of sequence pattern concepts is denoted as $(A_1, d_1) \leq_{\mathcal{K}} (A_2, d_2)$ iff $A_1 \subseteq A_2$ and $d_2 \sqsubseteq d_1$ where $A_1, A_2 \subseteq \mathcal{A}$ and $d_1, d_2 \in D$. Finally, the set of all sequence pattern concepts with the order $\leq_{\mathcal{K}}$ defines a sequence pattern concept lattice denoted as $(\mathfrak{B}(\mathcal{K}), \leq_{\mathcal{K}})$.

Within a sequence pattern concept we have that $d^{\square\square} = d$, i.e. the set of sequences d is closed. This should not be confused with the notion of *set of closed sequences* given in Definition 4 which is a property of the elements inside d. Actually, d is a set of common closed subsequences (SCCS) when $d^+ = d$. With this new constraint, the correct denomination for d would be a *closed set of common closed subsequences*. To avoid confusions, hereafter we will denominate a set d such that $d^+ = d$ and $d^{\square\square} = d$ as a *maximal SCCS*.

A sequence pattern concept represents a set of sequences A in the database and their maximal SCCS, i.e. closed subsequences of elements in A are contained in d. Consider from the previous example that $\{e|a|c|b\}^{\square} = \{\alpha^3, \alpha^4\}$. Now, we can calculate $\{\alpha^3, \alpha^4\}^{\square} = \{e|a|c|b, e|f|c|b, e|b|c, f|b|c\}$. From this it follows that $\{e|a|c|b\}$ is not a maximal SCCS and that $(\{\alpha^3, \alpha^4\}, \{e|a|c|b, e|f|c|b, e|b|c, f|b|c\})$ is a sequence pattern concept. Figure 1 shows the sequence pattern concept lattice built from entries in Table 1. The bottom concept in the figure is a *fake* element in the lattice since there is no definition for the meet between pattern concepts. We have included it to show that it represents the concept with empty extent, i.e. the "subsequences of no sequences". The indefinition of the meet between formal concepts also means that we can only *join* sequence pattern concepts starting from object concepts $\gamma(\alpha) = (\{\alpha\}, \delta(\alpha))$. Consequently our mining approach is a *rare* sequence mining technique.

The concept lattice structure allows proving a characterization of an important property of sequences, namely we can show that all sequences in an intent are closed w.r.t. the database $\mathcal{A}$.

Proposition 2. *Given a sequence pattern concept (A, d) any sequence $\beta \in d$ is not only closed w.r.t. A but it is also closed w.r.t. $\mathcal{A}$.*

Proof. Consider two sequence pattern concepts $(A_1, d_1), (A_2, d_2) \in \mathfrak{B}(\mathcal{K})$ such as $(A_1, d_1) <_{\mathcal{K}} (A_2, d_2)$ (then $A_1 \subset A_2$ and $d_2 \sqsubseteq d_1$). Particularly, the latter relation indicates that $\forall \beta^j \in d_2 \; \exists \beta^i \in d_1$ s.t. $\beta^j \preceq \beta^i$ and $d_1 \neq d_2$ (Definition 7). In the case that $\beta^j \prec \beta^i$, the fact that $A_1 \subset A_2 \implies |A_1| < |A_2|$ secures that both sequences have different supports making β^j closed. In the case that $\beta^j = \beta^i$, there are two different cases. Either a third concept $(A_3, d_3) \in \mathfrak{B}(\mathcal{K})$ exists such that $(A_2, d_2) <_{\mathcal{K}} (A_3, d_3)$ where $\beta^j \in d_2$ and $\beta^j \notin d_3$ and thus, we fall in the previous case and β^j is closed, or $\beta^j \in A^{\square}$ and β^j is closed with support $|A|$.

Indeed, since $(A^{\square}) = (A^{\square})^+$, any $\beta \in A^{\square}$ is closed w.r.t. A, even *when we may not know the support for that particular sequence*. Moreover, the support of any closed sequence β is given by the cardinality of the extent of the concept (A_1, d_1) where $\beta \in d_1$ s.t. $\nexists (A_2, d_2)$ where $(A_1, d_1) <_{\mathcal{K}} (A_2, d_2)$ and $\beta \in d_2$.

For example, consider Fig. 1 and the concept labelled as $\{\alpha^1, \alpha^2\}$ containing sequence $a|c|c$. While we can be sure that $a|c|c$ is a closed subsequence w.r.t. A, the support is not 2. Actually, $a|c|c$ exists also in the intent of the concept labelled $\{\alpha^1, \alpha^2, \alpha^4\}$. Since it does not exists in the suprema (the only superconcept of the latter), we can conclude that the support of $a|c|c$ is 3.

Finally, we briefly mention the pattern complexity filtering capabilities of our approach. For a relation $(A_1, d_1) <_{\mathcal{K}} (A_2, d_2)$, sequences in d_2 will be shorter and with fewer items per itemset than those in d_1. Given thresholds for size and lenght, the sequence search space can be pruned similarly to support pruning.

4   Implementing the Similarity Operator ⊓

So far we have taken for granted the ability to calculate the set of common closed subsequences (SCCS) of two sequence sets (i.e. $d_1 \sqcap d_2$). In what follows, we will describe and discuss our approach to achieve this.

4.1   Rationale

Sequences are composed of two parts, firstly the elements in the sequence and secondly, the order they have. Considering the latter, we can think about how

Table 1. An example sequence database.

sid	Sequence
α^1	$a|abc|ac|d|cf$
α^2	$ad|c|bc|ae$
α^3	$ef|ab|df|c|b$
α^4	$e|g|af|c|b|c$

Fig. 1. Extents and Intents of the sequence pattern structure built from Table 1.

a "common subsequence" (an element in the SCCS) β of two sequences β^1, β^2 relates to them. Given the definition of subsequence (Definition 1), since $\beta \preceq \beta^1$ we have a sequence of integers $i_1^1 < i_2^1 < i_3^1 < \ldots < i_m^1$ such that $\beta_j \subseteq \beta_{i_j}^1$ with $i_j^1 \in [1, n]$ and $j \in [1, m]$ with $size(\beta) = m$, $size(\beta^1) = n$, and $m < n$. A similar sequence of integers can be derived from the relation $\beta \preceq \beta^2$, namely $i_1^2 < i_2^2 < i_3^2 < \ldots < i_m^2$. Since both sequence of integers have the same length, we can arrange them in a sequence of tuples $P = \langle (i_1^1, i_1^2), (i_2^1, i_2^2), \ldots, (i_m^1, i_m^2) \rangle$. Furthermore, given that $\beta_j \subseteq \beta_{i_j}^1$ and $\beta_j \subseteq \beta_{k_j}^2$, it is only natural to define $\beta_j = \beta_{i_j}^1 \cap \beta_{k_j}^2$ (notice that defining $\beta_j \subset \beta_{i_j}^1 \cap \beta_{k_j}^2$ would render β not a closed sequence). Consequently, tuple P is a possible characterization of β in terms of its parents supersequences β^1 and β^2. Actually, given any set of sequences A, a common subsequence of them $\beta \in A^\square$ has one or more characterizations P.

Generally, we call $P := \langle t_1, t_2, \ldots, t_m \rangle$ an alignment with size m of sequences in A where $r = |A|$, n_k is the size of the k-th sequence in A with $k \in [1, r]$, $t_j = (i_j^k)$, and $j \in [1, m]$ ($t_j \in \{1, 2, \ldots, n\}^r$). Furthermore, we will require that with $k \in [1, r]$ and $j \in [1, (m-1)]$ we have that: (i) ($\exists i_1^k \in t_1$) s.t. $i_1^k = 1$, (ii) ($\exists i_m^k \in t_m$) s.t. $i_m^k = n_k$, (iii) ($\exists i_{j+1}^k \in t_{j+1}$) s.t. $i_{j+1}^k = i_j^k + 1$, and (iv) ($\forall i_{j+1}^k \in t_{j+1}$) $i_j^k < i_{j+1}^k$. Conditions (i), (ii) and (iii) secure that alignments always begin with the first element of at least one sequence in A, always end with the last element of a least one sequence in A, and that they are *maximal*, e.g. $\langle (1,1), (3,3) \rangle$ is not a maximal alignment if $\langle (1,1), (2,2), (3,3) \rangle$ exists. Condition (iv) secures that alignments only consider incremental tuples, e.g. $\langle (1,2), (2,1) \rangle$ is not an alignment. $\mathcal{P}_n^r$ is the space of all maximal alignments of sequences in A.

It is possible to show that $|A^\square| \leq |\mathcal{P}_n^r|$ provided that conditions (i), (ii), (iii) and (iv) hold for alignments in $\mathcal{P}_n^r$ and that $|M| = n^r$. This is demonstrated by showing that there exists a scenario (however unlikely) where for each different closed subsequence there exists a unique alignment, i.e. ($\forall \beta \in A^\square$)($\exists! \langle t_1, t_2, \ldots, t_m \rangle \in \mathcal{P}_n^r$) s.t. $\beta_j = \bigcap \alpha_{t_j}^i = \pi(t_j)$ where $\pi : \mathbb{N}^r \to \mathbb{N}$ is pairing function encoding tuples in $\{1, 2, \ldots, n\}^r$ into $M = \{\pi(t_j) \mid t_j \in \{1, 2, \ldots, n\}^r\}$. Thus, given a set of sequences A we only need to compute *all possible alignments* in order to obtain their *complete set of common closed subsequences* $A^\square$ (SCCS).

Notably, this relation allows calculating the number of possible closed subsequences between two sequences of arbitrary length[1]. Table 2 shows this number with $n \in [1, 10]$. For example, for two sequences of size 10 we have a maximum of 26797 alignments/closed subsequences (requiring $|M| = 2^{10} = 1024$ items).

Enumerating all possible alignments may look an extremely naive strategy considering how the space of alignment grows w.r.t. the number of sequences in A and their size, however the way we encode alignments makes the representation of subsequences more compact than simply listing them. In addition, the sparsity of sequence element intersections (most $\beta_j = \beta_{i_j}^1 \cap \beta_{k_j}^2$ are empty) provides a very efficient pruning method. In the following, we show how to encode alignments in $\mathcal{P}_n^r$ as paths of a directed acyclic graph (DAG).

---

[1] https://oeis.org/A171155.

Definition 9. DAG OF ALIGNMENTS. *We call* $\mathcal{G}_n^r = (V, E, \lambda)$ *a directed acyclic graph of alignments* $\mathcal{P}_n^r$ *(DAG of alignments) for a set of r sequences* $\mathbf{A} \subseteq \mathcal{A}$ *of size n, if the set of vertices, edges and labelling function* $\lambda : V \to \wp(\mathsf{M})$ *are:*

$$V = \{t_j = \langle i_j^k \rangle \mid t_j \in \{1, 2, \dots, n\}^r\} \cup \{0_r, (n+1)_r\}$$

$$E = \{(t_a, t_b) \iff (\exists i_b^k \in t_b)\, i_b^k = i_a^k + 1 \text{ and } (\forall i_b^k \in t_b)\, i_a^k < i_b^k\}$$

$$\lambda(t_j) = \bigcap_{i \in [1,r]} \alpha_{i_j^i}^i$$

$0_r, (n+1)_r$ *are r-tuples filled with 0 and $n+1$ values respectively. We denominate 0_r the source vertex (or source node) and $(n+1)_r$ the sink vertex (or sink node).*

Notice that conditions (iii) and (iv) for alignments in $\mathcal{P}_n^r$ described above are encoded in the definition of edges E in $\mathcal{G}_n^r$. The inclusion of the source and sink nodes ensure that all paths between them corresponds to alignments in $\mathcal{P}_n^r$ that also respect conditions (i) and (ii). Thus, the DAG $\mathcal{G}_n^r$ is designed so that all possible maximal alignments among r sequences of size n correspond to *paths between the source and sink nodes*. In addition, all possible common subsequences can be read from the graph by means of the labelling function λ. Trivially, for a set of sequences with different sizes, a DAG of alignments can be built with n corresponding to the maximal sequence size in the set. Figure 2 shows the DAG $\mathcal{G}_4^2$ without labels, where the number of paths between $(0, 0)$ and $(5, 5)$ is 27 (as predicted by Table 2).

4.2 Calculating $\delta(\alpha^1) \sqcap \delta(\alpha^2)$

Consider sequences $\alpha^1 = ab|cd|e|f$, $\alpha^2 = b|ac|d|ef$ and their itemset intersections shown in Table 3 (empty cells indicate an empty intersection) Since both sequences have size 4, we can use the DAG $\mathcal{G}_4^2$ shown in Fig. 2 to obtain all possible alignments between these sequences. Then, we can derive associated subsequences using the labelling function given by Table 3. In these subsequences the SCCS must be included.

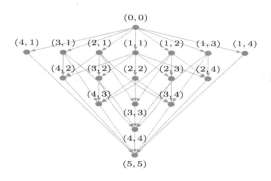

Fig. 2. DAG for the alignments of two sequences of size 4

Table 2. Number of maximal alignments for $n \in [1, 10]$.

n	$|\mathcal{P}_n^2|$
1	1
2	3
3	9
4	27
5	83
6	259
7	817
8	2599
9	8323
10	26797

In this case more than half the itemset intersections are empty which makes the task of deriving paths easier. Figure 3 shows a subgraph of $\mathcal{G}_4^2$, where vertex (i,j) is labelled as $\lambda((i,j)) = \alpha_i^1 \cap \alpha_j^2$. Vertices $v \in V$ for which $\lambda(v) = \emptyset$ were removed from Fig. 3. We have also connected $(2,2)$ with $(4,4)$ since vertex $(3,3)$ is among those vertices removed.

A path between $(0,0)$ and $(5,5)$ is $\langle(1,1),(2,2),(4,4)\rangle$ (removing the source and sink nodes). Thus, a subsequence can be derived using the labels of the vertices in the path, i.e. $\beta = \lambda((1,1))|\lambda((2,2))|\lambda((4,4))$ or $\beta = b|c|f$. All the subsequences of α^1 and α^2 are shown in Table 4 from which we can derive that the SCCS of α^1 and α^2 is $\delta(\alpha^1) \sqcap \delta(\alpha^2) = \{b|c|e, b|c|f, b|d|e, b|d|f, a|d|e, a|d|f\}$.

This result illustrates the usefulness of the DAG of alignments in Fig. 2. Indeed, instead of enumerating all possible sequences (like in usual subsequence mining algorithms) we just need to intersect the itemsets of two sequences and later interpret them using the corresponding DAG.

To extend the example, let us consider a third sequence $\alpha^3 = ab|c|d|ef$ and the SCCS of α^1, α^2 and α^3. We need to calculate the table of intersections that in this case corresponds to a *cube of intersections* where the cell (i,j,k) in the cube contains $\alpha_i^1 \cap \alpha_j^2 \cap \alpha_k^3$. While the construction of this cube requires $4^3 = 64$ intersections, using the results in Table 4 we need only to calculate 24 intersections (those with empty intersections in Table 4 can be disregarded). Using the DAG of alignments $\mathcal{G}_4^3$, we can obtain the SCCS such that $\delta(\alpha^1) \sqcap \delta(\alpha^2) \sqcap \delta(\alpha^3) = \{b|c|e, b|c|f, b|d|e, b|d|f, a|d|e, a|d|f\}$.

4.3   Discussion

In the previous example, one important thing to notice is that we did not need to interpret the sequences from the intersection table of $\mathbf{d} = \{\alpha^1, \alpha^2\}^\square$ in order to

Table 3. Intersection table for sequences $\alpha^1 = ab|cd|e|f$ and $\alpha^2 = b|ac|d|ef$.

	α_1^2	α_2^2	α_3^2	α_4^2
α_1^1	b	a		
α_2^1		c	d	
α_3^1			e	
α_4^1				f

Table 4. List of paths in the DAG of Fig. 3 and subsequences derived.

Path	Sequence
$\langle(1,1),(2,2),(4,4)\rangle$	$b|c|f$
$\langle(1,1),(2,2),(3,4)\rangle$	$b|c|e$
$\langle(1,1),(2,3),(4,4)\rangle$	$b|d|f$
$\langle(1,1),(2,3),(3,4)\rangle$	$b|d|e$
$\langle(1,2),(2,3),(3,4)\rangle$	$a|d|e$
$\langle(1,2),(2,3),(4,4)\rangle$	$a|d|f$

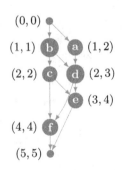

Fig. 3. DAG for sequences $\alpha^1 = ab|cd|e|f, \alpha^2 = b|ac|d|ef$.

calculate $d \sqcap \delta(\alpha^3)$. This information is encoded in the table itself in the form of index tuples of the itemsets we had intersected and can be recovered at any time using the corresponding DAG. Nevertheless, we are not spared from interpreting the sequences out of the table of intersections. This is because we still need the sequences to establish the order between two sequence sets. We have not found a property that would allow us to compare sequence sets from their intersection tables or from their DAG of alignments. Actually, this may not be possible. In this work we use the DAG of alignments just as a tool to explore and store the space of common closed subsequences. Implementation details of the DAG (such as the canonical order in the node tuples) were left out of the scope of this article for the sake of brevity, as well as some other elements of the search space exploration, e.g. the DAG allows filtering out sequences below a threshold of size and length.

5   Related Work

Sequence mining is usually based on what is called "prefix enumeration" or "pattern growth". This is, given a vocabulary M and a lexicographical order in it, we proceed by enumerating prefixes (using two main operations, namely *i-extension* and *s-extension* of sequences) and counting in a sequence database $\mathcal{A}$ how many elements contain them. The seminal paper in sequence mining is [1] introduces an Apriori-based algorithm to cut down the search space in a similar way to how it is done for standard itemset mining. A very popular extension of this idea was implemented in the PrefixSpan algorithm [10] using the notion of "projected databases" to decrease the number of comparisons in the database of sequences to calculate pattern support. PrefixSpan inspired different extensions such as CloSpan [6] to mine closed subsequences, OrderSpan [5] to mine *partially-ordered subsequences*, and CCSpan [14] to mine subsequences with a restriction of "contiguity" in the items they contain.

A different approach was taken in [13] using a "vertical database format" in an algorithm called SPADE (PrefixSpan-based algorithms are considered based on a "horizontal database format"). A rather interesting element of [13] is the characterization of the sequence search space as a *hyperlattice* which can be factored allowing parallel computation of patterns. A similar approach was adopted in BIDE [11] and ClaSP [7] for closed subsequence mining based on a vertical database format.

Partially-ordered subsequences is an idea introduced in [4] based on the notion that the subsumption order among sequences allows integrating different subsequences into a single pattern by means of a directed acyclic graph, which may provide a richer representation to an end user. This is closely related to our own model for sequence mining. Indeed, our own representation of sequence sets is through the DAG of alignments which are themselves partially-ordered patterns. This work inspired Frecpo [10] and OrderSpan [5] for mining closed partially-ordered subsequence patterns.

A comprehensive review on the sequence pattern mining techniques can be found in [9]. To the author's knowledge, our work is the first attempt to approach the problem of sequence mining in a general unrestricted manner using the framework of formal concept analysis. In this regard, an interesting work is presented in [2] where an FCA-based technique for mining Gradual patterns is introduced. Gradual patterns consider a total order between attribute values generating sequences of objects. The technique introduced allows characterizing sequences derived from these gradual patterns. However, much like BIDE, sequences are restricted to strings of elements, not sequences of itemsets. In [3], a different sequence mining framework based on pattern structures is introduced for dealing with medical records. These sequences contain complex elements composed of taxonomical elements and attribute sets. The authors decided to simplify the problem by considering subsequences containing just contiguous elements.

6   Experiments and Discussion

In this section, we present an experimental evaluation on our sequence mining approach over a series of synthetic datasets generated using the IBM Quest Dataset Generator software hosted in the SPMF Website[2]. Our approach was implemented in Python and embedded into a system named *Pypingen* available through a subversion repository[3].

6.1   Datasets and Experimental Setup

Pypingen was applied over 34 different datasets in 4 four groups. Each group allows showing how our algorithm handle different settings. The first group, denoted as $\mathcal{D}_{nseq_i}$, $i \in [1,7]$, contains datasets of different size $|\mathcal{A}|$, i.e. the number of sequences inside a dataset. The second group, $\mathcal{D}_{nitems_i}$, $i \in [1,12]$, contains datasets where the size of the itemset $|\mathsf{M}|$ varies. The third group, $\mathcal{D}_{size_i}$, $i \in [1,9]$, contains datasets of sequences that have different average sizes $mean(size)$, i.e. number of itemsets in a given sequence. Finally, the last group, $\mathcal{D}_{length_i}$, $i \in [1,6]$, consists of datasets where sequences have a varying average number of items by itemset, i.e. a variation in the *length* of sequences $mean(length/size)$ for a fixed size. Table 5 displays the different settings for each synthetic dataset generated. The variation in the parameters is a consequence of considering outputs that took *less than 15 min* to calculate. Experiments were performed on a 3.10 GHz processor with 8 GB main memory running Ubuntu 14.04.1 LTS.

6.2   Performance Study - Comparison

All 34 datasets were also processed using "closed sequence pattern mining" algorithms ClaSP, CM-ClaSP, CloSpan and BIDE implemented in the SPMF software[4] (written in Java) under the same conditions of the execution of

---

[2] Open-source data mining library - http://www.philippe-fournier-viger.com/spmf/.

[3] http://gforge.inria.fr/projects/pypingen.

[4] Version 0.97d / 0.97e - 2015-12-06.

Table 5. Characteristics of the datasets where $|\mathcal{A}|$ is the number of sequences in the dataset, $|\mathsf{M}|$ the cardinality of the set of items or symbols, $mean(size)$ the average size of the sequences of the dataset and $mean(length/size)$ the average number of items in the itemsets of the sequences.

Group 1 $\mathcal{D}_{nseq_i}$	
$|\mathsf{M}|$	50
$mean(size)$	5
$mean(lenght/size)$	4
$|\mathcal{A}|$	$\{9, 27, 40, 54, 69, 84, 101\}$
Group 2 $\mathcal{D}_{nitems_i}$	
$|\mathsf{M}|$	$\{10, 20, 30, 40, 50, 100, 150, 200, 250, 300, 350, 400\}$
$mean(size)$	5
$mean(lenght/size)$	4
$|\mathcal{A}|$	36
Group 3 $\mathcal{D}_{size_i}$	
$|\mathsf{M}|$	50
$mean(size)$	$\{2, 3, 4, 5, 6, 7, 8, 9, 10\}$
$mean(lenght/size)$	4
$|\mathcal{A}|$	21
Group 4 $\mathcal{D}_{lenght_i}$	
$|\mathsf{M}|$	50
$mean(size)$	5
$mean(lenght/size)$	$\{1, 2, 3, 4, 5, 6\}$
$|\mathcal{A}|$	36

Pypingen. Algorithms were given a threshold value of 0 to obtain all closed sequence patterns. From all four algorithms, only CloSpan and BIDE were able to obtain results in less than 15 min. However, we do not report on these results for two reasons. Firstly, with respect to CloSpan, it was only able to obtain results for just a single dataset, namely $\mathcal{D}_{length_1}$ which is the simplest. Secondly, with respect to BIDE, even when it was able to obtain results in less time, these results were incorrect and incomplete. This is due to the fact that even when BIDE is catalogued as a "closed sequence miner", its formalization actually corresponds to that of a "closed string miner", that is, sequences where each itemset has cardinality 1. This simplification of the problem greatly reduces the search space and makes any comparison unfair. More importantly, in datasets where this condition is not met such as those in our evaluation, it produces incorrect (in the form of non-closed sequences) and incomplete results.

Execution times for *Pypingen* can be visualized in Fig. 4 (execution times are given in logarithmic scale). The curves presented are consistent with the description of the algorithm. Runtimes when varying the size of the sequences (Fig. 4c) grow exponentially due to the fact that the search space grows w.r.t. n^r where $|\mathcal{A}| = r$ and n is the maximum size of the sequences in the dataset. When varying the sequence length (Fig. 4d), the search space does not change (it remains at n^r which in this case is 5^{36}). However, the larger the itemset cardinalities, the less empty intersections we can expect in the intersection table, and thus, the more closed subsequences for a sequence set.

Figure 4a is very interesting because it is more related to scalability. We can see that as the number of sequences grows, the time grows polynomially and not exponentially, which is desirable for a mining technique. Finally, Fig. 4b shows that the execution time gets lower as the cardinality of itemset M gets larger. Indeed, since the remainder dataset parameters are left unchanged, the increase in the number of possible items leads to sparser datasets where most intersections are empty.

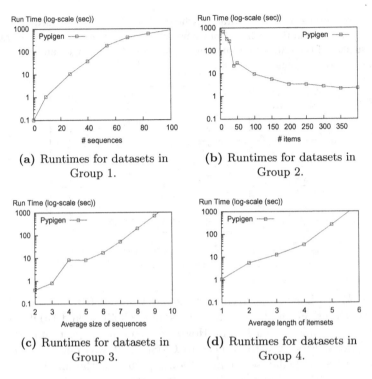

(a) Runtimes for datasets in Group 1.

(b) Runtimes for datasets in Group 2.

(c) Runtimes for datasets in Group 3.

(d) Runtimes for datasets in Group 4.

Fig. 4. Runtimes results for each dataset group.

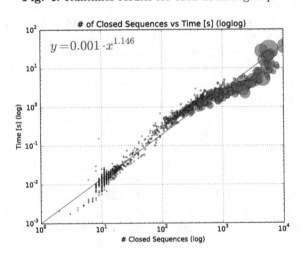

Fig. 5. Number of closed subsequences vs execution time for 1200 datasets.

Figure 5 presents the relation between the number of closed subsequences and the execution time to calculate them using *Pypingen* over 1200 different synthetic datasets (each circle represents a dataset) created for 60 different configurations of vocabulary size (M), average size of sequences and average length of itemsets. All datasets contained only 10 sequences ($|\mathcal{A}| = 10$). Axes in the figure are provided in logarithmic scale, while the curve is the best fit to the data w.r.t. the coefficient of determination $R^2 = 0.97$ (a quadratic curve has $R^2 = 0.78$ while an exponential has $R^2 = 0.29$). We can appreciate that the curve is a power function with an exponent very close to 1 indicating a quasilinear relation between the number of closed sequences in a dataset and execution time required to calculate them, which is a desirable property for a pattern mining algorithm. Nevertheless, the performance of the algorithm slowly degrades as the number of patterns in a dataset increases. The radius of each point in the figure represents the ratio between the number of closed sequences and the number of sequence patterns found for that dataset.

6.3   Discussion

Algorithms CloSpan, ClaSP, CM-ClaSP and BIDE are *frequent* sequence miners, while *Pypingen* is a *rare* sequence miner. This means that the only "fair" comparison requires for all algorithms to explore the entire search space and mine all possible closed sequences. Under this circumstances, *Pypingen* clearly outperforms all of them as they are not able to achieve this task for all but one dataset in our experiments.

Arguably, our comparison can be understood as "fair" considering that all the algorithms are set to the worst case scenario. Frequent mining algorithms are designed under the assumption that results are in the "upper" part of the search space. Thus, asking them to explore the entire search space goes against the very assumption behind their designs and it becomes *unfair* to evaluate their performance under the worst of the conditions. On the other hand, *Pypingen* was designed to begin its search from the "bottom" part of the search space and thus, it is also an evaluation in the worst case scenario. The conclusion that follows is that, compared with the state-of-the-art algorithms, *Pypingen* is able to explore a much larger region of the search space.

It is clear nonetheless that given a large enough dataset *Pypingen* will not be able to explore the entire search space and thus, it will be unable to mine frequent subsequences. Under these circumstances, frequent miners have the advantage of obtaining a set of results out of reach for a rare pattern miner such as *Pypingen*. In this light, we highlight the *benefits* of our approach. Like frequent miners use a threshold for support, we can provide a threshold for sequence size to mine *large sequences*. Using a proper value for this threshold, we can obtain all sequences for this restriction and particularly, those that are most frequent. Mining this type of *frequent large sequences* is not possible for frequent miners. Another threshold that *Pypingen* can easily support is the minimum cardinality per itemset of a sequence pattern, adding in the capabilities of restricting the complexity of the desired patterns.

Finally, one of the main features of *Pypingen* is its flexibility. Consider the intersection table that we use to generate the DAG of alignments such as the one shown in Table 3. In here, the intersection operation may be easily replaced by an abstract similarity operator that holds under idempotence, commutativity and associativity to support sequences of elements other than itemsets, in the same way that it is done when defining pattern structures. Such adaptation would allow supporting sequences of complex elements such as intervals, taxonomical elements or heterogeneous representations [8]. This kind of sequences of complex elements would be hard to mine in current sequence mining algorithms designed for prefix enumeration.

7   Conclusions

In this article we have introduced the main notions behind *rare sequence pattern* mining using the framework of formal concept analysis, particularly the "pattern structures" extension to deal with complex object descriptions. Indeed, we have modelled patterns as sequence sets where the similarity operator allows formalizing an algorithm for extracting closed subsequences.

We have shown how the similarity operator can be implemented using a directed acyclic graph (DAG) of sequence alignments where a set of sequences can be easily encoded. The DAG of alignments also allows for an easy interpretation of sequence patterns as well as for pruning procedures that allows cutting down the search space.

We have implemented our approach in a system called *Pypingen* and showed that w.r.t. state-of-the-art closed sequence mining algorithms, it performs better in the worst of the cases, this is, when obtaining the full set of closed subsequences from a database.

Finally, we have discussed the implications of our system and its possible extensions regarding the mining of sequences composed by elements other than itemsets.

References

1. Agrawal, R., Srikant, R.: Mining sequential patterns. In: Proceedings of the Eleventh International Conference on Data Engineering, ICDE 1995, pp. 3–14. IEEE Computer Society, Washington, D.C. (1995)
2. Ayouni, S., Laurent, A., Yahia, S.B., Poncelet, P.: Mining closed gradual patterns. In: Rutkowski, L., Scherer, R., Tadeusiewicz, R., Zadeh, L.A., Zurada, J.M. (eds.) ICAISC 2010. LNCS, vol. 6113, pp. 267–274. Springer, Heidelberg (2010). doi:10. 1007/978-3-642-13208-7_34
3. Buzmakov, A., Egho, E., Jay, N., Kuznetsov, S.O., Napoli, A., Raïssi, C.: On projections of sequential pattern structures (with an application on care trajectories). In: Ojeda-Aciego, M., Outrata, J. (eds.) The Tenth International Conference on Concept Lattices and their Applications - CLA 2013, La Rochelle, France, pp. 199–208. Université de La Rochelle (2013)
4. Casas-Garriga, G.: Summarizing sequential data with closed partial orders. In: Proceedings of the 2005 SIAM International Conference on Data Mining, SDM 2005, Newport Beach, CA, USA, 21–23 April 2005, pp. 380–391. SIAM (2005)

5. Fabrègue, M., Braud, A., Bringay, S., Le Ber, F., Teisseire, M.: Mining closed partially ordered patterns, a new optimized algorithm. Knowl.-Based Syst. **79**, 68–79 (2015)
6. Ganter, B., Kuznetsov, S.O.: Pattern structures and their projections. In: Delugach, H.S., Stumme, G. (eds.) ICCS-ConceptStruct 2001. LNCS, vol. 2120, pp. 129–142. Springer, Heidelberg (2001). doi:10.1007/3-540-44583-8_10
7. Gomariz, A., Campos, M., Marin, R., Goethals, B.: ClaSP: an efficient algorithm for mining frequent closed sequences. In: Pei, J., Tseng, V.S., Cao, L., Motoda, H., Xu, G. (eds.) PAKDD 2013. LNCS, vol. 7818, pp. 50–61. Springer, Heidelberg (2013). doi:10.1007/978-3-642-37453-1_5
8. Kaytoue, M., Codocedo, V., Buzmakov, A., Baixeries, J., Kuznetsov, S.O., Napoli, A.: Pattern structures and concept lattices for data mining and knowledge processing. In: Bifet, A., May, M., Zadrozny, B., Gavalda, R., Pedreschi, D., Bonchi, F., Cardoso, J., Spiliopoulou, M. (eds.) ECML PKDD 2015. LNCS, vol. 9286, pp. 227–231. Springer, Cham (2015). doi:10.1007/978-3-319-23461-8_19
9. Mooney, C.H., Roddick, J.F.: Sequential pattern mining - approaches and algorithms. ACM Comput. Surv. **45**(2), 19:1–19:39 (2013)
10. Pei, J., Han, J., Mortazavi-Asl, B., Pinto, H., Chen, Q.C.Q., Dayal, U., Hsu, M.: PrefixSpan: mining sequential patterns efficiently by prefix-projected pattern growth. In: Proceedings 17th International Conference on Data Engineering, pp. 215–224 (2001)
11. Wang, J., Han, J.: BIDE: efficient mining of frequent closed sequences. In: 20th International Conference on Data Engineering, Proceedings, pp. 79–90 (2004)
12. Yan, X., Han, J., Afshar, R.: CloSpan: mining: closed sequential patterns in large datasets. In: Proceedings of the 2003 SIAM International Conference on Data Mining, pp. 166–177 (2003)
13. Zaki, M.J.: SPADE: an efficient algorithm for mining frequent sequences. Mach. Learn. **42**(1–2), 31–60 (2001)
14. Zhang, J., Wang, Y., Yang, D.: CCSpan: mining closed contiguous sequential patterns. Knowl.-Based Syst. **89**, 1–13 (2015)

An Investigation of User Behavior in Educational Platforms Using Temporal Concept Analysis

Sanda-Maria Dragoş$^{(\boxtimes)}$, Christian Săcărea, and Diana-Florina Şotropa

Babeş-Bolyai University, Cluj-Napoca, Romania
sanda@cs.ubbcluj.ro, csacarea@math.ubbcluj.ro, diana.halita@ubbcluj.ro

Abstract. In this paper, we focus on the problem of investigating user behavior using conceptual structures distilled from weblogs of an educational e-platform. We define a set of so-called attractors as sets of scales in conceptual time systems and compute user life tracks in order to highlight different types of behaviors. These life tracks can give valuable feedback to the instructor how his students are using the online educational resources, analyzing their behavior and extracting as much knowledge as possible from the log access files. This might also be helpful to analyze the usability of the online educational content, eventually for improving the structure of the platform and to develop new educational instruments.

Keywords: Life tracks · Temporal Concept Analysis · Web logs analysis · Conceptual structures · User behavior · Attractors

1 Introduction

Weblogs are log files which are created and maintained by a web server. Every activity on the website, including views of documents, images and other objects is logged. It contains raw information about the site's visitors: activity statistics, accessed files, referring pages, paths through the site, time stamps for every activity, as well as browsers, operating systems and many more. Hence, weblogs are raw data strings, comprising in a concise format all relevant information related to user's activity on a website.

Of particular interest are weblogs of online educational platforms, since they contain the entire knowledge on how students are using the available educational resources. While in traditional education the instructor has a limited insight on how his students are using the provided educational resources, obtaining the necessary information via indirect feedback (assessments, examinations or interaction with students), weblogs might provide a complete landscape of their online activity. Moreover, the feedback might be used in web personalisation or in developing new recommendation systems or educational tools.

Using the metaphor of conceptual landscapes of knowledge introduced by Wille in [1] and methods of Formal Concept Analysis in order to find relevant

---

Diana Şotropa was supported by Tora Trading Services private scholarship.

K. Bertet et al. (Eds.): ICFCA 2017, LNAI 10308, pp. 122–137, 2017.
DOI: 10.1007/978-3-319-59271-8_8

conceptual structures in weblogs lies at hand. The challenge to find and investigate relevant conceptual structures from weblogs is not minor since they are structured according to some particularly defined rules, having their own logic and containing in a rather hidden way the knowledge we are looking for. Nevertheless, this research shows once more the power of FCA methods and tools in distilling valuable knowledge from various data formats.

This research is driven by the necessity of understanding *how* students are using certain educational contents and *why* are they behaving in a way or another. There is no direct interaction with the students, but merely a silent observation of *what they are doing*, at which particular time and in what particular order. By this, we want to detect relevant behavioral patterns distilling them directly from the stored weblogs.

In order to understand how and why are students using an educational system, creating their user profiles is a necessity. This can be done by gathering information specific to each visitor, either explicitly or implicitly: users interests on the information presented on the platform or users behavior while navigating the platform. This kind of information might be exploited in order to customize the content and the structure of a web site to serve the visitors specific needs.

We exemplify the methods we have proposed on a locally developed e-learning platform called PULSE [2]. This portal, which is personalized for every user that enters it, was mainly designed to be used for presenting theoretical support for the studied subjects and automatically setting assignments and recording evaluations for individual work and tests. The system was progressively built and enhanced according to its users needs. This gave us the necessary access to improve not only the educational content on PULSE but also its design, since we wanted to have an informed learning management system that continually educates itself about the requirements of its users as a result of the feedback offered by various pattern mining tools and thus to evaluate the effectiveness of PULSE.

An educational platform can offer the possibility for educators to distribute information to students, produce content material, prepare assignments and tests, manage distance classes and enable collaborative learning with file storage areas, news services, etc. [3].

The traditional development of e-learning courses is a laborious activity because the instructor has to choose the contents that will be shown, decide on the structure of the contents, and determine the most appropriate content elements for each type of potential user of the course. Due to the complexity of these decisions, a one-shot design is hardly feasible, even when carefully done. Instead, most cases will probably need evaluation and possibly modification of course content, structure and navigation based on students' usage information. To facilitate this, data analysis methods and tools are used to observe students' behaviour and to assist instructors in detecting possible errors and shortcomings and in incorporating possible improvements.

The application of data mining in e-learning systems is an iterative cycle [3]: collect data, preprocess the data, apply data mining and interpret, evaluate and deploy the results.

A detailed discussion of various methods in behavioral pattern mining in web based educational systems is made in [4]. Quantitative methods are providing an overall perspective, wherefrom the focus can be set on specific data segments. FCA grounded data analysis methods are then used to provide a more detailed perspective. Weblogs are considered as many-valued contexts and particularly built conceptual scales are used to decode the contained relevant knowledge. Other contributions like [3] are focusing on behavioral pattern analysis in well-established learning platforms.

This research builds upon previous work on investigating conceptual structures distilled from weblogs using FCA. An initial analysis on some conceptual structures is performed in [5], while [6] describes how user dynamics can be investigated using ToscanaJ [7] and Triadic FCA (3FCA). The idea of applying 3FCA to the study of web usage behavior has been continued in [8], where a circular visualization tool called CIRCOS[1] has been employed to represent patterns of user dynamics and in [9] where the notion of *attractor* has been introduced. Paper [10] employs 3FCA to investigate correlations between similar page chains and the time when a certain pattern occurs. Tetradic FCA (4FCA) is then used to compare web usage patterns wrt. temporal development and occurence. Paper [11] investigates trend-setters in online educational platforms using polyadic FCA (4FCA and 5FCA) and Answer Set Programming, while [12] describes how new educational strategies correlate with students' academic performance and the use of online learning resources.

In this paper, we apply the methods of Temporal Concept Analysis (TCA) to define and discuss a particular set of *attractors*: navigational, habitual, and critical and to investigate how users adhere to them. Briefly defined, attractors can be considered as a category of so-called *qualitative patterns* (a complementary category to the more quantitative patterns detected by usual web usage tools) which are distilled from weblogs. While accessing the educational platform, users more or less *adhere* to them and this is considerably influencing their navigation behavior. A major category are *educational attractors*, reflecting an intended structure of the learning resources as it has been structured by the educator. We make use of the various time stamps in weblogs to study the relationship between these educational attractors and the temporal development of users behavior. We are also interested to study their trajectories through the educational content of the PULSE platform. We use TCA to define life tracks and to study them wrt the educational attractor. This gave us the possibility to understand how these users are experiencing the educational content and to correlate their behavior with academic performance.

2   Related Work

Browsing web pages became an essential aspect of our everyday life. The World Wide Web developed to a huge information service center, having an explosive growth of information. Navigation on the Internet has a social dimension [13]

---

[1] www.circos.ca.

and becomes a very effective mechanism for acquiring knowledge [14]. Improving web communication is an essential task, in order to adapt the content and to make it more feasible with the expectations of the visitors, and, on the other hand, to improve both content and design of the web pages. Behavioral aspects become more and more important, and web usage tools have been developed in order to analyze the web usage behavior. Recent investigations try to unveil the way how people interact with the Internet, its impact on our daily life, the cultural transformations which come along with it [15].

Initially emerged from the corporate environment, but then rapidly adapted to any other needs, web analytics and its newer research field, web usage mining, offer many communication analysis techniques [16]. Statistics and/or data mining techniques are used to discover and extract useful information from weblogs [17], discover interesting usage patterns, cluster the users into groups according to their navigational behavior or discover potential correlations between Web pages and user groups. Other techniques include predictions about the user's behavior by discovering of interesting web usage patterns [18].

Web analytics are mainly used on e-commerce sites to provide a rough insight about the usage of the analysed web site. However, such instruments are not precise enough for the educational content [19]. Web usage mining [20] constitutes an important feedback for website optimization, for web personalization [21] and predictions [3]. Surveys [22,23] present educational data mining works in order to show recent educational data mining (EDM) advances.

Authors of [24] define blended learning as a hybrid of traditional and online learning, where the online component becomes a natural extension of traditional learning. There are a variety of blended learning classes in universities. The study [25] considers on one hand the case of the discussion-based blended learning course, which involves active learner's participation in online forums, and on the other hand the case of the lecture-based blended learning course, which involves submitting tasks or downloading materials as main online activities. The results show that the data collected in first case can be processed in order to predict linear relations between online activities and student performance (i.e., total score). However, the same analysis model was not appropriate for prediction for the data in the second case.

It is found in [3] that a single algorithm with the best classification and accuracy in all cases are not possible, even though highly complicated and advanced data-mining technique are used. Thus, offline information such as classroom attendance, punctuality, participation, attention and predisposition were suggested to increase the efficiency of such algorithms.

3   Preliminaries

In an e-learning platform, a huge amount of data is stored from multiple sources. They provide a database that stores all system information: personal information about the users (profile), academic results and users' interaction data. The weblog system we have used records all PULSE accesses into a MySQL database.

The data fields from the collected information are: the time stamp (date and time) of the request, the IP address of the, originating Web page request, screen resolution, full request-URI (the file, the name, and the method used to retrieve it), including the domain, the requested URL, and any applicable query parameters, full unmodified user-agent string (which encodes the operating system and the browser used by the user), referrer (which is the URL the client was visiting before requesting that URL), login id, cookie id, along with information regarding students attendance and performance [26]. During two academic years, over 130,000 entries have been logged in the database. The educational content uploaded on PULSE was regarding a compulsory course for which students have to use the e-learning platform. These systems can generate, when intensively used, vast quantities of data, which can be analyzed using several techniques, from purely quantitative approaches to more and more sophisticated knowledge discovery methods, where FCA can prove its efectiveness (Table 1).

Table 1. The structure of a log record

Ip	Access file	Referrer	Screen res	Time stamp	User agent	Login	Cookie Id
14704	/~sanda...	http://...	1600×900	1360348736	Mozilla/5.0 ...	sanda	1499640

Web log data are preprocessed in order to identify users, sessions, page views, etc. Data are cleaned from accesses of robots and spider crawlers, since such entries do not provide useful information about the site's usability. Also, web log data may need to be cleaned from entries involving pages that returned an error. In some cases such information might be useful, but in others such data should be eliminated from a log file. We aggregated time stamps into larger granules, i.e., weeks and access files into classes. The sessions are actual HTTP sessions, while the user is identified by his/her login ID. The logged data is integrated with the records about student performance (per laboratory work, tests and exams) and attendance, and with the correlation between the contents between different site sections regarding theoretical support (i.e., lectures, explained test papers, and theoretical support for labs) and laboratory work.

The most important issue is the user and session identification. More accurate approaches for a priori identification of unique visitors is the requirement for user registration. Assuming a user is identified, the next step is to perform session identification. The algorithm of identifying sessions in an educational environment might be different than in others due to the fact that the learning process usually takes time and it does not comply to usual heuristics [19].

We consider the web log data as a many-valued context and use the ToscanaJ suite[2] to conceptually scale the many-valued attributes, to build new scales, and to use them to discover and then investigate interesting patterns.

We will briefly recall some definitions introduced by Wille in [1] regarding many-valued contexts and conceptual scaling.

---

[2] https://sourceforge.net/projects/toscanaj/.

Definition 1. *Many-valued contexts*

A many-valued context (G, M, W, I) *consists of sets* G, M, *and* W *and a ternary relation* I *between* G, M *and* W *(i.e.,* $I \subseteq G \times M \times W$*) for which it holds that* $(g, m, w) \in I$ *and* $(g, m, v) \in I$ *always implies* $w = v$.

The triple $(g, m, w) \in I$ is read as "the attribute m has the value w for the object g". The many-valued attributes can be regarded as partial maps from G in W. Therefore, it seems reasonable to write $m(g) = w$ instead of $(g, m, w) \in I$.

In order to derive the conceptual structure of a many-valued context, we need to scale every many-valued attribute. This process is called *conceptual scaling* and it is always driven by the semantics of the attribute values.

Definition 2. *Conceptual scales*

A scale *for the attribute* m *of a many-valued context is a formal context* $S_m := (G_m, M_m, I_m)$ *with* $m(G) \subseteq G_m$. *The objects of a scale are called* **scale values**, *the attributes are called* **scale attributes**.

Every context can be used as a scale. Formally there is no difference between a scale and a context. However, we will use the term "scale" only for contexts which have a clear conceptual structure and which bear meaning.

The set of scales can then be used to navigate within the conceptual structure of the many-valued context (and the subsequent scaled context). Some scales are predefined (like nominally, ordinally, etc.), while for more complex views, we need to define particular scales.

Conceptual time systems have been introduced by Wolff in [27] in order to investigate conceptual structures of data enhanced with a time layer. Basically, conceptual time systems are many-valued contexts, comprising a time part and an event part, which are subject of conceptual scaling, unveiling the temporal development of the analyzed data, object trajectories and life tracks. We briefly recall some basic definitions.

Definition 3. *Conceptual Time System*

Let G be an arbitrary set, (G, M, W, I_T) and (G, E, V, I_E) *many-valued contexts. Let* $\{S_m \mid m \in M\}$ *be a set of scales for* (G, M, W, I_T), *and* $\{S_e \mid e \in E\}$ *a set of scales for* (G, E, V, I_E). *We denote by* $T := ((G, M, W, I_T), (S_m \mid m \in M))$ *and* $C := ((G, E, V, I_e), (S_e \mid e \in E))$ *the correspondent scaled many-valued contexts (on the same object set* G*). The pair* (T, C) *is called a* conceptual time system on G. T *is called the* time part *and* C *the* event part *of* (T, C).

Definition 4. *Conceptual Time Systems with a Time Relation*

Let (T, C) *be a conceptual time system on* G *and* $R \in G \times G$. *The triple* (T, C, R) *is called a* conceptual time system on G with a time relation.

Definition 5. *Transitions in Conceptual Time Systems with a Time Relation*

Let (T, C, R) *be a conceptual time system on* G *with a time relation. Then any pair* $(g, h) \in R$ *is called an* R-transition on G. *The element* g *is called the* start *and* h *the* end *of* (g, h).

Definition 6. *Conceptual Time Systems with Actual Objects and Time Relation*

Let P be a set of objects, G a set of points in time and $\Pi \subseteq P \times G$ a set of actual objects. Let (T, C) be a conceptual time system on Π and $R \subseteq \Pi \times \Pi$. Then the tuple (P, G, Π, T, C, R) is called a conceptual time system on $\Pi \subseteq P \times G$ with actual objects and a time relation R (shortly a CTSOT).

For each object $p \in P$ we can be define the set $p^{\Pi} = \{g \in G \mid (p, g) \in \Pi\}$. Then the set $R_P = \{(g, h) \mid ((p, g), (p, h)) \in R\}$ is called the set of R transitions of p and the relational structure (p^{Π}, R_P) is called the time structure of p.

Definition 7. *Life track of an object*

Let (P, G, Π, T, C, R) be a CTSOT and $p \in P$. Then, for any mapping $f \colon \{p\} \times p^{\Pi} \to X$, the set $f = \{((p, g), f(p, g)) \mid g \in p^{\Pi}\}$ is called the f-life track of p.

4   Investigating User Behaviors

Attractors have been first introduced in [9]. As the name suggests, an attractor is a conceptual structure having a clear meaning and which is considerably influencing the users behavior in the educational platform. By this, attractors prove to be special categories of scales which need to be related to specific time granules, when the attractor occurs. Moreover, an attractor reveals to be a specific behavioral pattern. Students, while browsing the e-learning platform, *adhere* to some attractors or not, showing thus particular browsing habits. Some attractors are *intended*, i.e., encoded in the conceptual structure by the educator itself, while some others are not intended and are occuring at some particular points in time.

Once more we need to highlight that the scope of this investigation was to observe what are users doing, at which time and to understand why are they developing a certain behavior over time, as well as how this behavior is changing in the students group. All this *silent observation* is done by conceptual scaling of raw web log data, and since we are particularly interested in points in time and users behavior over time, using conceptual time systems and thus TCA lies at hand.

Based on users habits on navigating on the platform, we classified them into three main classes. *Early birds* are users that have accessed the provided material before it was expected. *Common users* are users that behave as expected, i.e., they visit the provided material within the expected time interval. *Late rise users* are users that visit the provided material later than expected. These behaviors are not clearly partitioning the set of users. For example, we can find users which somehow lie in between these classes but most of them can be characterized by one of these classes.

4.1   Navigational Attractors

The collection of navigational attractors reflects the conceptual structure of the educational platform as it was intended by the instructor.

Definition 8. *A navigational attractor is a so-called intended conceptual struc-*
ture which reflects the educational purpose of the instructor and suggest users to
follow specific navigational patterns in order to visit a certain collection of pages
within the e-learning platform at a specific point in time. The collection of nav-
igational attractors is formed by a set of conceptual scales on time granules. By
this, each navigational attractor is comprising the event part of the conceptual
time system at the specific time granule.

Navigational attractors have a dual origin. On the one hand, they reflect how
the instructor is structuring the educational content at a specific time granule,
here one week, since students are expected to perform some tasks weekly. On the
other, they are crystallizing from users habits of visiting the educational platform
and they might not have a one-to-one correspondent in the actual navigational
structure of the platform. They are called *navigational attractors* because they
actually persuade the user to visit the educational content of some week in a
specific way. Distilling information while analyzing these attractors might give
valuable information about users' habits, as well as the usage of the educational
content over time.

For instance, Figs. 1, 2 and 3 are all displaying the concept lattice of the
navigational attractor of the ninth week of the semester. As we can see, the
instructor has structured the content for the ninth week in Lab Assignment,
Lab Theory, a Test Paper and an example, and requests the visit of the content
of Lecture 7 in order to fulfil the assignments. Then, we compute life tracks
for every single user, classify them in early-birds, common users and late risers
and then superimpose all these life tracks on the structure of the navigational
attractors. For instance, some early bird user (Fig. 1) visits Lab Assignment 9
(S4, S5) in the 4th and 5th week of the semester (while expected to do this
in the 9th), in the 6th week it is already visiting both Lab Assignment 9 and

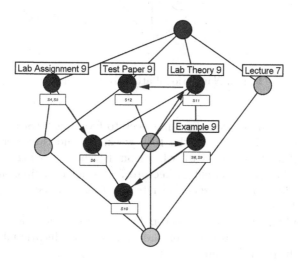

Fig. 1. Navigational attractor - early bird user

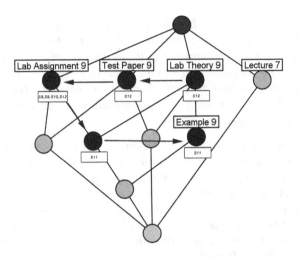

Fig. 2. Navigational attractor - common user

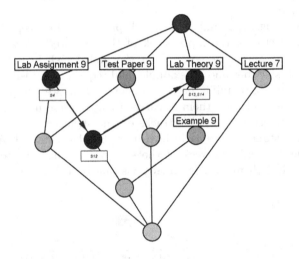

Fig. 3. Navigational attractor - late user

Lab Theory 9 (S6), moving in the same week to Example 9 (S6) (and revisiting at deadline, i.e., week 9/S9). In the 10th week, we find this user in the most bottom node of his lifetrack (S10), wherefrom he/she is moving to Lab Theory 9 in week 11 (S11). The content of Lecture 7 was not consulted at all. Similar interpretations for common users and late rise users, respectively, can be read off from Figs. 2 and 3.

One can build navigational attractors for every week, but they become interesting in investigating users behavior and usage of the educational content only after some weeks from the course start.

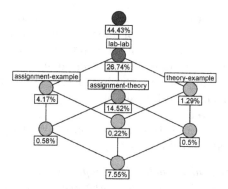

Fig. 4. Habitual attractor - users make branches in their chain of visited pages to/from the classes of pages labeled in the nodes

4.2   Habitual Attractors

Another class of attractors are so-called *unintended attractors*, a subclass of which are the habitual attractors. Opposite to intended attractors, like navigational, the unintended attractors are crystallizing behavioral patterns of users which actually shows how students are using the resources, independently to the intention of the educator. The question is why are students behaving like this while navigating the e-learning platform? Analyzing habitual attractors might give hints on how students would like to have the information structured, or what they are actually looking for while consulting the educational resources and what they expect or wish to be included in the websites structure.

Investigating this class of behavioral patterns started with the observation of so-called *branching behavior*. The browsing history is captured in page chains (i.e., sequences of visited pages where the accessed page becomes the referrer for the next one). When the referrer is not the same as the last page accessed it means that the user opened a new browser tab or window, and we called that part of our session chain a new *branch*.

Definition 9. *A habitual attractor is a conceptual scale on a given time granule, which reflects the branching behavior and reveals clues about restructuring information on the e-learning platform or some other unintended patterns showed off by the users.*

Figure 4 shows the structure of some habitual attractor. This type of attractor gives clues to what users consider to be interesting. By analysing the branching behavior, we have observed that more than half of the separate branches contain pages regarding lab related content.

4.3   Critical Attractors

Let T be a set of time granules (i.e., in our case the weeks of academic year). Similar to the navigational attractor, we can define a new habitual attractor for each time granule. Specific events (like examination periods or deadlines), are generating some inflexion points in the students behavior and usage habits of the e-learning platform. It is for sure of interest to investigate how these specific events are reflected in the structure of the habitual attractors.

Let CP be the set of events (called *critical points*) which might arise independently or might be imposed by the instructor. The critical points can be detected and thus correlated with peaks of accesses on the educational platform around certain critical events.

Let (T, C) be a conceptual time system on G. We will consider the set of empirically determined critical points.

Definition 10. *A critical attractor is a habitual attractor on a critical time granule from CP and is defined as a conceptual scale which reflects users behavior around critical events.*

Figure 5 shows that out of all pages which students visit during the examination period, the one presenting educational content are more visited. The main characteristic for this type of attractor is a flash-like visit: students stay only a few seconds and/or they refresh allot to see new content.

Even if the analysed event is a critical point in time, we have observed that within the educational content the accesses are quite balanced as depicted in Fig. 6.

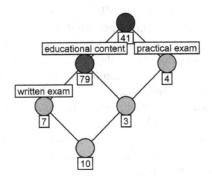

Fig. 5. Critical attractor - users visit educational content on examination period

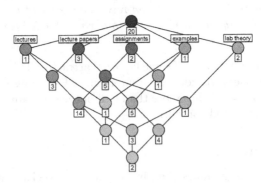

Fig. 6. Critical attractor - what educational content is visited?

5   Discussion: Students Life Tracks

Let (T, C) be a conceptual time system on G. We resume our discussion to life tracks in navigational attractors and the information it might provide to the instructor which observes how his students are using the online educational resources of his course. We set the time granulation on one week, which is a quite fair "time resolution" for our purposes. Then we consider for every week the corresponding navigational attractor and then build users life tracks superimposing them on the concept lattice of the navigational attractor.

As we have seen before, there are multiple stages in which a user can be over time. Most users are starting at the attribute concept level. This is quite normal for a first encounter with the course content, but it might also signalize - if continued over more navigational attractors - a quite superficial approach to the learning process, which needs to be corrected by the instructor by active means. Some others can be found on the top of the lattice because they visited some pages from the platform just out of curiosity. More deeper a life track goes in the concept lattice, the more content is visited and one can expect that the user is learning more and more specific skills related to the subject treated in a certain week. This also reflects how seriously users are approaching a specific subject imposed by the instructor, its structure being unveiled by the corresponding navigational attractor.

For the considered conceptual scale, the more down node is reached by a user the better is the overview on the learning topic. A naturally question is arising: what can the instructor do suggest, recommend or even develop new educational means in order to modify the trajectories of his students through the learning content?

We can also remark that both early birds and late rise users are significantly biased wrt to their time granule. While in the first case, curiosity or willingness to learn is the main motivation factor, in the second, there are some other motivations, sometimes because students are being re-enrolled in the course. In any case, life tracks and different categories of attractors proves to be a quite valuable instrument for the instructor to evaluate the behavior of his students and the way how they are approaching the learning process.

Summarizing, in educational platforms, life tracks offer valuable insights about the usage of the platform at the full potential. This feedback might be used in order to improve the platform, to design recommendation systems for students or to develop new educational tools.

6   Example: Formalizing the Life Track of a Student

Let $P = \{student_1, student_2, \ldots, student_x\}$ be the set of students who are enrolled on the e-learning platform and $G = \{w_1, w_2, \ldots, w_y\}$ be the set of points in time. Then we can define $\Pi \subseteq P \times G$ as the set of actual objects.

For each student $p \in P$ we define

$$p^{\Pi} = \{g \in G | (p, g) \in \Pi\} = \{w_{p_1}, \ldots, w_{p_q}\} \tag{1}$$

as the set of weeks in which student p had activity, i.e. visited the platform and

$$R_p = \{(g,h)|((p,g),(p,h)) \in R\} \tag{2}$$

as the set of pairs of weeks in which student p had activity.

We define for each navigational attractor a time interval in which the student is expected to adhere to it. For a particular attractor $n \in \{1, \ldots, y-2\}$, the expected interval of adhering to the attractor is w_n, w_{n+1}, or w_{n+2}.

Considering each student p and each attractor n, we will define the set of R transitions for each class of users:

$$Early_p = \{g \in G|(p,g) \in \Pi, g < n\} \tag{3}$$

$$Common_p = \{g \in G|(p,g) \in \Pi, n \leq g \leq n+2\} \tag{4}$$

$$Late_p = \{g \in G|(p,g) \in \Pi, g > n+2\} \tag{5}$$

In this example, we select a subset of common users (i.e., $P = \{D, H, I\}$), their corresponding set of points in time (i.e. $G = \{w8, w9, w10, w11, w12, w13, w14\}$) and the concept lattice of the navigational attractor of the ninth week of the semester. The results are presented in Fig. 7.

$\Pi = \{(D,w8); (D,w9); (D,w10); (D,w11); (D,w12); (H,w8); (H,w10);$
$(H,w11); (H,w13); (H,w14); (I,w10); (I,w11); (I,w12); (I,w13)\};$
$D^\Pi = \{w8, w9, w10, w11, w12\};$
$H^\Pi = \{w8, w10, w11, w13, w14\};$
$I^\Pi = \{w10, w11, w12, w13\}.$

We generate the sets R_D, R_H and R_I and we represent the visiting process and the transitions between classes of pages of the platform, where $\{H1, H2, H3, H4\}$ represent the set of transitions for user H through educational content.

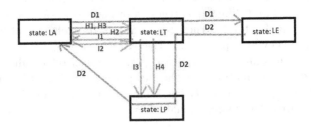

Fig. 7. Common user - transitions through the platform

7   Conclusions and Further Research

Temporal Concept Analysis has been specifically developed to deal with data having a clear temporal structure. Nevertheless, there are surprisingly few real world applications of TCA. This paper proves the effectiveness of combining conceptual scale building and TCA to investigate students behavior in online educational platforms. These methods can be eventually extended to any data set enhanced with a temporal layer, where users behavior is of interest. We consider these methods as a proof of concept on how powerful FCA can be. We have also proved that by using FCA, TCA, etc. one is able to extract lots of very valuable knowledge directly from raw web logs, knowledge which couldn't be extracted herefrom by any other methods besides FCA.

By taking advantage of the user's navigational behavior, characteristics and interests, as revealed through analyzing weblogs, one may customize the content and structure of a Web site to the specific and individual needs of the user. Such information can be further analyzed in association with the content of a Web site, resulting in improvement of the system performance, users' retention, and/or site modification.

We intend to continue this research, considering also pattern structures, relational FCA and other FCA varieties, as well as to build efficient tools supporting FCA grounded data analysis on various data formats.

References

1. Wille, R.: Conceptual landscapes of knowledge: a pragmatic paradigm for knowledge processing. In: Gaul, W., Locarek-Junge, H. (eds.) Classification in the Information Age. Studies in Classification, Data Analysis, and Knowledge Organization, pp. 344–356. Springer, Heidelberg (1999)
2. Dragos, S.: PULSE extended. In: The Fourth International Conference on Internet and Web Applications and Services, pp. 510–515. IEEE Computer Society, Venice/Mestre, May 2009
3. Romero, C., Espejo, P.G., Zafra, A., Romero, J.R., Ventura, S.: Web usage mining for predicting final marks of students that use moodle courses. Comput. Appl. Eng. Educ. **21**(1), 135–146 (2013)
4. Dragos, S., Halita, D., Sacarea, C.: Behavioral pattern mining in web based educational systems. In: Rozic, N., Begusic, D., Saric, M., Solic, P. (eds.) 23rd International Conference on Software, Telecommunications and Computer Networks, SoftCOM 2015, Split, Croatia, 16–18 September 2015, pp. 215–219. IEEE (2015). http://ieeexplore.ieee.org/xpl/mostRecentIssue.jsp?punumber=7303284
5. Dragos, S., Sacarea, C.: Analysing the usage of pulse portal with formal concept analysis. Studia Universitatis Babes-Bolyai Series Informatica **LVII**(3), 65–75 (2012)
6. Dragos, S., Halita, D., Sacarea, C., Troanca, D.: An FCA grounded study of user dynamics through log exploration. Studia Universitatis Babes-Bolyai Series Informatica **2**, 82–97 (2014)
7. Becker, P., Correia, J.H.: The ToscanaJ suite for implementing conceptual information systems. In: Ganter, B., Stumme, G., Wille, R. (eds.) Formal Concept Analysis. LNCS, vol. 3626, pp. 324–348. Springer, Heidelberg (2005). doi:10.1007/11528784_17
8. Dragoş, S., Haliţă, D., Săcărea, C., Troancă, D.: Applying triadic FCA in studying web usage behaviors. In: Buchmann, R., Kifor, C.V., Yu, J. (eds.) KSEM 2014. LNCS, vol. 8793, pp. 73–80. Springer, Cham (2014). doi:10.1007/978-3-319-12096-6_7

9. Dragoş, S., Haliţă, D., Săcărea, C.: Attractors in web based educational systems a conceptual knowledge processing grounded approach. In: Zhang, S., Wirsing, M., Zhang, Z. (eds.) KSEM 2015. LNCS, vol. 9403, pp. 190–195. Springer, Cham (2015). doi:10.1007/978-3-319-25159-2_18

10. Dragoş, S.-M., Haliţă, D.-F., Săcărea, C.: Distilling conceptual structures from weblog data using polyadic FCA. In: Haemmerlé, O., Stapleton, G., Faron Zucker, C. (eds.) ICCS 2016. LNCS, vol. 9717, pp. 151–159. Springer, Cham (2016). doi:10.1007/978-3-319-40985-6_12

11. Dragos, S., Halita, D., Troanca, D.: Investigating trend-setters in e-learning systems using polyadic formal concept analysis and answer set programming. In: The 4th International Workshop on Artificial Intelligence for Knowledge Management (AI4KM), New York, USA, pp. 42–48, July 2016

12. Dragos, S., Halita, D., Sacarea, C.: Analysing the effect of changing the educational methods by using FCA. In: 24th International Conference on Software, Telecommunications and Computer Networks, SoftCOM 2016, Split, Croatia, 22–24 September 2016, pp. 1–5. IEEE (2016). http://ieeexplore.ieee.org/xpl/mostRecentIssue.jsp?punumber=7754733

13. Dieberger, A.: Supporting social navigation on the world wide web. Int. J. Hum.-Comput. Stud. 46(6), 805–825 (1997)

14. Beydoun, G., Kultchitsky, R., Manasseh, G.: Evolving semantic web with social navigation. Expert Syst. Appl. 32(2), 265–276 (2007)

15. Gonçalves, B., Ramasco, J.J.: Human dynamics revealed through web analytics. CoRR, vol. abs/0803.4018 (2008). http://arxiv.org/abs/0803.4018

16. Norguet, J., Tshibasu-Kabeya, B., Bontempi, G., Zimányi, E.: A page-classification approach to web usage semantic analysis. Eng. Lett. 14(1), 120–126 (2007)

17. Kosala, R., Blockeel, H.: Web mining research: a survey. SIGKDD Explor. 2(1), 1–15 (2000)

18. Srivastava, J., Cooley, R., Deshpande, M., Tan, P.: Web usage mining: discovery and applications of usage patterns from web data. SIGKDD Explor. 1(2), 12–23 (2000)

19. Macfadyen, L.P., Dawson, S.: Numbers are not enough. Why e-learning analytics failed to inform an institutional strategic plan. Educ. Technol. Soc. 15(3), 149–163 (2012)

20. Spiliopoulou, M., Faulstich, L.C.: WUM: a tool for web utilization analysis. In: Atzeni, P., Mendelzon, A., Mecca, G. (eds.) WebDB 1998. LNCS, vol. 1590, pp. 184–203. Springer, Heidelberg (1999). doi:10.1007/10704656_12

21. Romero, C., Ventura, S., Zafra, A., de Bra, P.: Applying web usage mining for personalizing hyperlinks in web-based adaptive educational systems. Comput. Educ. 53(3), 828–840 (2009)

22. Peña-Ayala, A.: Educational data mining: a survey and a data mining-based analysisof recent works. Expert Syst. Appl. 41(4), 1432–1462 (2014)

23. Romero, C., Ventura, S.: Data mining in education. Wiley Interdisc. Rev.: Data Min. Knowl. Discov. 3(1), 12–27 (2013)

24. Liebowitz, J., Frank, M.: Knowledge Management and e-Learning. CRC Press, Boca Raton (2010)

25. Jo, I.-H., Park, Y., Kim, J., Song, J.: Analysis of online behavior and prediction of learning performance in blended learning environments. Educ. Technol. Int. 15(2), 71–88 (2014)

26. Dragoş, S.-M.: Why Google analytics cannot be used for educational web content. In: 2011 7th International Conference on Next Generation Web Services Practices (NWeSP), pp. 113–118. IEEE (2011)
27. Wolff, K.E.: Temporal concept analysis. In: ICCS-2001 International Workshop on Concept Lattices-Based Theory, Methods and Tools for Knowledge Discovery in Databases, pp. 91–107. Stanford University, Palo Alto (2001)

Hierarchies of Weighted Closed Partially-Ordered Patterns for Enhancing Sequential Data Analysis

Cristina Nica[✉], Agnès Braud, and Florence Le Ber

ICube, University of Strasbourg, CNRS, ENGEES, Strasbourg, France
{cristina.nica,florence.leber}@engees.unistra.fr, agnes.braud@unistra.fr
http://icube-sdc.unistra.fr

Abstract. Discovering sequential patterns in sequence databases is an important data mining task. Recently, hierarchies of closed partially-ordered patterns (cpo-patterns), built directly using Relational Concept Analysis (RCA), have been proposed to simplify the interpretation step by highlighting how cpo-patterns relate to each other. However, there are practical cases (e.g. choosing interesting navigation paths in the obtained hierarchies) when these hierarchies are still insufficient for the expert. To address these cases, we propose to extract hierarchies of more informative cpo-patterns, namely weighted cpo-patterns (wcpo-patterns), by extending the RCA-based approach. These wcpo-patterns capture and explicitly show not only the order on itemsets but also their different influence on the analysed sequences. We illustrate how the proposed wcpo-patterns can enhance sequential data analysis on a toy example.

1 Introduction

Searching for sequential patterns [1] is a well-known data mining task whose aim is to find regularities and tendencies in sequential data that can be interpreted and assessed by experts. Various algorithms have therefore been proposed [9] and many of them focus on extracting efficiently concise representations of sequential patterns (e.g. closed sequential patterns [15]). To obtain a more compact set of such sequential patterns, efficient algorithms for directly mining closed partially-ordered patterns (cpo-patterns, [2]) were proposed in [5,12]. Precisely, a cpo-pattern summarises a set of closed sequential patterns, which coexist in the same sequences, and it has a graphical representation that facilitates the interpretation step. However, regardless of the fewer number of obtained cpo-patterns, the interpretation step remains difficult since these cpo-patterns are unorganized.

In [10], Relational Concept Analysis (RCA, [13]) is used to directly extract hierarchies of cpo-patterns that help the interpretation step by highlighting the relationships between cpo-patterns. Indeed, RCA classifies sets of objects described by attributes and relations, allowing the discovery of hierarchies of patterns. Nica et al. have proposed to extract cpo-patterns by navigating only the intents of the interrelated concepts from the RCA result, i.e. a family of concept lattices, beginning with concept intents from the *main lattice*. The extracted

© Springer International Publishing AG 2017
K. Bertet et al. (Eds.): ICFCA 2017, LNAI 10308, pp. 138–154, 2017.
DOI: 10.1007/978-3-319-59271-8_9

hierarchies of cpo-patterns help in understanding the obtained knowledge and provide a quick way to navigate to interesting cpo-patterns.

Nevertheless, cpo-patterns still do not capture all the particularities hidden in the analysed sequential data. A cpo-pattern considers only the order on itemsets in its supporting sequences, and, besides, the itemsets are treated uniformly even if they can have different roles in these sequences. In fact, previous studies showed that exploiting the time information from the analysed sequences, such as capturing time-intervals between adjacent itemsets [4] in the mined sequential patterns, leads to more valuable knowledge. In addition, there are practical cases (e.g. choosing among cpo-patterns that have the same frequency in the analysed data) when the hierarchical order on the extracted cpo-patterns given by the lattice is still insufficient for the expert. In contrast to existing works, here, we propose to study and measure the repetitive occurrences of *preceded itemsets* in a cpo-pattern, i.e. itemsets with specific predecessors; this measure may show the non-accidental occurrence of such itemsets in the considered sequences.

To address the aforementioned limitations, this paper focuses on extracting hierarchies of more informative cpo-patterns, namely *weighted cpo-patterns* (wcpo-patterns), that capture and explicitly show the different weightiness of itemsets. These hierarchies can be directly obtained by extending the RCA-based extraction method presented in [10]. Briefly, we suggest as well to navigate the extents of the interrelated concept that reveal the different weightiness of preceded itemsets in the analysed sequential data. Accordingly, by exploiting the RCA result, we extract hierarchies of wcpo-patterns that better characterise the analysed sequential data.

Our paper is structured as follows. In Sect. 2 we give the theoretical background of our work. Section 3 introduces a running medical example and details how to explore it using RCA. Section 4 formally defines our proposal for mining directly wcpo-patterns. Then, we illustrate how the proposed wcpo-patterns can enhance the sequential data analysis in Sect. 5. Finally, we present an overview of the related work in Sect. 6, and conclude the paper in Sect. 7.

2   Preliminaries

Our approach relies both on sequential patterns and formal concept analysis domains.

2.1   Sequences, Sequential Patterns and PO-Patterns

Let $\mathcal{I} = \{I_1, I_2, ..., I_m\}$ be a set of *items*. An *itemset* IS is a non empty, unordered, set of items, $IS = (I_{j_1}...I_{j_k})$ where $I_{j_i} \in \mathcal{I}$. Let $\mathcal{IS}$ be the set of all itemsets built from $\mathcal{I}$. A *sequence* S is a non empty ordered list of itemsets, $S = \langle IS_1 IS_2...IS_p \rangle$ where $IS_j \in \mathcal{IS}$. The sequence S is a *subsequence* of another sequence $S' = \langle IS'_1 IS'_2...IS'_q \rangle$, denoted as $S \preceq_s S'$, if $p \leq q$ and if there are integers $j_1 < j_2 < ... < j_k < ... < j_p$ such that

$IS_1 \subseteq IS'_{j_1}, IS_2 \subseteq IS'_{j_2}, ..., IS_p \subseteq IS'_{j_p}$. An item can occur only once in an itemset, but can occur several times in different itemsets of the same sequence.

Sequential patterns have been defined by [1] as frequent subsequences found in a sequential dataset. A sequential pattern is associated to a support, i.e. the number of sequences containing the pattern, that has to be greater than or equal to a minimum support, denoted by θ. Formally, the support of a sequential pattern M extracted from a sequential dataset $\mathcal{D_S}$ is defined as $Support(M) = |\{S \in \mathcal{D_S}|M \preceq_s S\}|$. For instance, $M_1 = \langle(d)(bc)\rangle$ and $M_2 = \langle(d)(e)\rangle$ are two sequential patterns found in Fig. 1(a) sequence database for $\theta = 2$.

Sequence Id	Sequence
S_1	$\langle(d)(bce)\rangle$
S_2	$\langle(ad)(bc)(e)\rangle$

(a) $\mathcal{D_S}$ (b) $\mathcal{G}_1$

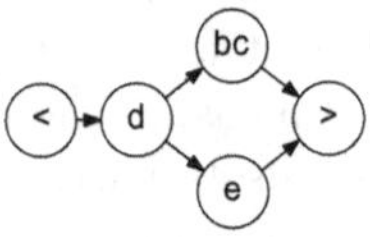

Fig. 1. (a) $\mathcal{D_S}$ a sequence database; (b) $\mathcal{G}_1$ cpo-pattern

Partially-ordered patterns, *po-patterns*, have been introduced by [2], to synthesise sets of sequential patterns. Formally, a *po-pattern* is a directed acyclic graph $\mathcal{G} = (\mathcal{V}, \mathcal{E}, l)$. $\mathcal{V}$ is the set of vertices, $\mathcal{E} \subseteq \mathcal{V} \times \mathcal{V}$ is the set of directed edges, and l is the labelling function mapping each vertex to an itemset. With such a structure, we can determine a strict partial order on vertices u and v such that $u \neq v : u < v$ if there is a directed path from u to v. However, if there is no directed path from u to v, these elements are not comparable. Each path of the graph represents a sequential pattern, and the set of paths in $\mathcal{G}$ is denoted by $\mathcal{P_G}$. A po-pattern is associated to the set of sequences $\mathcal{S_G}$ that contain all paths of $\mathcal{P_G}$. The support of a po-pattern is defined as $Support(\mathcal{G}) = |\mathcal{S_G}| = |\{S \in \mathcal{D_S}|\forall M \in \mathcal{P_G}, M \preceq_s S\}|$. Furthermore, let $\mathcal{G}$ and $\mathcal{G}'$ be two po-patterns with $\mathcal{P_G}$ and $\mathcal{P_{G'}}$ their sets of paths. $\mathcal{G}'$ is a sub po-pattern of $\mathcal{G}$, denoted $\mathcal{G}' \preceq_g \mathcal{G}$, if $\forall M' \in \mathcal{P_{G'}}, \exists M \in \mathcal{P_G}$ such that $M' \preceq_s M$. A po-pattern $\mathcal{G}$ is *closed*, denoted *cpo-pattern*, if there exists no po-pattern $\mathcal{G}'$ such that $\mathcal{G} \prec_g \mathcal{G}'$ with $\mathcal{S_G} = \mathcal{S_{G'}}$. For example, Fig. 1(b) shows $\mathcal{G}_1$ cpo-pattern that synthesises M_1 and M_2 sequential patterns that coexist exactly in the same sequences S_1 and S_2.

2.2 FCA and RCA

Formal Concept Analysis (FCA, [6]) considers an object-attribute context which is a set of objects described by attributes, and builds from it a concept lattice used to analyse the objects. Concisely, an object-attribute context K is a 3-tuple (G, M, I), where G is a set of objects, M a set of attributes, and $I \subseteq G \times M$ an incidence relation. $C = (X, Y)$ where $X = \{g \in G|\forall m \in Y, (g, m) \in I\}$ and $Y = \{m \in M|\forall g \in X, (g, m) \in I\}$ is a formal concept built from K. X and Y

are respectively the extent and the intent of the concept. Let $\mathcal{C}_K$ be the set of all formal concepts that can be built on K. Let $C_1 = (X_1, Y_1)$ and $C_2 = (X_2, Y_2)$ be two concepts from $\mathcal{C}_K$, the concept generalisation order $\preceq_K$ is here defined by $C_1 \preceq_K C_2$ iff $X_1 \subseteq X_2$ ($\Leftrightarrow Y_2 \subseteq Y_1$). $\mathcal{L}_K = (\mathcal{C}_K, \preceq_K)$ is the concept lattice built from K. We denote by $\top(\mathcal{L}_K)$ the concept from $\mathcal{L}_K$ whose extent has all the objects, and by $\bot(\mathcal{L}_K)$ the concept from $\mathcal{L}_K$ whose intent has all the attributes.

RCA extends the purpose of FCA to relational data. RCA applies iteratively FCA on a Relational Context Family (RCF). An RCF is a pair $(\mathcal{K}, \mathcal{R})$, where $\mathcal{K}$ is a set of object-attribute contexts and $\mathcal{R}$ is a set of object-object contexts. $\mathcal{K}$ contains n object-attribute contexts $K_i = (G_i, M_i, I_i), i \in \{1, ..., n\}$. $\mathcal{R}$ contains m object-object contexts $R_j = (G_k, G_l, r_j), j \in \{1, ..., m\}$, where $r_j \subseteq G_k \times G_l$ is a binary relation with $k, l \in \{1, ..., n\}$, $G_k = dom(r_j)$ the domain of the relation and $G_l = ran(r_j)$ the range of the relation. G_k and G_l are the sets of objects of the object-attribute contexts K_k and K_l, respectively. RCA relies on a relational scaling mechanism that is used to transform a relation r_j into a set of *relational attributes* that extends the object-attribute context describing the set of objects $dom(r_j)$. A relational attribute $\exists r_j(C)$, where $\exists$ is the existential quantifier, and $C = (X, Y)$ is a concept whose extent contains objects from the $ran(r_j)$, describes an object $g \in dom(r_j)$ if $r_j(g) \cap X \neq \emptyset$. Other quantifiers can be found in [13]. RCA process consists in applying FCA first on each object-attribute context of an RCF, and then iteratively on each object-attribute context extended by the relational attributes created using the concepts from the previous step. The RCA result is obtained when the families of lattices of two consecutive steps are isomorphic and the object-attribute contexts are unchanged.

3   Relational Analysis of Sequential Data

3.1   Running Example

Patterns hidden in sequential medical data about patients and their medical histories can provide valuable medical knowledge for physicians. Here, we propose to study the symptoms (e.g. fever and cough) that indicate the presence of viruses (e.g. influenza) in patients. The symptoms and viruses are detected by medical examinations and viral tests, respectively. In Fig. 2 is a medical sequence, i.e. a chronologically ordered set of medical examinations with a viral test at the end, all undergone by the same patient. The medical examinations are itemsets of symptoms, while the viral test is the *target 1-itemset* (set of only one item) that contains the studied *object of interest* (here, the influenza virus).

Fig. 2. Medical sequence

Table 1 illustrates a medical toy example of sequential data. We consider only pertinent medical sequences to recognize influenza outbreaks (e.g. only medical examinations undergone by patients within two weeks before a viral test). There are three items as follows: two symptoms FEVER and COUGH, and one virus Influenza. The symptoms can be moderate or high, while the influenza virus can be of type A or B. Thus, in this example we deal with *qualitative sequential data*. For instance, *S3* sequence consists in one medical examination undergone by a patient who experienced moderate cough and moderate fever before being diagnosed with influenza A virus.

Table 1. Medical toy sequential dataset.

Sequence Id	Sequence
S1	$\langle(\text{FEVER}_{\text{moderate}})(\text{FEVER}_{\text{moderate}} \text{ COUGH}_{\text{high}})(\text{FEVER}_{\text{high}} \text{ COUGH}_{\text{high}})(\text{FEVER}_{\text{moderate}})(\text{Influenza}_\text{A})\rangle$
S2	$\langle(\text{FEVER}_{\text{moderate}})(\text{FEVER}_{\text{high}} \text{ COUGH}_{\text{high}})(\text{Influenza}_\text{A})\rangle$
S3	$\langle(\text{COUGH}_{\text{moderate}} \text{ FEVER}_{\text{moderate}})(\text{Influenza}_\text{A})\rangle$
S4	$\langle(\text{FEVER}_{\text{moderate}})(\text{FEVER}_{\text{moderate}} \text{ COUGH}_{\text{high}})(\text{FEVER}_{\text{high}} \text{ COUGH}_{\text{high}})(\text{Influenza}_\text{A})\rangle$
S5	$\langle(\text{FEVER}_{\text{moderate}})(\text{FEVER}_{\text{high}})(\text{FEVER}_{\text{high}})(\text{COUGH}_{\text{high}})(\text{Influenza}_\text{A})\rangle$
S6	$\langle(\text{FEVER}_{\text{moderate}})(\text{COUGH}_{\text{high}})(\text{FEVER}_{\text{high}})(\text{FEVER}_{\text{high}})(\text{Influenza}_\text{B})\rangle$
S7	$\langle(\text{FEVER}_{\text{moderate}})(\text{COUGH}_{\text{high}} \text{ FEVER}_{\text{high}})(\text{COUGH}_{\text{high}})(\text{Influenza}_\text{B})\rangle$
S8	$\langle(\text{COUGH}_{\text{moderate}} \text{ FEVER}_{\text{moderate}})(\text{Influenza}_\text{B})\rangle$
S9	$\langle(\text{FEVER}_{\text{moderate}})(\text{FEVER}_{\text{high}})(\text{COUGH}_{\text{high}})(\text{Influenza}_\text{B})\rangle$
S10	$\langle(\text{FEVER}_{\text{moderate}})(\text{FEVER}_{\text{high}} \text{ COUGH}_{\text{high}})(\text{Influenza}_\text{B})\rangle$

The physicians try to assess the cough and fever symptoms felt by patients to better understand how to early recognize outbreaks of influenza A/B virus, and, besides, to distinguish between influenza A and influenza B outbreaks. To this end, we build qualitative sequential sub-datasets from Table 1, based on the diagnosed type of influenza. Hence, there are two sub-datasets referred to as $\mathcal{D}_{SfluA}$ (sequences from *S1* to *S5*) and $\mathcal{D}_{SfluB}$ (sequences from *S6* to *S10*).

3.2 Preprocessing Qualitative Sequential Data

In this section and in Sect. 3.3, we briefly recall and improve the temporal relational analysis step proposed in [10], and we exemplify it with only $\mathcal{D}_{SfluA}$. Exploiting the relational character of our medical qualitative sequential data given in Table 1, we define a *temporal model* for $\mathcal{D}_{SfluA}$ composed of four sets of objects, as follows: viruses (V), symptoms (S), viral tests (VT), and medical examinations (ME). The viral tests are linked to viruses by a qualitative binary relation *has virus A*. Similarly, medical examinations are linked to symptoms by qualitative relations *has symptom* (mS or hS) differentiated by the type of identified symptoms, e.g. *moderate* or *high*. Viral tests/medical examinations and medical examinations are linked by a temporal binary relation *is preceded by* (ipb) that associates a viral test/medical examination to a medical examination if the viral test/medical examination is preceded in time by the medical examination. There is no temporal binary relation between viral tests since our aim is to study the symptoms that prognosticate influenza A virus.

As explained in Sect. 3.1, the set of viruses contains only one object (item) Influenza and the set of symptoms contains two objects COUGH and FEVER. In order to build the set of viral tests and the one of medical examinations (since an itemset can correspond to several viral tests or medical examinations in the analysed sequences, each occurrence of the itemset should be uniquely identified), we re-modelled our medical sequences of itemsets as the medical sequences of unique identifiers (UIDs) given in Table 2.

Table 2. $\mathcal{D}'_{SfluA}$ of unique identifiers.

Sequence
$\langle$IS1_Seq1 IS2_Seq1 IS3_Seq1 IS4_Seq1 Seq1$\rangle$
$\langle$IS1_Seq2 IS2_Seq2 Seq2$\rangle$
$\langle$IS1_Seq3 Seq3$\rangle$
$\langle$IS1_Seq4 IS2_Seq4 IS3_Seq4 Seq4$\rangle$
$\langle$IS1_Seq5 IS2_Seq5 IS3_Seq5 IS4_Seq5 Seq5$\rangle$

Formally, let $\mathcal{D}_S$ be a sequential dataset and $S \in \mathcal{D}_S$ a sequence of itemsets. We model S as $\langle$IS1_Seq IS2_Seq...ISp_Seq Seq$\rangle$, that is, a sequence of UIDs. Let $\mathcal{D}'_S$ be the set of all such sequences of UIDs derived from $\mathcal{D}_S$ sequences. Seq$i$ is the UID of the *target 1-itemset* and it uniquely identifies the sequence S. We define $G_m = \{$Seq$i\}_{i\in[1,n]}$, where $n = |\mathcal{D}'_S|$, as the set of all UIDs of the target 1-itemsets in $\mathcal{D}'_S$. The function $getS : G_m \rightarrow \mathcal{D}_S$ maps a target 1-itemset UID to the corresponding sequence of itemsets. ISj_Seq$i$ is the UID of an itemset and specifies Seq$i$ sequence that owns the itemset. We define $G_t = \{$ISj_Seq$i\}_{i\in[1,n];j\in[1,l_i]}$, where l_i is the number of itemsets (except the target 1-itemset) in Seq$i$ sequence, as the set of all itemset UIDs, excluding G_m, in $\mathcal{D}'_S$. The function $getSeq : G_t \rightarrow G_m$ maps an itemset UID to the sequence that owns it. The function $getIS : G_m \cup G_t \rightarrow \mathcal{IS}$ maps an itemset/target 1-itemset UID to the corresponding itemset.

For instance, $\mathcal{D}'_{SfluA}$ (Table 2) is a sequence database of UIDs, where G_m is the set of all viral test UIDs, while G_t is the set of all medical examination UIDs. The third sequence $\langle$IS1_Seq3 Seq3$\rangle$ is derived from $getS($Seq3$) = S3$ (Table 1). Seq3 uniquely identifies the viral test $getIS($Seq3$) = ($Influenza$_A)$ in $S3$. IS1_Seq3 is owned by $getSeq($IS1_Seq3$) = $ Seq3 and it uniquely identifies the medical examination $getIS($IS1_Seq3$) = ($COUGH$_{moderate}$ FEVER$_{moderate})$ in $S3$.

3.3 Exploring Qualitative Sequential Data Using RCA

Firstly, the RCA input (RCF) – an excerpt is depicted in Table 3 – is built following the temporal modelling described in Sect. 3.2. Tables KS (symptoms), KVT

Table 3. RCF excerpt composed of object-attribute contexts (KS, KVT and KME), and object-object contexts (RVT-ipb-ME and RhS).

KS

KS	FEVER	COUGH
FEVER	×	
COUGH		×

KVT

KVT
Seq1
Seq2
Seq3
Seq4
Seq5

KME

KME
IS1_Seq1
IS2_Seq1
IS3_Seq1
IS4_Seq1
IS1_Seq2
IS2_Seq2
IS1_Seq3
IS1_Seq4
IS2_Seq4
IS3_Seq4
IS1_Seq5
IS2_Seq5
IS3_Seq5
IS4_Seq5

RVT-ipb-ME

RVT-ipb-ME	IS1_Seq1	IS2_Seq1	IS3_Seq1	IS4_Seq1	IS1_Seq2	IS2_Seq2	IS1_Seq3	IS1_Seq4	IS2_Seq4	IS3_Seq4	IS1_Seq5	IS2_Seq5	IS3_Seq5	IS4_Seq5
Seq1	×	×	×	×										
Seq2					×	×								
Seq3							×							
Seq4								×	×	×				
Seq5											×	×	×	×

RhS

RhS	FEVER	COUGH
IS1_Seq1		×
IS2_Seq1		×
IS3_Seq1	×	×
IS4_Seq1		
IS1_Seq2		
IS2_Seq2	×	×
IS1_Seq3		
IS1_Seq4		
IS2_Seq4		×
IS3_Seq4	×	×
IS1_Seq5		
IS2_Seq5	×	
IS3_Seq5	×	
IS4_Seq5		×

(viral tests) and KME (medical examinations) represent object-attribute contexts. Let us note that, KME has no column since a medical examination is described only using *has symptom* qualitative relations, and the rows represent the UIDs of medical examinations from Table 2. There is no object-attribute context of viruses since we focus on a specific virus and thus all viral tests detect influenza A. Tables RVT-ipb-ME (viral test *ipb* medical examination), RME-ipb-ME (medical examination *ipb* medical examination), RmS (medical examination detects a moderate symptom) and RhS (medical examination detects a high symptom) represent object-object contexts. For example, RVT-ipb-ME has viral tests as rows and medical examinations as columns. A cross indicates a link between objects, e.g. the cell identified by the viral test Seq3 and the medical examination IS1_Seq3 contains a cross since both are undergone by the same patient and the medical examination precedes the viral test, as shown in Table 2.

Secondly, RCA is applied[1] on the aforementioned RCF and the family of concept lattices given in Fig. 3 is obtained after four iterations. There is a concept lattice for each object-attribute context as follows: $\mathcal{L}_{KVT}$ (viral tests), $\mathcal{L}_{KS}$ (symptoms) and $\mathcal{L}_{KME}$ (medical examinations). $\mathcal{L}_{KVT}$ is considered as the *main lattice* since it contains the target 1-itemsets. $\mathcal{L}_{KME}$ is considered as the *temporal lattice* since it describes the temporal links between the itemsets. $\mathcal{L}_{KME}$ and $\mathcal{L}_{KVT}$ are modified during the iterative steps due to the qualitative and temporal relations that have respectively as domain the set of objects of KME and KVT. Each concept is represented by a box structured from top to bottom as follows: concept name, simplified intent, simplified extent. The representation of each lattice is simplified as every attribute/object is top-down/bottom-up inherited. The navigation amongst these lattices follows the concepts used to build relational attributes, e.g. the relational attribute ∃RVT-ipb-ME(CKME_6), which is a temporal one since it introduces the temporal relation *is preceded by*, of the concept intent CKVT_5 in $\mathcal{L}_{KVT}$ lattice allows us to navigate from CKVT_5 to CKME_6 concept in $\mathcal{L}_{KME}$ lattice.

---

[1] Using RCAExplore tool (http://dolques.free.fr/rcaexplore).

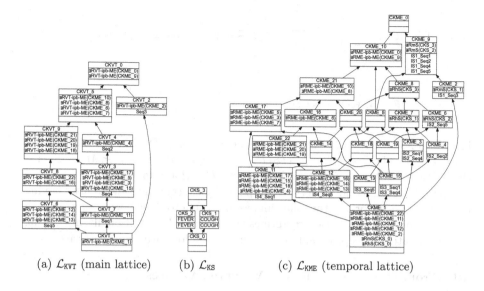

(a) $\mathcal{L}_{\mathrm{KVT}}$ (main lattice)   (b) $\mathcal{L}_{\mathrm{KS}}$ (c) $\mathcal{L}_{\mathrm{KME}}$ (temporal lattice)

Fig. 3. $\mathcal{L}_{\mathrm{KVT}}$ lattice of viral tests, $\mathcal{L}_{\mathrm{KS}}$ lattice of symptoms and $\mathcal{L}_{\mathrm{KME}}$ lattice of medical examinations obtained by applying RCA on the RCF given in Table 3

4   Extracting WCPO-Patterns from the RCA Result

In [10], Nica et al. have shown how to directly extract hierarchies of cpo-patterns by navigating interrelated concept intents from the RCA result (which is built as explained in Sect. 3) beginning with concept intents from the main lattice. For each navigated concept intent, a vertex (itemset) is derived from all qualitative relational attributes, while an edge is derived from each temporal relational attribute. When a cpo-pattern is extracted the order on itemsets in the analysed sequences is exploited, while the itemsets themselves are considered uniformly.

Figure 4, on the left hand side, illustrates a set of concepts whose intents are navigated starting with CKVT_4 from the main lattice $\mathcal{L}_{\mathrm{KVT}}$ (Fig. 3(a)); on the right hand side, is depicted the extracted cpo-pattern, which is contained in each sequence of CKVT_4 extent (*S1*, *S2* and *S4*). The last vertex of the cpo-pattern is derived from the first navigated concept intent. It is noted that the cpo-pattern preserves the order on itemsets in these sequences. However, the cpo-pattern can be misleading if the itemsets have different numbers of occurrences in these sequences. For instance, the cpo-pattern given in Fig. 4 does not encapsulate that in our sequences there are 3 occurrences of (FEVER$_{\mathrm{high}}$ COUGH$_{\mathrm{high}}$) itemset when each occurrence is preceded by (FEVER$_{\mathrm{moderate}}$) itemset, and 5 occurrences of (FEVER$_{\mathrm{moderate}}$) with no constraint on its order.

Formally, let $\mathcal{G} = (\mathcal{V}, \mathcal{E}, l)$ be a cpo-pattern, and $\mathcal{S}_{\mathcal{G}}$ the set of sequences that support $\mathcal{G}$. Let $v_i \in \mathcal{V}$ be a vertex of $\mathcal{G}$, and $\mathcal{V}_i = \{v \in \mathcal{V} | v \le v_i\}$ the set of predecessors of v_i in $\mathcal{G}$ (including v_i). Furthermore, $\mathcal{E}_i = \{(v_k, v_l) \in \mathcal{E} | v_k \in \mathcal{V}_i \text{ and } v_l \in \mathcal{V}_i\}$. $\mathcal{G}_i = (\mathcal{V}_i, \mathcal{E}_i, l)$ is a sub-graph of $\mathcal{G}$, associated to vertex v_i. Let

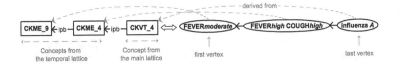

Fig. 4. From a set of navigated concepts to a cpo-pattern

introduce $IS_i \supseteq l(v_i)$ an itemset in a sequence $S \in S_\mathcal{G}$. IS_i is a *preceded itemset* w.r.t. $v_i \in \mathcal{V}$, iff $\exists S_i \preceq_s S, S_i = \langle IS_1 IS_2...IS_p IS_i \rangle$ and $\forall M \in \mathcal{P}_{\mathcal{G}_i}, M \preceq_s S_i$ (i.e. there exists a subsequence of S, ending with IS_i, that supports $\mathcal{G}_i$).

Actually, each sequence of $S_\mathcal{G}$ can repeatedly contain the same itemset having the same predecessors. In the following, we are going to show how to capture such information (that reveal the weightiness of vertices) by additionally navigating the interrelated concept extents to obtain more informative cpo-patterns.

4.1 From Uniform Vertices to Weighted Vertices

Our purpose is to formalise an approach for determining the weightiness of vertices (derived from concepts of the temporal lattice) that correspond to itemsets with specific predecessors, namely *preceded itemsets*. To this end, as explained in Sect. 2.1, let $\mathcal{D}_S$ be a sequence dataset re-modelled as $\mathcal{D}_S'$ of UIDs. G_m is the set of all target 1-itemset UIDs in $\mathcal{D}_S'$, while G_t is the set of all other itemset UIDs.

Let $\mathcal{L}_{K_m} = (\mathcal{C}_{K_m}, \preceq_{K_m})$ be the main lattice (e.g. the lattice of viral tests $\mathcal{L}_{KVT}$) whose set of main concepts $\mathcal{C}_{K_m}$ is derived from the formal context $K_m = (G_m, M_m, I_m)$. G_m is the domain of a temporal relation *is preceded by*, denoted by $ipb_1 \subseteq G_m \times G_t$ (e.g. viral test ipb_1 medical examination). A main concept $C_m \in \mathcal{C}_{K_m}$ is a pair (X_m, Y_m) such that:

- **the intent** Y_m consists of temporal relational attributes that are navigated to reveal $\mathcal{G}_{C_m} = (\mathcal{V}_m, \mathcal{E}_m, l_m)$ cpo-pattern whose last vertex v_m is the one derived from C_m; v_m is labelled with the *target 1-itemset*;
- **the extent** X_m gathers all UIDs in G_m of the sequences that contain all paths in $\mathcal{G}_{C_m}$; $S_{\mathcal{G}_{C_m}} = \{getS(\text{Seq}) \in \mathcal{D}_S | \text{Seq} \in X_m\}$.

Note that the range of ipb_1 temporal relation is G_t, and thus the set of vertices $\mathcal{V}_m$ contains one or more vertices v_t derived from temporal concepts, and v_m vertex. Indeed, let $\mathcal{L}_{K_t} = (\mathcal{C}_{K_t}, \preceq_{K_t})$ be the temporal lattice (e.g. the lattice of medical examinations $\mathcal{L}_{KME}$) whose set of temporal concepts $\mathcal{C}_{K_t}$ is derived from $K_t = (G_t, M_t, I_t)$ context. G_t is both the domain and the range of a second temporal relation *is preceded by*, denoted by $ipb_2 \subseteq G_t \times G_t$ (e.g. medical examination ipb_2 medical examination). A temporal concept $C_t \in \mathcal{C}_{K_t}$ is a pair (X_t, Y_t) such that:

- **the intent** Y_t contains temporal relational attributes that are navigated to reveal $\mathcal{G}_{C_m} = (\mathcal{V}_m, \mathcal{E}_m, l_m)$ cpo-pattern whose vertex v_t is derived from C_t; v_t vertex is labelled with $l_m(v_t)$ itemset;

– **the extent** X_t gathers all UIDs in G_t that identify itemsets containing the itemset $l_m(v_t)$ and respect the temporal order with the UIDs pointed by temporal relational attributes of Y_t. We introduce $X_{t|m} = \{\text{IS\_Seq} \in X_t | getSeq(\text{IS\_Seq}) \in X_m\}$.

Proposition 1. $X_{t|m}$ *is the set of all UIDs that identify a preceded itemset w.r.t.* $v_t \in \mathcal{V}_m$. *Furthermore,* X_t *is the set of all UIDs that identify a preceded itemset w.r.t.* $v_t^k \in \mathcal{V}_m^k$, *with* $k \in \{1, ..., |\mathcal{L}_{K_m}|\}$.

Proof. Let IS_i be a preceded itemset w.r.t. $v_t \in \mathcal{V}_m$. Then $IS_i \supseteq l_m(v_t)$ and $\exists S \in \mathcal{D}_\mathcal{S}, \exists S_i \preceq_s S$ such that IS_i is the last itemset in S_i and S_i supports the sub-graph of v_t predecessors in $\mathcal{G}_{C_m}$ while S supports $\mathcal{G}_{C_m}$. Let Seq be the UID of S: $\text{Seq} \in X_m$. Furthermore, the UID of IS_i, i.e. IS_i_Seq, owns all temporal relational attributes of Y_t and is thus included in $X_{t|m}$.

Let C_t be a temporal concept revealing a vertex $v_t \in \mathcal{V}_m$. Let $\text{IS}_i\_\text{Seq} \in X_{t|m}$ be the UID of $IS_i = getIS(\text{IS}_i\_\text{Seq}) \supseteq l_m(v_t)$, and $S \in \mathcal{D}_\mathcal{S}$ the sequence referred by $getSeq(\text{IS}_i\_\text{Seq}) \in X_m$. $IS_i \in S$, S supports the graph $\mathcal{G}_{C_m}$. We can define $S_i \preceq_s S$ the subsequence of S ending with IS_i. Let $\mathcal{G}_t$ be the sub-graph of v_t predecessors in $\mathcal{G}_{C_m}$: $\forall M \in \mathcal{P}_{\mathcal{G}_t}, M \preceq_s S_i$. Thus IS_i is a preceded itemset w.r.t. $v_t \in \mathcal{V}_m$. □

Definition 1 (Weighted CPO-pattern). *Given a main concept* C_m, *the vertex* v_m *derived from* C_m, *the associated cpo-pattern* $\mathcal{G}_{C_m} = (\mathcal{V}, \mathcal{E}, l)$, *and a function* $w : (\mathcal{V} - \{v_m\}) \to \mathbb{R}_{\geq 0}^n$, *where* n *is constant. A weighted cpo-pattern is a quadruple* $(\mathcal{V}, \mathcal{E}, l, w)$, *i.e. the cpo-pattern* $\mathcal{G}_{C_m}$ *with the function* w *that maps each vertex to a* n-*tuple of real positive numbers (vertex measures of weightiness).*

We propose three vertex measures of weightiness that represent: the *persistency* of the corresponding preceded itemset in the subset of sequences of $\mathcal{D}_\mathcal{S}$ (how many repetitions of it are in that subset); the *overall weight* of the preceded itemset (how often it occurs) in $\mathcal{D}_\mathcal{S}$; the *specificity* of the preceded itemset in the subset of sequences of $\mathcal{D}_\mathcal{S}$ (the extent to which it belongs only to that subset).

In the following, we consider a main concept $C_m = (X_m, Y_m)$, the associated cpo-pattern $\mathcal{G}_{C_m} = (\mathcal{V}, \mathcal{E}, l)$, and a vertex $v_t \in \mathcal{V}$ derived from a temporal concept $C_t = (X_t, Y_t)$.

Definition 2 (Vertex Persistency). *The* persistency *of* v_t, *denoted by* ϖ_{v_t}, *is the total number of repetitions (repetitive occurrences in the same sequence) of preceded itemsets w.r.t.* v_t.

$$\varpi_{v_t} = \frac{|X_{t|m}| - |X_m|}{|X_m|} \tag{1}$$

Persistency of a vertex measures the persistence of the corresponding preceded itemset in a subset of the analysed dataset. We consider that the preceded itemset characterizes the subset of the analysed data if it is not accidental, i.e. the preceded itemset occurs repeatedly in the subset.

Definition 3 (Vertex Overall Weight). *The* overall weight *of* v_t, *denoted by* ω_{v_t}, *is the total number of occurrences of preceded itemset w.r.t.* $v_t^i \in \mathcal{G}_{C_m}^i$, $i \in \{1, ..., |\mathcal{L}_{K_m}|\}$.

$$\omega_{v_t} = |X_t| \tag{2}$$

Overall Weight of a vertex measures how numerous is the corresponding preceded itemset in all analysed sequences. Therefore, the overall weight provides an overview of the number of occurrences of the preceded itemset in the analysed dataset and it can be a reference point used in decision-making by the expert. Using the overall weight of a vertex v_t, the *overall frequency* of v_t in $\mathcal{D}_S$ can be computed as $\varphi_{v_t} = \frac{|X_t|}{|G_t|}$.

Definition 4 (Vertex Specificity). *The* specificity *of* v_t, *denoted by* ς_{v_t}, *is the relative number of preceded itemsets w.r.t.* v_t.

$$\varsigma_{v_t} = \frac{|X_{t|m}|}{|X_t|} 100 \in (0\%, 100\%] \tag{3}$$

Specificity of a vertex measures the extent to which the corresponding preceded itemset is specific for a subset of the analysed data. We consider that the vertex is likely to be more interesting for low values of the specificity, that is, if the preceded itemset characterises the current subset and other sequences from the analysed dataset as well.

Using these three measures, a vertex derived from a temporal concept can be mapped to a 3-tuple such as $(\varpi_{v_t}, \omega_{v_t}, \varsigma_{v_t})$.

4.2   Application to the Running Example

To illustrate our method, let us examine the set of interrelated concepts navigated to extract $\mathcal{G}_{\text{CKVT\_5}}$ cpo-pattern associated to CKVT_5 main concept from $\mathcal{L}_{\text{KVT}}$ (Fig. 3(a)). More precisely, we propose to investigate the navigated concept extents. To this end, Fig. 5 illustrates $\mathcal{G}_{\text{CKVT\_5}}$ cpo-pattern, whose vertices are annotated with 3-tuples $(\varpi_{v_t}, \varphi_{v_t}, \varsigma_{v_t})$, and the navigated concept extents.

The vertex labelled with (Influenza$_\text{A}$) target 1-itemset, is derived from CKVT_5 intent. CKVT_5 extent comprises the sequences in $\mathcal{D}_{SfluA}$ (Table 1) containing all the paths in $\mathcal{G}_{\text{CKVT\_5}}$, i.e. $\mathcal{S}_{\mathcal{G}_{\text{CKVT\_5}}} = \{S1, S2, S4, S5\}$. There are 4 distinct (Influenza$_\text{A}$) target 1-itemsets in $\mathcal{D}_{SfluA}$ that are preceded by the itemsets (FEVER$_\text{high}$), (COUGH$_\text{high}$) and (FEVER$_\text{moderate}$) in the order they appear in $\mathcal{G}_{\text{CKVT\_5}}$. The vertex labelled with (FEVER$_\text{high}$) preceded itemset is derived from CKME_6 temporal concept intent and it is denoted by $v_{\text{CKME\_6}}$. The CKME_6 extent gathers the 5 itemsets (each UID represents an itemset) in $\mathcal{D}_{SfluA}$ that contain FEVER$_\text{high}$ item and that are preceded by the (FEVER$_\text{moderate}$) itemset. Therefore, the overall weight of $v_{\text{CKME\_6}}$ is $\omega_{v_{\text{CKME\_6}}} = 5$. Since all the itemsets in the CKME_6 extent are owned by the sequences in $\mathcal{S}_{\mathcal{G}_{\text{CKVT\_5}}}$, the $v_{\text{CKME\_6}}$ specificity is $\varsigma_{v_{\text{CKME\_6}}} = \frac{5}{5} 100 = 100\%$. In addition, we observe that CKME_6 extent contains the group of itemsets {IS2_Seq5, IS3_Seq5} that occur in the same sequence $getSeq(\text{IS2\_Seq5}) = getSeq(\text{IS3\_Seq5}) = \text{Seq5}$ s.t. $getS(\text{Seq5}) \in \mathcal{S}_{\mathcal{G}_{\text{CKVT\_5}}}$. Then,

CKME_6 extent contains only one repetition identified by IS3_Seq5, and thus v_{CKME_6} persistency is $\varpi_{v_{CKME\_6}} = 0.25$. Similarly, the vertex v_{CKME_7} labelled with (COUGH$_{high}$) preceded itemset is derived from CKME_7 temporal concept intent and has the overall weight $\omega_{v_{CKME\_7}} = 6$ and the specificity $\varsigma_{v_{CKME\_7}} = \frac{6}{6}100 = 100\%$. Since there are two groups of two itemsets that occur in Seq1 and Seq4, respectively, the persistency of v_{CKME_7} is $\varpi_{v_{CKME\_7}} = 0.5$. The vertex v_{CKME_9} labelled with (FEVER$_{moderate}$) preceded itemset is derived from CKME_9 temporal concept intent. CKME_9 extent comprises the 8 itemsets in $\mathcal{D}_{\mathcal{S}_{fluA}}$ (Table 1) that contain FEVER$_{moderate}$ item, i.e. the overall weight of v_{CKME_9} is $\omega_{v_{CKME\_9}} = 8$. The itemset IS1_Seq3 (gray colored in Fig. 5) is owned by $getS(\mathsf{Seq3}) \notin \mathcal{S}_{\mathcal{G}_{CKVT\_5}}$ and thus, the v_{CKME_9} specificity is $\varsigma_{v_{CKME\_9}} = \frac{7}{8}100 = 87.5\%$. Since the CKME_9 extent contains three repetitions identified by IS2_Seq1, IS4_Seq1, and IS2_Seq4, the v_{CKME_9} persistency is $\varpi_{v_{CKME\_9}} = 0.75$.

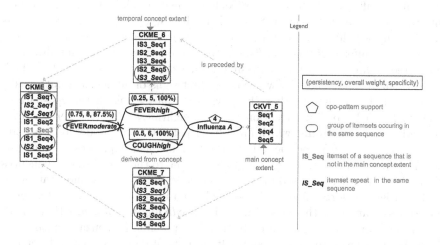

Fig. 5. Extraction of the wcpo-pattern associated to CKVT_5 concept (Fig. 3(a))

5   Enhancing Sequential Data Analysis Using WCPO-Patterns

Using RCA as explained in [10], hierarchies of cpo-patterns are obtained in order to improve the interpretation step by highlighting how the extracted cpo-patterns relate to each other. However, there are practical cases (three such cases are discussed in Sects. 5.1, 5.2 and 5.3) when the order between the extracted cpo-patterns is insufficient for the expert. To improve these practical cases, we propose to use hierarchies of wcpo-patterns and to exploit the vertex measures of weightiness introduced in Sect. 4.1.

Henceforth, we use our running example to illustrate three practical cases that take advantage of the proposed wcpo-patterns when a physician tries to interpret the extracted medical knowledge. As these examples demonstrate, for the sake of our approach illustration, the wcpo-patterns can lead to more informative medical knowledge since the different importance of vertices or paths are considered. It is worth mentioning that the persistency, overall weight and the specificity of a vertex can be considered simultaneously or not depending on the motivation behind the analysis step.

5.1   Practical Case: Ranking CPO-Pattern Vertices and Paths

In a cpo-pattern, its vertices/paths are considered uniformly. The expert can easily be misled by this assumption into thinking that all vertices/paths in a cpo-pattern have the same impact on the object of interest. To illustrate that, let us suppose that a physician tries to interpret the cpo-pattern given in Fig. 5 by disregarding the weightiness of vertices. The physician finds that often before outbreaks of influenza A the patients feel high cough and high fever in any order, but after feeling moderate fever. Since the medical knowledge (cpo-pattern) was mined with very high support (4 out of 5 analysed medical sequences), the physician can infer *with high confidence* that "moderate fever should always be considered as an early sign of a possible influenza A outbreak" and that "high fever and high cough should always be the first signs of influenza A outbreak".

However, let us assume that the physician analyses again the cpo-pattern given in Fig. 5 by paying attention to the weightiness of vertices. High fever and high cough symptoms, which are felt by patients after moderate fever, are $\varsigma_{v_{CKME\_6}} = \varsigma_{v_{CKME\_7}} = 100\%$ specific only to the four medical sequences. In contrast, moderate fever symptom is $\varsigma_{v_{CKME\_9}} = 87.5\%$ specific to the four medical sequences and besides, to other analysed medical sequences. Consequently, moderate fever felt by patients before influenza A outbreaks can be a global available tendency in the dataset and thus the physician can infer *with higher confidence* that "moderate fever should always be an early sign of a possible influenza A outbreak". Since the high fever and high cough felt by patients after moderate fever is a tendency available only in a subset of the analysed dataset, the physician can conclude *with less confidence* that "high fever and high cough should always be the first signs of influenza A outbreak".

In addition, the physician can deduce that high cough is more persistent than high fever in the four medical sequences, i.e. $\varpi_{v_{CKME\_7}} > \varpi_{v_{CKME\_6}}$. Therefore, the high cough is more likely to be the first sign of influenza A outbreak, while high fever, for example, can be caused by a bacterial infection. Similarly, this assumption holds if the overall weights are analysed. Relying on the persistency of high cough, the physician can rank the paths, i.e. FEVER$_{moderate}$ ← COUGH$_{high}$ ← Influenza$_A$ path is more pertinent to recognize influenza A outbreak.

5.2   Practical Case: Selecting Interesting Navigation Paths in CPO-Pattern Hierarchies

Usually the extracted hierarchies of cpo-patterns are very large, and even if the relationships between cpo-patterns are highlighted, and the support measure can be considered, their navigation is still not an easy task for the expert. For instance, let us suppose that a physician tries to navigate the hierarchy of cpo-patterns shown in Fig. 6 while ignoring the weightiness of vertices. This figure depicts an excerpt (with five cpo-patterns from (a) to (e)) from the hierarchy of wcpo-patterns obtained adding new medical sequences to the RCF given in Table 3. The physician begins the navigation from the simple cpo-patterns (having only one vertex) (a) and (b). Thus, the physician has an overview of the common tendencies of the analysed $\mathcal{D}_{SfluA}$ and minimises the chance of overlooking interesting cpo-patterns. It is noted that both cpo-patterns were mined with the same support, and apparently they mark out two interesting navigation paths in the hierarchy. Nevertheless, when the physician considers the persistency of vertices, their different importance are highlighted. The physician can easily infer that high cough is more probable to be a sign of influenza A outbreak, i.e. high cough is more persistent ($\varpi = 2$) than high fever ($\varpi = 1$). Accordingly, the physician selects the navigation path that consists in the descendant wcpo-patterns of (b) and the analysis continues by applying the same ranking criterion.

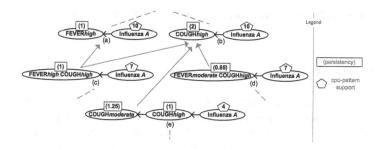

Fig. 6. Excerpt from the hierarchy of wcpo-patterns obtained by adding new medical examinations and viral tests to the RCF given in Table 3

5.3 Practical Case: Distinguishing the Best Represented Sub-Dataset by a CPO-Pattern

There are cases when it is useful to find out discriminant tendencies for different types of the studied object of interest. Here, in our running example, the physician is interested in distinguishing between outbreaks of influenza A and B by assessing the symptoms felt by patients. Usually, the physician determines that the same extracted cpo-pattern belongs rather to $\mathcal{D}_{SfluA}$ or $\mathcal{D}_{SfluB}$ (given in Table 1) by relying on support measure. However, there are cases when a cpo-pattern is found with equal support in both datasets.

For example, let us consider that the physician tries to understand if the cpo-pattern given in Fig. 7 represents a discriminant tendency for influenza A or influenza B outbreak. The cpo-patterns given in Fig. 7(a) and (b) were extracted from $\mathcal{D}_{SfluA}$ and from $\mathcal{D}_{SfluB}$, respectively. Both cpo-patterns were mined with the same support and thus it is impossible to distinguish between them by disregarding the weightiness of vertices. In contrast, when the physician considers, for instance, the persistencies of vertices, it is easily noted that high cough and moderate fever are more persistent in the subset of $\mathcal{D}_{SfluA}$, while high fever has the same persistence in both subsets (one of $\mathcal{D}_{SfluA}$ and one of $\mathcal{D}_{SfluB}$). Accordingly, the physician can conclude that the cpo-pattern is a distinguishing characteristic of influenza A outbreak since two out of three vertices are more significant in Fig. 7(a). Moreover, this inference is drawn with more confidence by additionally considering the overall weights of the vertices, i.e. mainly, high cough and moderate fever are more numerous in $\mathcal{D}_{SfluA}$ than in $\mathcal{D}_{SfluB}$.

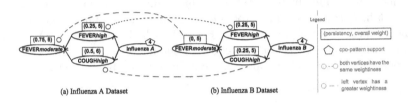

Fig. 7. Distinguishing between outbreaks of influenza A and B

6   Related Work

Traditional pattern mining algorithms in sequence databases, such as those surveyed in [9] for sequential pattern mining and closed sequential pattern mining, or FRECPO [12] and ORDERSPAN [5] for cpo-pattern mining, consider only the order on itemsets in concerned sequences and treat all the itemsets uniformly. To capture more particularities hidden in the analysed data, Srikant and Agrawal [14] propose to extract more informative sequential patterns by adding time constraints in advance, and thus a pattern is extracted only if it admits a max-gap and a min-gap between adjacent itemsets. Pei et al. [11] push various constraints, e.g. time-interval and gap information between items, into the mining process to limit the mining results. Chen et al. [4] propose to extract time-interval sequential patterns that reveal the time interval between successive items and besides these time-intervals are explicitly shown in the patterns. To capture the time interval between all pairs of items in the extracted patterns, Hu et al. [7] introduce the multi-time-interval sequential pattern. Chang [3] propose to find weighted sequential patterns by pushing a time-interval weight measure (the weight of a sequence derived from the time-interval of the sequence itemsets) into the mining process. Besides, in [8,16] more informative sequential patterns are obtained pushing pre-assigned quantitative information, which are recorded in the analysed database, into the mining process. In contrast, our RCA-based approach focuses more on enhancing the interpretation step by extracting hierarchies of wcpo-patterns. We capture for each vertex (preceded itemset) in a cpo-pattern its weightiness (e.g. specificity), in the analysed sequences and we show them explicitly in the extracted wcpo-pattern. Consequently, the expert is guided by (i) the relationships between wcpo-patterns that are revealed by the obtained hierarchies, (ii) the weightiness of vertices and (iii) the more informative order (partial order) of itemsets.

7 Conclusion

This work presents an approach for enhancing sequential data analysis within the framework of RCA. To this end, we propose to extract more informative patterns, namely weighted cpo-patterns, that capture and explicitly show not only the order on itemsets (as do traditional cpo-patterns) but also their different influence on the analysed sequence database through measures such as persistency, specificity and overall weight. Moreover, thanks to the hierarchical RCA results, we directly obtain the relationships between these wcpo-patterns that guide the interpretation step and help in better understanding the extracted knowledge. In this paper, we have formally defined our approach and we have illustrated it on a toy example.

In the future, we plan to study the properties of the proposed measures, and to make a comparison with existing measures of interest. In addition, a possible extension of our work is to consider time-intervals between itemsets, or to add quantitative information recorded in the analysed sequence database to obtain more valuable knowledge.

References

1. Agrawal, R., Srikant, R.: Mining sequential patterns. In: International Conference on Data Engineering, pp. 3–14 (1995)
2. Casas-Garriga, G.: Summarizing sequential data with closed partial orders. In: 2005 SIAM International Conference on Data Mining, pp. 380–391 (2005)
3. Chang, J.H.: Mining weighted sequential patterns in a sequence database with a time-interval weight. Know.-Based Syst. **24**(1), 1–9 (2011)
4. Chen, Y.L., Chiang, M.C., Ko, M.T.: Discovering time-interval sequential patterns in sequence databases. Expert Syst. Appl. **25**(3), 343–354 (2003)
5. Fabrègue, M., Braud, A., Bringay, S., Le Ber, F., Teisseire, M.: Mining closed partially ordered patterns, a new optimized algorithm. Know.-Based Syst. **79**, 68–79 (2015)
6. Ganter, B., Wille, R.: Formal Concept Analysis: Mathematical Foundations. Springer, Heidelberg (1999)
7. Hu, Y.H., Huang, T.C.K., Yang, H.R., Chen, Y.L.: On mining multi-time-interval sequential patterns. Data Knowl. Eng. **68**(10), 1112–1127 (2009)
8. Kim, C., Lim, J.H., Ng, R.T., Shim, K.: SQUIRE: sequential pattern mining with quantities. J. Syst. Softw. **80**(10), 1726–1745 (2007)
9. Mabroukeh, N.R., Ezeife, C.I.: A taxonomy of sequential pattern mining algorithms. ACM Comput. Surv. **43**(1), 3:1–3:41 (2010)
10. Nica, C., Braud, A., Dolques, X., Huchard, M., Le Ber, F.: Extracting hierarchies of closed partially-ordered patterns using relational concept analysis. In: Haemmerlé, O., Stapleton, G., Faron Zucker, C. (eds.) ICCS 2016. LNCS (LNAI), vol. 9717, pp. 17–30. Springer, Cham (2016). doi:10.1007/978-3-319-40985-6_2

11. Pei, J., Han, J., Wang, W.: Mining sequential patterns with constraints in large databases. In: Proceedings of the 11th International Conference on Information and Knowledge Management, CIKM 2002, pp. 18–25. ACM (2002)
12. Pei, J., Wang, H., Liu, J., Wang, K., Wang, J., Yu, P.S.: Discovering frequent closed partial orders from strings. IEEE Trans. Knowl. Data Eng. **18**(11), 1467–1481 (2006)
13. Rouane-Hacene, M., Huchard, M., Napoli, A., Valtchev, P.: Relational concept analysis: mining concept lattices from multi-relational data. Ann. Math. Artif. Intell. **67**(1), 81–108 (2013)
14. Srikant, R., Agrawal, R.: Mining sequential patterns: generalizations and performance improvements. In: Apers, P., Bouzeghoub, M., Gardarin, G. (eds.) EDBT 1996. LNCS, vol. 1057, pp. 1–17. Springer, Heidelberg (1996). doi:10.1007/BFb0014140
15. Yan, X., Han, J., Afshar, R.: CloSpan: mining closed sequential patterns in large datasets. In: SDM, pp. 166–177 (2003)
16. Yun, U.: A new framework for detecting weighted sequential patterns in large sequence databases. Know.-Based Syst. **21**(2), 110–122 (2008)

First Notes on Maximum Entropy Entailment for Quantified Implications

Francesco Kriegel$^{(\boxtimes)}$ (iD)

Institute of Theoretical Computer Science,
Technische Universität Dresden, Dresden, Germany
francesco.kriegel@tu-dresden.de

Abstract. Entropy is a measure for the *uninformativeness* or *randomness* of a data set, i.e., the higher the entropy is, the lower is the amount of information. In the field of propositional logic it has proven to constitute a suitable measure to be maximized when dealing with models of probabilistic propositional theories. More specifically, it was shown that the model of a probabilistic propositional theory with maximal entropy allows for the deduction of other formulae which are somehow expected by humans, i.e., allows for some kind of common sense reasoning.

  In order to pull the technique of *maximum entropy entailment* to the field of Formal Concept Analysis, we define the notion of entropy of a formal context with respect to the frequency of its object intents, and then define maximum entropy entailment for quantified implication sets, i.e., for sets of partial implications where each implication has an assigned degree of confidence. Furthermore, then this entailment technique is utilized to define so-called *maximum entropy implicational bases (ME-bases)*, and a first general example of such a ME-base is provided.

Keywords: Maximum entropy · Formal context · Partial implication · Formal Concept Analysis · Implicational base · Uncertain knowledge

1 Introduction

Entropy is a measure for describing the amount of *randomness* or *uninforma-tiveness* of a particular system, and has variants in different fields, e.g., in *thermodynamics* (where it was defined first), in *statistical mechanics*, but also in *probability theory* (where it describes a lack of *predicatability*), and in *information theory* as well (where it is also called *Shannon entropy*, and describes the amount of information contained in a message sent through a channel). In particular, one of the first works on entropy in information theory was published by Shannon and Weaver in 1949, cf. [10]. Later, there were several researchers who adapted the notion of entropy to *probabilistic logic*. For example, a motivating and wide introduction can be found in the book *The Uncertain Reasoner's Companion - a Mathematical Perspective* [9] written by Paris. A thorough justification for reasoning under maximum entropy semantics can be found therein on pages 76 ff. Furthermore, there is ongoing and active research on maximum entropy

© Springer International Publishing AG 2017
K. Bertet et al. (Eds.): ICFCA 2017, LNAI 10308, pp. 155–167, 2017.
DOI: 10.1007/978-3-319-59271-8_10

semantics, in particular for the case of *probabilistic first-order-logic*, which is
e.g. driven by Kern-Isberner, cf. [3]. Both Paris and Kern-Isberner show that reasoning under maximum entropy semantics somehow implements common sense
reasoning, and yield conclusions which are expected by humans when dealing
with probabilistic theories or data sets.

This paper shall give a short overview and some first notes on utilizing the
measure of entropy in the field of *Formal Concept Analysis*, and on a possibility for defining maximum entropy semantics for sets of partial implications
equipped with probabilities (which are called *quantified implication sets* herein).
Additionally, this paper presents a definition of an implicational base under such
a maximum entropy semantics (*ME-base*), and provides a first general example
of such a base which is induced by Luxenburger's base, cf. [8]. However, there
is a specific example of a data set showing that this first ME-base is neither
non-redundant nor minimal, and thus leaves possibilities for future research.

2   Formal Concept Analysis

The field of *Formal Concept Analysis* was invented for at least two reasons:
Firstly, it should formalize the philosophical notion of a *concept* – which is
characterized by its *extent*, i.e., the objects it describes, and its *intent*, i.e.,
the properties it satisfies – in a formal and mathematical way, cf. [5, p. 58].
Secondly, it was meant to be a new approach to the field of *lattice theory*, and
in particular each complete lattice is isomorphic to a concept lattice (the lattice
that is induced by the set of all formal concepts of a formal context). In this
section, all necessary definitions for understanding of the subsequent sections
are presented. For a more sophisticated overview on Formal Context Analysis,
the interested reader is rather referred to the standard book [5] by Ganter and
Wille.

Definition 1 (Formal Context). *A formal context is a triple* $\mathbb{K} := (G, M, I)$
which consists of a set G of objects, *a set M of* attributes, *and an* incidence
*relation $I \subseteq G \times M$. In case $(g, m) \in I$, we say that the object g has the attribute
m in $\mathbb{K}$, and we may also denote this infix as $g\,I\,m$. Furthermore, $\mathbb{K}$ induces two
derivation operators $\cdot^I \colon \wp(G) \to \wp(M)$ and $\cdot^I \colon \wp(M) \to \wp(G)$, respectively,
as follows:*

$$A^I := \{\, m \in M \mid \forall\, g \in A \colon g\,I\,m \,\},$$
$$and \quad B^I := \{\, g \in G \mid \forall\, m \in B \colon g\,I\,m \,\}.$$

It is well-known [5] that both derivation operators form a so-called *Galois
connection* between the powersets $\wp(G)$ and $\wp(M)$, i.e., the following statements
hold true for all subsets $A, A_1, A_2 \subseteq G$ and $B, B_1, B_2 \subseteq M$:

1. $A \subseteq B^I \Leftrightarrow B \subseteq A^I \Leftrightarrow A \times B \subseteq I$
2. $A \subseteq A^{II}$
3. $A^I = A^{III}$
4. $A_1 \subseteq A_2 \Rightarrow A_2^I \subseteq A_1^I$

5. $B \subseteq B^{II}$
6. $B^I = B^{III}$
7. $B_1 \subseteq B_2 \Rightarrow B_2^I \subseteq B_1^I$

An attribute set $B \subseteq M$ with $B = B^{II}$ is called an *intent* of $\mathbb{K}$, and we shall denote the set of all intents of $\mathbb{K}$ as $\mathsf{Int}(\mathbb{K})$.

Let $\mathbb{K} = (G, M, I)$ be a finite formal context. For an attribute set $X \subseteq M$, its *frequency* in $\mathbb{K}$ is defined as

$$\mathsf{freq}_{\mathbb{K}}(X) := \frac{|\{g \in G \mid \{g\}^I = X\}|}{|G|},$$

and its *support* is given by

$$\mathsf{supp}_{\mathbb{K}}(X) := \frac{|X^I|}{|G|}.$$

It is easy to verify that $\mathsf{freq}_{\mathbb{K}}$ is a discrete probability measure on $\wp(M)$, i.e., $\sum_{X \subseteq M} \mathsf{freq}_{\mathbb{K}}(X) = 1$.

Definition 2 (Implication). *An* implication *over M is of the form $X \to Y$ where $X, Y \subseteq M$, and we call X the* premise, *and Y the* conclusion.[1] *It is* valid *in a formal context $\mathbb{K} := (G, M, I)$, denoted as $\mathbb{K} \models X \to Y$, if each object that possesses all attributes in X also has all attributes in Y, i.e., if $X^I \subseteq Y^I$. Furthermore, the* confidence *of $X \to Y$ in $\mathbb{K}$ is defined as*

$$\mathsf{conf}_{\mathbb{K}}(X \to Y) := \frac{|(X \cup Y)^I|}{|X^I|},$$

i.e., as the fraction of the number of objects satisfying both premise and conclusion compared to the number of objects satisfying only the premise.

An implication set *$\mathcal{L}$ is a set of implications. Then, $\mathcal{L}$ is* valid *in $\mathbb{K}$ if all implications in $\mathcal{L}$ are valid in $\mathbb{K}$, and we symbolize this by $\mathbb{K} \models \mathcal{L}$. If an implication $X \to Y$ is valid in all formal contexts in which $\mathcal{L}$ is valid, then we say that $\mathcal{L}$ entails $X \to Y$ and denote this as $\mathcal{L} \models X \to Y$.*

The confidence of an implication measures the degree of validity in a given formal context. It is readily verified that an implication $X \to Y$ is valid in a formal context $\mathbb{K}$ if, and only if, its confidence in $\mathbb{K}$ equals 1. (Then $X^I \cap Y^I = (X \cup Y)^I = X^I$, and so $X^I \subseteq Y^I$.)

---

[1] In the field of machine learning, an implication is also called *association rule*, a premise is called *antedecent*, and a conclusion is called *consequent*.

In [6], the *canonical base* of a formal context $\mathbb{K}$ was introduced as the implication set

$$\mathsf{Can}(\mathbb{K}) := \{\, P \to P^{II} \mid P \in \mathsf{PsInt}(\mathbb{K}) \,\},$$

where $\mathsf{PsInt}(\mathbb{K})$ consists of all *pseudo-intents* of $\mathbb{K}$, i.e., those sets $P \subseteq M$ that are no intents $(P \neq P^{II})$, but contain all intents Q^{II} for pseudo-intents $Q \subsetneq P$. In particular, the above mentioned canonical base is an *implicational base* for $\mathbb{K}$, i.e., for all implications $X \to Y$, it holds true that $X \to Y$ is valid in $\mathbb{K}$ if, and only if, $\mathsf{Can}(\mathbb{K})$ entails $X \to Y$. Furthermore, it is in fact a *minimal* implicational base, i.e., there is no implicational base of $\mathbb{K}$ containing fewer implications.

We continue by defining three different notions of equivalence of formal contexts, one for each of the measures freq, supp, and conf, that are introduced above. As a not very surprising result, it is afterwards shown that these three kinds of equivalence are indeed the same, i.e., two formal contexts are equivalent with respect to one of the measures if, and only if, they are equivalent with respect to all of the three measures.

Definition 3 (Equivalence of Formal Contexts). *Let $\mathbb{K}_1$ and $\mathbb{K}_2$ be two formal contexts with a common attribute set M.*

1. *$\mathbb{K}_1$ and $\mathbb{K}_2$ are* frequency-equivalent *if $\mathsf{freq}_{\mathbb{K}_1}(X) = \mathsf{freq}_{\mathbb{K}_2}(X)$ for all $X \subseteq M$.*
2. *$\mathbb{K}_1$ and $\mathbb{K}_2$ are* support-equivalent *if $\mathsf{supp}_{\mathbb{K}_1}(X) = \mathsf{supp}_{\mathbb{K}_2}(X)$ for all $X \subseteq M$.*
3. *$\mathbb{K}_1$ and $\mathbb{K}_2$ are* confidence-equivalent *if $\mathsf{conf}_{\mathbb{K}_1}(X \to Y) = \mathsf{conf}_{\mathbb{K}_2}(X \to Y)$ for all $X, Y \subseteq M$.*

Lemma 4. *Let $\mathbb{K}_1$ and $\mathbb{K}_2$ be two formal contexts with a common attribute set M. Then the following statements are equivalent:*

1. *$\mathbb{K}_1$ and $\mathbb{K}_2$ are frequency-equivalent.*
2. *$\mathbb{K}_1$ and $\mathbb{K}_2$ are support-equivalent.*
3. *$\mathbb{K}_1$ and $\mathbb{K}_2$ are confidence-equivalent.*

Proof. 1. $\Rightarrow$ 2. We may express $\mathsf{supp}_{\mathbb{K}}$ in terms of $\mathsf{freq}_{\mathbb{K}}$ as follows:

$$\mathsf{supp}_{\mathbb{K}}(X) = \sum_{X \subseteq Y} \mathsf{freq}_{\mathbb{K}}(Y).$$

It follows that frequency-equivalence implies support-equivalence.

2. $\Rightarrow$ 3. Obviously, $\mathsf{conf}_{\mathbb{K}}$ can be expressed by means of $\mathsf{supp}_{\mathbb{K}}$ by

$$\mathsf{conf}_{\mathbb{K}}(X \to Y) = \frac{\mathsf{supp}_{\mathbb{K}}(X \cup Y)}{\mathsf{supp}_{\mathbb{K}}(X)}.$$

Thus, support-equivalence implies confidence-equivalence.

3. $\Rightarrow$ 1. Note that for arbitrary formal contexts $\mathbb{K} := (G, M, I)$,

$$\begin{aligned}
\mathsf{conf}_\mathbb{K}(X \to Y) &= \frac{|(X \cup Y)^I|}{|X^I|} \\
&= \frac{|\{\, g \in G \mid X \cup Y \subseteq \{g\}^I \,\}|}{|\{\, g \in G \mid X \subseteq \{g\}^I \,\}|} \\
&= \frac{\sum_{X \cup Y \subseteq Z} |\{\, g \in G \mid Z = \{g\}^I \,\}|}{\sum_{X \subseteq Z} |\{\, g \in G \mid Z = \{g\}^I \,\}|} \\
&= \frac{\sum_{X \cup Y \subseteq Z} \mathsf{freq}_\mathbb{K}(Z)}{\sum_{X \subseteq Z} \mathsf{freq}_\mathbb{K}(Z)}.
\end{aligned}$$

Thus, it follows that

$$\mathsf{conf}_\mathbb{K}(\emptyset \to X) = \sum_{X \subseteq Z} \mathsf{freq}_\mathbb{K}(Z),$$

and hence

$$\mathsf{freq}_\mathbb{K}(X) = \mathsf{conf}_\mathbb{K}(\emptyset \to X) - \sum_{X \subsetneq Z} \mathsf{freq}_\mathbb{K}(Z)$$

yields a recursion with $\mathsf{freq}_\mathbb{K}(M) = \mathsf{conf}_\mathbb{K}(\emptyset \to M)$. Consequently, confidence-equivalence implies frequency-equivalence.   □

3 Quantified Implication Sets

In [8], Luxenburger introduced the notion of a *stochastic context* as a means to capture statistical knowledge between classes (of objects) and attributes. Formally, a stochastic context is a triple (G, M, i) consisting of a set G of *classes*, a set M of *attributes*, and an *incidence function* $i \colon G \times M \to [0, 1]$. Then, the value $p = i(g, m)$ indicates that the (statistical) probability of an object in the class g having the attribute m is p. For introducing semantics, the notion of a *realizer* was defined: A formal context $(H, G \cup M, J)$ *realizes* a stochastic context (G, M, i) if, for all classes $g \in G$ and all attributes $m \in M$, the implication $g \to m$ has a confidence $i(g, m)$. Then, an implication is *certainly valid* (*possibly valid*) in (G, M, i) if it is valid in all (some) realizers of (G, M, i).

It turns out that this type of knowledge representation is too restricted. On the one hand, it is not possible to denote unknown incidence values,[2] and on the other hand, we may not express dependencies between attributes or sets of attributes. As a solution, we could consider i as a partial function with type $G \times \wp(M) \dashrightarrow [0, 1]$, but then we would not be able to express relations between the different classes, e.g., when it is known that two classes are disjoint, or overlap to a certain degree. Generalizing further, we could utilize an incidence function $i \colon \wp(G \cup M) \times \wp(G \cup M) \dashrightarrow [0, 1]$, which leads us directly to sets of implications that are annotated with a value between $[0, 1]$. Then, we do not

---

[2] Of course, this may be easily solved by regarding i as a partial function.

have to distinguish between the classes and attributes, and simply say that a (generalized) stochastic context over M is a partial function $\mathcal{L}\colon \mathsf{Imp}(M) \dashrightarrow [0,1]$.

Definition 5 (Quantified Implication). *A quantified implication over M is of the form* $\mathsf{d} \bowtie p.\, X \to Y$ *where* $\bowtie \in \{<, \leq, =, \neq, \geq, >\}$, $p \in [0,1]$, *and* $X, Y \subseteq M$. *It is* valid *in a formal context* $\mathbb{K} := (G, M, I)$, *symbolized as* $\mathbb{K} \models \mathsf{d} \bowtie p.\, X \to Y$, *if* $\mathrm{conf}_{\mathbb{K}}(X \to Y) \bowtie p$. *The set of all quantified implications over M is denoted as* $\mathsf{dImp}(M)$.

A quantified implication set *is a set of quantified implications. A formal context* $\mathbb{K} := (G, M, I)$ *is a* realizer *of a quantified implication set* $\mathcal{L}$, *and we shall denote this as* $\mathbb{K} \models \mathcal{L}$, *if all implications in* $\mathcal{L}$ *are valid in* $\mathbb{K}$.

In the following text, we will only consider quantified implications of the form $\mathsf{d} = p.\, X \to Y$. Consequently, a quantified implication set $\mathcal{L}$ may also be viewed as a partial mapping $\mathcal{L}\colon \mathsf{Imp}(M) \dashrightarrow [0,1]$ where $\mathcal{L}(X \to Y) := p$ if, and only if, $\mathsf{d} = p.\, X \to Y \in \mathcal{L}$. Of course, this requires that there are no two different quantified implications $\mathsf{d} = p.\, X \to Y$ and $\mathsf{d} = q.\, X \to Y$ where $p \neq q$. We will not distinguish between both notational variants, and use that one which is better readable. Furthermore, by $\mathsf{dom}(\mathcal{L})$ we denote the *domain* of a quantified implication set $\mathcal{L}\colon \mathsf{Imp}(M) \dashrightarrow [0,1]$, i.e., the set of all implications to which $\mathcal{L}$ assigns a probability.

Clearly, quantified implication sets generalize the stochastic contexts of Luxenburger, since if (G, M, i) is a stochastic context, then $\mathcal{L}\colon (g \to m) \mapsto i(g, m)$ is a quantified implication set over $G \cup M$, and both have the same realizers.

As an alternative approach to defining semantics for the quantified implications, it is also possible to use probabilistic formal contexts which were introduced in [7]. In particular, we could define that a probabilistic formal context $\mathbb{K} := (G, M, W, I, \mathbb{P})$ *realizes* a quantified implication $\mathsf{d} \bowtie p.\, X \to Y$ if in $\mathbb{K}$ it holds true that $\mathbb{P}(X \to Y) \bowtie p$. However, we do not want to follow this approach here.

As a next step, we define the notion of entailment for quantified implication sets. For this purpose, let $\mathcal{L} \cup \{\mathsf{d} = p.\, X \to Y\} \subseteq \mathsf{dImp}(M)$. Then $\mathcal{L}$ *certainly entails* $\mathsf{d} = p.\, X \to Y$ if each realizer of $\mathcal{L}$ is a realizer of $\mathsf{d} = p.\, X \to Y$, too, and we shall denote this by $\mathcal{L} \models_{\forall} \mathsf{d} = p.\, X \to Y$. Dually, $\mathcal{L}$ *possibly entails* $\mathsf{d} = p.\, X \to Y$ if there is a common realizer of $\mathcal{L}$ and $\mathsf{d} = p.\, X \to Y$, or if $\mathcal{L}$ is *inconsistent*, i.e., if $\mathcal{L}$ has no realizers at all, and this is symbolized as $\mathcal{L} \models_{\exists} \mathsf{d} = p.\, X \to Y$. The certain entailment $\models_{\forall}$ was investigated by Borchmann in [4], where a suitable system of linear equations must be solved in order to decide an entailment question $\mathcal{L} \models_{\forall} \mathsf{d} = p.\, X \to Y$. Furthermore, Borchmann showed that this entailment can be solved in polynomial space. In the next section, we will define the notion of *maximum entropy entailment* $\models_{\mathsf{ME}}$, which is stronger than $\models_{\exists}$, but weaker than $\models_{\forall}$.

4   Maximum Entropy Entailment

In [9], Paris motivates and introduces an inference process called *maximum entropy reasoning* for probabilistic propositional logic. Of course, there is a strong correspondence between propositional logic and Formal Concept Analysis, and we want to use this fact to apply the maximum entropy reasoning to Formal Concept Analysis.

Definition 6 (Entropy). *Let* $\mathbb{K} := (G, M, I)$ *be a formal context. Then, the entropy of* $\mathbb{K}$ *is defined as*[3]

$$\mathbf{H}(\mathbb{K}) := - \sum_{X \subseteq M} \mathsf{freq}_{\mathbb{K}}(X) \cdot \log(\mathsf{freq}_{\mathbb{K}}(X)).$$

For a quantified implication set $\mathcal{L}$*, a* maximum entropy realizer *(ME-realizer) of* $\mathcal{L}$ *is a realizer of* $\mathcal{L}$ *that has maximal entropy among all realizers of* $\mathcal{L}$*. It can be shown that modulo frequency-equivalence, maximum entropy realizers are unique, and hence we shall denote the maximum entropy realizer of* $\mathcal{L}$ *by* $\mathcal{L}^*$*. The* maximum entropy confidence *(ME-confidence) of an implication* $X \to Y$ *with respect to* $\mathcal{L}$ *is defined as the confidence of* $X \to Y$ *in the ME-realizer* $\mathcal{L}^*$*, i.e.,* $\mathsf{conf}_{\mathcal{L}}(X \to Y) := \mathsf{conf}_{\mathcal{L}^*}(X \to Y).$

Furthermore, for a quantified implication set $\mathcal{L} \cup \{\mathsf{d} \bowtie p. \, X \to Y\}$*, we say that* $\mathcal{L}$ *entails* $\mathsf{d} \bowtie p. \, X \to Y$ *with respect to maximum entropy reasoning (or abbreviated: ME-entails) if* $\mathsf{d} \bowtie p. \, X \to Y$ *is valid in the ME-realizer* $\mathcal{L}^*$*, or if* $\mathcal{L}$ *is inconsistent, i.e., if* $\mathcal{L}$ *has no realizers at all. We shall denote this as* $\mathcal{L} \models_{\mathsf{ME}} \mathsf{d} \bowtie p. \, X \to Y.$

For the case of one event and its converse, Fig. 1 shows a plot of the corresponding entropy function, while for the case of two events, Fig. 2 presents a suitable plot.

It is easy to see that $\mathcal{L} \models_{\forall} \mathsf{d} \bowtie p. \, X \to Y$ implies $\mathcal{L} \models_{\mathsf{ME}} \mathsf{d} \bowtie p. \, X \to Y$, and the latter implies $\mathcal{L} \models_{\exists} \mathsf{d} \bowtie p. \, X \to Y$, i.e., *ME-entailment* $\models_{\mathsf{ME}}$ is weaker than certain entailment $\models_{\forall}$, but stronger than possible entailment $\models_{\exists}$. A number of reasons that support the claim of ME-entailment being more natural than the other entailment types are presented by Paris in [9], where it is shown that reasoning under maximum entropy semantics somehow captures human expectation and common sense reasoning.

In the remaining part of this section, the problem of deciding entailment under maximum entropy reasoning is considered. We hence assume that a quantified implication set $\mathcal{L}$ is given, and its ME-realizer is to be computed. For this purpose, we show how $\mathcal{L}$ induces a system of linear equations, for which the entropy function has to be maximized, i.e., we define an optimization problem the solution of which describes the ME-realizer $\mathcal{L}^*$ (modulo frequency-equivalence).

---

[3] We use the logarithm with base 2 here, since we are dealing with data sets or informations, respectively, which are encoded as bits. However, using another base would not cause any problem, since this would only distort the entropy by a multiplicative factor.

$$f(x) := -(x \cdot \log_2(x) + (1 - x) \cdot \log_2(1 - x))$$

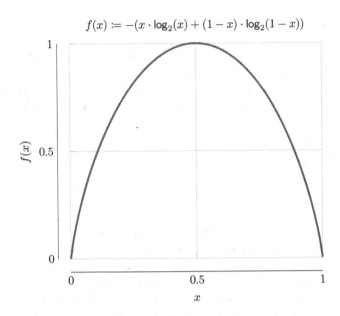

Fig. 1. A plot of the entropy function for one event.

$$f(x,y) := -(x \cdot \log_2(x) + (1 - x) \cdot \log_2(1 - x) + y \cdot \log_2(y) + (1 - y) \cdot \log_2(1 - y))$$

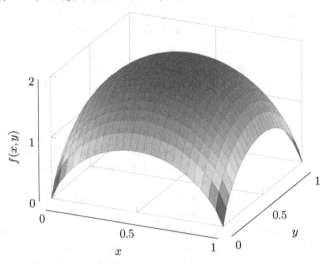

Fig. 2. A plot of the entropy function for two events.

For each quantified implication $\mathsf{d} = p.\, X \to Y$ of $\mathcal{L}$, a realizer $\mathbb{K}$ of $\mathcal{L}$ must satisfy $\mathsf{conf}_\mathbb{K}(X \to Y) = p$, i.e.,

$$p = \frac{\mathsf{supp}_\mathbb{K}(X \cup Y)}{\mathsf{supp}_\mathbb{K}(X)} = \frac{\sum_{X \cup Y \subseteq Z} \frac{|\{g \in G \mid \{g\}^I = Z\}|}{|G|}}{\sum_{X \subseteq Z} \frac{|\{g \in G \mid \{g\}^I = Z\}|}{|G|}} = \frac{\sum_{X \cup Y \subseteq Z} \mathsf{freq}_\mathbb{K}(Z)}{\sum_{X \subseteq Z} \mathsf{freq}_\mathbb{K}(Z)}.$$

This is equivalent to the condition $\sum_{X \cup Y \subseteq Z} \mathsf{freq}_\mathbb{K}(Z) - p \cdot \sum_{X \subseteq Z} \mathsf{freq}_\mathbb{K}(Z) = 0$. Hence, we formulate a matrix equation

$$\mathbf{A} \cdot \mathbf{x} = \mathbf{b},$$

where $\mathbf{A} \colon \mathsf{dom}(\mathcal{L}) \cup \{*\} \times \wp(M) \to [0,1]$, $\mathbf{x} \colon \wp(M) \to [0,1]$, and $\mathbf{b} \colon \mathsf{dom}(\mathcal{L}) \cup \{*\} \to [0,1]$, such that

$$\mathbf{A}(X \to Y, Z) := \begin{cases} 1 - p & \text{if } X \cup Y \subseteq Z, \\ -p & \text{else if } X \subseteq Z, \text{ and} \\ 0 & \text{otherwise.} \end{cases}$$

$$\mathbf{A}(*, Z) := 1$$

and $\mathbf{b}(X \to Y) := 0$ for all $X \to Y \in \mathsf{dom}(\mathcal{L})$, and $\mathbf{b}(*) := 1$. Then a solution $\mathbf{x}$ induces a class of formal contexts $\mathbb{K}$ such that for all $Z \subseteq M$, $\mathsf{freq}_\mathbb{K}(Z) = \mathbf{x}(Z) -$ we just have to determine the solution $\mathbf{x}$, for which the entropy $\mathbf{H}(\mathbf{x}) := -\sum_{X \subseteq M} \mathbf{x}(X) \cdot \log(\mathbf{x}(X))$ is maximal.

5   Maximum Entropy Bases

For the axiomatization of data sets containing assertional knowledge, implicational bases were intensively used to describe the deducible implications in a sound and complete manner. If only valid implications are of interest, then the *canonical base* of Guigues and Duquenne in [6] describes a minimal implicational base, i.e., it is of smallest size among all sound and complete implication sets for the input formal context. There are also some investigations on *direct bases*, for which entailment can be checked in one pass, i.e., each implication in a direct base must only be considered once when deciding entailment, cf., Adaricheva et al. [1]. Furthermore, several different bases for partial implications, i.e., for the implications in a formal context with a confidence less than 1, were introduced, and an overview as well as a comparison on different types are presented by Balcázar in [2], among which the most popular base is of course Luxenburger's base [8]. However, in this section a new notion of a base of partial implications of a formal context shall be defined, which utilizes the maximum entropy entailment from the preceeding sections.

Definition 7 (Maximum Entropy Implicational Base). *Let $\mathbb{K}$ be a formal context with attribute set M. Then, a quantified implication set $\mathcal{L}$ over M is a maximum entropy implicational base (ME-base) of $\mathbb{K}$ if for all $X \to Y \in \mathsf{Imp}(M)$,*

$$\mathsf{conf}_\mathbb{K}(X \to Y) = \mathsf{conf}_\mathcal{L}(X \to Y).$$

A ME-base $\mathcal{L}$ is non-redundant, *if none of its values can be removed, i.e., for all $X \to Y \in \mathsf{dom}(\mathcal{L})$ it holds that $\mathcal{L}(X \to Y) \neq \mathsf{conf}_{\mathcal{L} \setminus \{X \to Y\}}(X \to Y)$. Furthermore, a ME-base $\mathcal{L}$ of $\mathbb{K}$ is* minimal *if there is no ME-base with a smaller domain, i.e., if $|\mathsf{dom}(\mathcal{L})| \leq |\mathsf{dom}(\mathcal{L}')|$ for all ME-bases $\mathcal{L}'$ of $\mathbb{K}$.*[4]

---

[4] We denote by $\mathcal{L} \setminus \{X \to Y\}$ the quantified implication set which assigns no probability to $X \to Y$, but otherwise coincides with $\mathcal{L}$.

Lemma 8. *Let* $\mathbb{K}$ *be a formal context, and* $\mathcal{L}$ *be a quantified implication set. Then the following statements are equivalent:*

1. $\mathcal{L}$ *is a maximum entropy implicational base of* $\mathbb{K}$.
2. $\mathcal{L}^*$ *and* $\mathbb{K}$ *are confidence-equivalent.*
3. $\mathcal{L}^*$ *and* $\mathbb{K}$ *are support-equivalent.*
4. $\mathcal{L}^*$ *and* $\mathbb{K}$ *are frequency-equivalent.*

Proof. By the definition of realizers of quantified implication sets, we have $\mathcal{L}(X \to Y) = \mathrm{conf}_{\mathcal{L}^*}(X \to Y)$ for all implications $X \to Y \in \mathrm{dom}(\mathcal{L})$. Hence, $\mathcal{L}$ is a maximum entropy implicational base of $\mathbb{K}$ if, and only if, $\mathcal{L}^*$ and $\mathbb{K}$ are confidence-equivalent. The remaining equivalences are immediate consequences of Lemma 4. $\qquad\square$

As a corollary we infer that, since all realizers of the quantified implication set $\mathrm{conf}_{\mathbb{K}}$ are frequency-equivalent to $\mathbb{K}$, $\mathrm{conf}_{\mathbb{K}}$ is a maximal entropy base of $\mathbb{K}$. Furthermore, it follows that $\mathcal{L}$ with $\mathcal{L}(\emptyset \to X) := \mathrm{conf}_{\mathbb{K}}(\emptyset \to X)$ is a maximum entropy base of $\mathbb{K}$.

For an implication $X \to Y \in \mathrm{Imp}(M)$, it holds that $\mathrm{conf}_{\mathbb{K}}(X \to Y) = \mathrm{conf}_{\mathbb{K}}(X^{II} \to Y^{II})$.

Proposition 9. *Let* $\mathbb{K}$ *be a formal context, and define the quantified implication set* $\mathbb{K}^* \colon \mathrm{Imp}(M) \dashrightarrow [0,1]$ *where*

$$\mathbb{K}^*(P \to P^{II}) := 1 \qquad \text{for all } P \in \mathrm{PsInt}(\mathbb{K}), \text{ and}$$
$$\mathbb{K}^*(X \to Y) := \mathrm{conf}_{\mathbb{K}}(X \to Y) \quad \text{for all } X, Y \in \mathrm{Int}(\mathbb{K}) \text{ with } X \subsetneq Y$$
$$\text{where no } Z \in \mathrm{Int}(\mathbb{K}) \text{ with } X \subsetneq Z \subsetneq Y \text{ exists.}$$

Then $\mathbb{K}^*$ *is a maximum entropy implicational base for* $\mathbb{K}$.

Proof. We utilize Lemma 8, i.e., we prove that $\mathbb{K}$ and $\mathbb{K}^{**}$ are confidence-equivalent. Thus, consider an arbitrary implication $X \to Y \in \mathrm{Imp}(M)$ – we need to show that the confidences of $X \to Y$ in $\mathbb{K}$ and $\mathbb{K}^{**}$ are equal. Beforehand, it is useful to give a characterization or construction of the ME-realizer $\mathbb{K}^{**}$.

Consider an implication $X \to Y$ that is valid in $\mathbb{K}$, i.e., has a confidence of 1. Then it follows from the canonical base. Furthermore, all implications $P \to P^{II}$ where P is a pseudo-intent of $\mathbb{K}$ are valid in $\mathbb{K}^{**}$. Consequently, also $X \to Y$ must be valid in $\mathbb{K}^{**}$, i.e., has a confidence of 1 in $\mathbb{K}^{**}$. Consequently, $\mathbb{K}$ and $\mathbb{K}^{**}$ are confidence-equivalent for all valid implications of $\mathbb{K}$.

Now assume that $\mathrm{conf}_{\mathbb{K}}(X \to Y) = p < 1$. We consider the following four cases:

1. $\{X, Y\} \subseteq \mathrm{Int}(\mathbb{K})$ with $X \subseteq Y$, and there is no $Z \in \mathrm{Int}(\mathbb{K})$ such that $X \subsetneq Z \subsetneq Y$.
2. $\{X, Y\} \subseteq \mathrm{Int}(\mathbb{K})$ with $X \subseteq Y$, and there is a non-empty subset $\{Z_1, \ldots, Z_n\} \subseteq \mathrm{Int}(\mathbb{K})$ such that $X \subsetneq Z_1 \subsetneq \ldots \subsetneq Z_n \subsetneq Y$, and for all $i \in \{0, \ldots, n\}$, there is no $Z \in \mathrm{Int}(\mathbb{K})$ with $Z_i \subsetneq Z \subsetneq Z_{i+1}$ where $Z_0 := X$ and $Z_{n+1} := Y$.

3. $\{X, Y\} \subseteq \mathsf{Int}(\mathbb{K})$ with $X \not\subseteq Y$.
4. $\{X, Y\} \not\subseteq \mathsf{Int}(\mathbb{K})$.

We continue by proving the claim for each of the cases above.

1. is true by definition of $\mathbb{K}^*$.
2. Note that for sets $X \subseteq Y \subseteq Z \subseteq M$, and arbitrary formal contexts (G, M, I), it holds true that $\mathsf{conf}_{(G,M,I)}(X \rightarrow Y) \cdot \mathsf{conf}_{(G,M,I)}(Y \rightarrow Z) = \mathsf{conf}_{(G,M,I)}(X \rightarrow Z)$.
Then, by Case 1, each of the implications $Z_i \rightarrow Z_{i+1}$ where $i \in \{0, \ldots, n\}$ has the same confidence in $\mathbb{K}$ and $\mathbb{K}^{**}$. Consequently, we know that the following equations are valid:

$$\begin{aligned}
\mathsf{conf}_{\mathbb{K}}(X \rightarrow Y) &= \mathsf{conf}_{\mathbb{K}}(Z_0 \rightarrow Z_{n+1}) \\
&= \mathsf{conf}_{\mathbb{K}}(Z_0 \rightarrow Z_1) \cdot \ldots \cdot \mathsf{conf}_{\mathbb{K}}(Z_n \rightarrow Z_{n+1}) \\
&= \mathsf{conf}_{\mathbb{K}^{**}}(Z_0 \rightarrow Z_1) \cdot \ldots \cdot \mathsf{conf}_{\mathbb{K}^{**}}(Z_n \rightarrow Z_{n+1}) \\
&= \mathsf{conf}_{\mathbb{K}^{**}}(Z_0 \rightarrow Z_{n+1}) \\
&= \mathsf{conf}_{\mathbb{K}^{**}}(X \rightarrow Y).
\end{aligned}$$

3. The following equations are valid.

$$\begin{aligned}
\mathsf{conf}_{\mathbb{K}}(X \rightarrow Y) &= \mathsf{conf}_{\mathbb{K}}(X \rightarrow X \cup Y) \\
&= \mathsf{conf}_{\mathbb{K}}(X \rightarrow (X \cup Y)^{II}) \\
&= \mathsf{conf}_{\mathbb{K}^{**}}(X \rightarrow (X \cup Y)^{II}) \\
&= \mathsf{conf}_{\mathbb{K}^{**}}(X \rightarrow X \cup Y) \\
&= \mathsf{conf}_{\mathbb{K}^{**}}(X \rightarrow Y).
\end{aligned}$$

4. We know that both implications $X \rightarrow Y$ and $X^{II} \rightarrow Y^{II}$ have the same confidence in $\mathbb{K}$. Furthermore, $X \rightarrow Y$ is entailed by $\{X \rightarrow X^{II}, X^{II} \rightarrow Y^{II}, Y^{II} \rightarrow Y\}$, where $X \rightarrow X^{II}$ and $Y^{II} \rightarrow Y$ both have confidence 1 in $\mathbb{K}$, and hence also in $\mathbb{K}^{**}$. We prove that both implications $X \rightarrow Y$ and $X^{II} \rightarrow Y^{II}$ also have the same confidence in the ME-realizer $\mathbb{K}^{**}$.
Assume $X \subseteq Y$. Then we have that

$$\begin{aligned}
\mathsf{conf}_{\mathbb{K}^{**}}(X \rightarrow Y) &= \mathsf{conf}_{\mathbb{K}^{**}}(X \rightarrow Y) \cdot \mathsf{conf}_{\mathbb{K}^{**}}(Y \rightarrow Y^{II}) \\
&= \mathsf{conf}_{\mathbb{K}^{**}}(X \rightarrow Y^{II}) \\
&= \mathsf{conf}_{\mathbb{K}^{**}}(X \rightarrow X^{II}) \cdot \mathsf{conf}_{\mathbb{K}^{**}}(X^{II} \rightarrow Y^{II}) \\
&= \mathsf{conf}_{\mathbb{K}^{**}}(X^{II} \rightarrow Y^{II}).
\end{aligned}$$

Furthermore, let $X \not\subseteq Y$. The implication $(X \cup Y)^{II} \rightarrow X^{II} \cup Y^{II}$ is valid in $\mathbb{K}$, i.e., has confidence 1 both in $\mathbb{K}$ and in the maximum entropy realizer $\mathbb{K}^{**}$. We proceed as follows:

$$\text{conf}_{\mathbb{K}^{**}}(X \to Y) = \text{conf}_{\mathbb{K}^{**}}(X \to X \cup Y)$$
$$= \text{conf}_{\mathbb{K}^{**}}(X^{II} \to (X \cup Y)^{II})$$
$$= \text{conf}_{\mathbb{K}^{**}}(X^{II} \to (X \cup Y)^{II}) \cdot \text{conf}_{\mathbb{K}^{**}}((X \cup Y)^{II} \to X^{II} \cup Y^{II})$$
$$= \text{conf}_{\mathbb{K}^{**}}(X^{II} \to X^{II} \cup Y^{II})$$
$$= \text{conf}_{\mathbb{K}^{**}}(X^{II} \to Y^{II}).$$

We have shown that $\mathbb{K}$ and $\mathbb{K}^{**}$ are confidence-equivalent. Consequently, both are also frequency-equivalent, and thus must have the same intents. (The set of object intents consists of all those attribute sets with a frequency >0, hence both $\mathbb{K}$ and $\mathbb{K}^{**}$ have the same object intents.) $\qquad\square$

It should be emphasized that the ME-base from the preceding Proposition 9 is not a minimal ME-base – which is based on the example of a Boolean formal context $(M, M, \subseteq)$ for some $M := \wp(N)$: its Luxenburger base is

$$\{\, X \to X \cup \{m\} \mid X \subseteq M, \, m \notin X \,\}$$

where all contained implications possess a confidence of $\frac{1}{2}$. Of course, it can be easily verified that the empty implication set $\emptyset \subseteq \mathsf{dimp}(M)$ also constitutes a ME-base of $(M, M, \subseteq)$, since then $\mathsf{freq}_{\emptyset^*}$ is uniform in order to maximize the entropy, and hence each formal context being a ME-realizer of $\emptyset$ must be isomorphic to $(M, M, \subseteq)$ – in particular, it must be a subposition of some copies of the Boolean formal context $(M, M, \subseteq)$, modulo reordering of rows.

6   Conclusion

In this paper, the widely used measure of *entropy* is defined for the basic structure of a formal context in the field of *Formal Concept Analysis*. It is then utilized to define entailment for quantified implications under maximum entropy semantics, yielding a reasoning procedure which is more common for human thinking than the existing certain semantics and the possible semantics. Then some first steps towards implicational bases under maximum entropy semantics are presented. However, there is large room for future research. For example, possibilities of smaller ME-bases shall be investigated, and particularly and most importantly searching for a (canonical) minimal ME-base is a major goal. Of course, this would require a more sophisticated study of this idea.

Acknowledgements. The author gratefully thanks the anonymous reviewers for their constructive hints and helpful remarks.

References

1. Adaricheva, K.V., Nation, J.B., Rand, R.: Ordered direct implicational basis of a finite closure system. Discrete Appl. Math. **161**(6), 707–723 (2013)

2. Balcázar, J.L.: Minimum-size bases of association rules. In: Daelemans, W., Goethals, B., Morik, K. (eds.) ECML PKDD 2008. LNCS, vol. 5211, pp. 86–101. Springer, Heidelberg (2008). doi:10.1007/978-3-540-87479-9_24
3. Beierle, C., et al.: Extending and completing probabilistic knowledge and beliefs without bias. KI **29**(3), 255–262 (2015)
4. Borchmann, D.: Deciding entailment of implications with support and confidence in polynomial space. CoRR abs/1201.5719 (2012)
5. Ganter, B., Wille, R.: Formal Concept Analysis: Mathematical Foundations. Springer, Heidelberg (1999)
6. Guigues, J.-L., Duquenne, V.: Famille minimale d'implications informatives résultant d'un tableau de données binaires. Mathématiques et Sci. Hum. **95**, 5–18 (1986)
7. Kriegel, F.: Probabilistic implicational bases in FCA and probabilistic bases of GCIs in $\mathcal{EL}^{\perp}$. In: Yahia, S.B., Konecny, J. (eds.) Proceedings of the Twelfth International Conference on Concept Lattices and Their Applications, Clermont-Ferrand, France, 13–16 October 2015, vol. 1466. CEUR Workshop Proceedings. CEUR-WS.org, pp. 193–204 (2015)
8. Luxenburger, M.: Implikationen, Abhängigkeiten und Galois Abbildungen - Beiträge zur formalen Begriffsanalyse. Ph.D. thesis. Technische Hochschule Darmstadt (1993)
9. Paris, J.B.: The Uncertain Reasoner's Companion - A Mathematical Perspective. Cambridge Tracts in Theoretical Computer Science, vol. 39. Cambridge University Press, Cambridge (1994)
10. Shannon, C.E., Weaver, W.: The Mathematical Theory of Communication. University of Illinois Press, Urbana (1949)

Implications over Probabilistic Attributes

Francesco Kriegel$^{(\boxtimes)}$ (iD)

Institute of Theoretical Computer Science,
Technische Universität Dresden, Dresden, Germany
`francesco.kriegel@tu-dresden.de`

Abstract. We consider the task of acquisition of terminological knowledge from given assertional data. However, when evaluating data of real-world applications we often encounter situations where it is impractical to deduce only crisp knowledge, due to the presence of exceptions or errors. It is rather appropriate to allow for degrees of uncertainty within the derived knowledge. Consequently, suitable methods for knowledge acquisition in a probabilistic framework should be developed.

In particular, we consider data which is given as a probabilistic formal context, i.e., as a triadic incidence relation between objects, attributes, and worlds, which is furthermore equipped with a probability measure on the set of worlds. We define the notion of a probabilistic attribute as a probabilistically quantified set of attributes, and define the notion of validity of implications over probabilistic attributes in a probabilistic formal context. Finally, a technique for the axiomatization of such implications from probabilistic formal contexts is developed. This is done is a sound and complete manner, i.e., all derived implications are valid, and all valid implications are deducible from the derived implications. In case of finiteness of the input data to be analyzed, the constructed axiomatization is finite, too, and can be computed in finite time.

Keywords: Knowledge acquisition · Probabilistic formal context · Probabilistic attribute · Probabilistic implication · Knowledge base

1 Introduction

We consider data which is given as a probabilistic formal context, i.e., as a triadic incidence relation between objects, attributes, and worlds, which is furthermore equipped with a probability measure on the set of worlds. We define the notion of a probabilistic attribute as a probabilistically quantified set of attributes, and define the notion of validity of implications over probabilistic attributes in a probabilistic formal context. Finally, a technique for the axiomatization of such implications from probabilistic formal contexts is developed. This is done is a sound and complete manner, i.e., all derived implications are valid, and all valid implications are deducible from the derived implications. In case of finiteness of the input data to be analyzed, the constructed axiomatization is finite, too, and can be computed in finite time.

© Springer International Publishing AG 2017
K. Bertet et al. (Eds.): ICFCA 2017, LNAI 10308, pp. 168–183, 2017.
DOI: 10.1007/978-3-319-59271-8_11

This document is structured as follows. A brief introduction on *Formal Concept Analysis* is given in Sect. 2, and then the subsequent Sect. 3 presents basics on *Probabilistic Formal Concept Analysis*. Then, in Sect. 4 we define the notion of implications over probabilistic attributes, and infer some characterizing statements. The most important part of this document is the Sect. 5, in which we constructively develop a method for the axiomatization of probabilistic implications from probabilistic formal contexts; the section closes with a proof of soundness and completeness of the proposed knowledge base. Eventually, in Sect. 6 some closing remarks as well as future steps for extending and applying the results are given. In order to explain and motivate the definitions and the theoretical results, Sects. 3, 4 and 5 contain a running example.

2   Formal Concept Analysis

This section briefly introduces the standard notions of *Formal Concept Analysis* (abbr. *FCA*) [4]. A *formal context* $\mathbb{K} := (G, M, I)$ consists of a set G of *objects*, a set M of *attributes*, and an *incidence relation* $I \subseteq G \times M$. For a pair $(g, m) \in I$, we say that g *has* m. The *derivation operators* of $\mathbb{K}$ are the mappings $\cdot^I \colon \wp(G) \to \wp(M)$ and $\cdot^I \colon \wp(M) \to \wp(G)$ that are defined by

$$A^I := \{\, m \in M \mid \forall g \in A \colon (g, m) \in I \,\} \qquad \text{for object sets } A \subseteq G,$$

$$\text{and} \quad B^I := \{\, g \in G \mid \forall m \in B \colon (g, m) \in I \,\} \qquad \text{for attribute sets } B \subseteq M.$$

It is well-known [4] that both derivation operators constitute a so-called *Galois connection* between the powersets $\wp(G)$ and $\wp(M)$, i.e., the following statements hold true for all subsets $A, A_1, A_2 \subseteq G$ and $B, B_1, B_2 \subseteq M$:

1. $A \subseteq B^I \Leftrightarrow B \subseteq A^I \Leftrightarrow A \times B \subseteq I$
2. $A \subseteq A^{II}$
3. $A^I = A^{III}$
4. $A_1 \subseteq A_2 \Rightarrow A_2^I \subseteq A_1^I$
5. $B \subseteq B^{II}$
6. $B^I = B^{III}$
7. $B_1 \subseteq B_2 \Rightarrow B_2^I \subseteq B_1^I$

For obvious reasons, formal contexts can be represented as binary tables the rows of which are labeled with the objects, the columns of which are labeled with the attributes, and the occurrence of a cross $\times$ in the cell at row g and column m indicates that the object g has the attribute m.

An *intent* of $\mathbb{K}$ is an attribute set $B \subseteq M$ with $B = B^{II}$. The set of all intents of $\mathbb{K}$ is denoted by $\mathsf{Int}(\mathbb{K})$. An *implication* over M is an expression $X \to Y$ where $X, Y \subseteq M$. It is *valid* in $\mathbb{K}$, denoted as $\mathbb{K} \models X \to Y$, if $X^I \subseteq Y^I$, i.e., if each object of $\mathbb{K}$ that possesses all attributes in X also has all attributes in Y. An implication set $\mathcal{L}$ is *valid* in $\mathbb{K}$, denoted as $\mathbb{K} \models \mathcal{L}$, if all implications in $\mathcal{L}$ are valid in $\mathbb{K}$. Furthermore, the relation $\models$ is lifted to implication sets as follows: an implication set $\mathcal{L}$ *entails* an implication $X \to Y$, symbolized as $\mathcal{L} \models X \to Y$, if $X \to Y$ is valid in all formal contexts in which $\mathcal{L}$ is valid. More specifically, $\models$ is called the *semantic entailment relation*.

It was shown that entailment can also be decided *syntactically* by applying *deduction rules* to the implication set $\mathcal{L}$ without the requirement to consider all formal contexts in which $\mathcal{L}$ is valid. Recall that an implication $X \to Y$ is *syntactically entailed* by an implication set $\mathcal{L}$, denoted as $\mathcal{L} \vdash X \to Y$, if $X \to Y$ can be constructed from $\mathcal{L}$ by the application of *inference axioms*, cf. [11, p. 47], that are described as follows:

(F1) *Reflexivity:* $\qquad\qquad\qquad\qquad\qquad \emptyset \vdash X \to X$
(F2) *Augmentation:* $\qquad\qquad\qquad \{X \to Y\} \vdash X \cup Z \to Y$
(F3) *Additivity:* $\qquad\qquad \{X \to Y, X \to Z\} \vdash X \to Y \cup Z$
(F4) *Projectivity:* $\qquad\qquad\qquad \{X \to Y \cup Z\} \vdash X \to Y$
(F5) *Transitivity:* $\qquad\qquad\quad \{X \to Y, Y \to Z\} \vdash X \to Z$
(F6) *Pseudotransitivity:* $\quad \{X \to Y, Y \cup Z \to W\} \vdash X \cup Z \to W$

In the inference axioms above the symbols X, Y, Z, and W, denote arbitrary subsets of the considered set M of attributes. Formally, we define $\mathcal{L} \vdash X \to Y$ if there is a sequence of implications $X_0 \to Y_0, \dots, X_n \to Y_n$ such that the following conditions hold:

1. For each $i \in \{0, \dots, n\}$, there is a subset $\mathcal{L}_i \subseteq \mathcal{L} \cup \{X_0 \to Y_0, \dots, X_{i-1} \to Y_{i-1}\}$ such that $\mathcal{L}_i \vdash X_i \to Y_i$ matches one of the Axioms F1–F6.
2. $X_n \to Y_n = X \to Y$.

Often, the Axioms F1, F2, and F6, are referred to as *Armstrong's axioms*. These three axioms constitute a *complete* and *independent* set of inference axioms for implicational entailment, i.e., from it the other Axioms F3–F5 can be derived, and none of them is derivable from the others.

The semantic entailment and the syntactic entailment coincide, i.e., an implication $X \to Y$ is semantically entailed by an implication set $\mathcal{L}$ if, and only if, $\mathcal{L}$ syntactically entails $X \to Y$, cf. [11, Theorem 4.1 on p. 50], as well as [4, Proposition 21 on p. 81]. Consequently, we do not have distinguish between both entailment relations $\models$ and $\vdash$, when it is up to decide whether an implication follows from a set of implications.

A *model* of an implication set $\mathcal{L}$ is an attribute set $Z \subseteq M$ such that $X \subseteq Z$ implies $Y \subseteq Z$ for all $X \to Y \in \mathcal{L}$. By $X^{\mathcal{L}}$ we denote the smallest superset of X that is a model of $\mathcal{L}$.

The data encoded in a formal context can be visualized as a *line diagram* of the corresponding *concept lattice*, which we shall shortly describe. A *formal concept* of a formal context $\mathbb{K} := (G, M, I)$ is a pair (A, B) consisting of a set $A \subseteq G$ of objects as well as a set $B \subseteq M$ of attributes such that $A^I = B$ and $B^I = A$. We then also refer to A as the *extent*, and to B as the *intent*, respectively, of (A, B). In the denotation of $\mathbb{K}$ as a cross table, those formal concepts are the maximal rectangles full of crosses (modulo reordering of rows and columns). Then, the set of all formal concepts of $\mathbb{K}$ is denoted as $\mathfrak{B}(\mathbb{K})$, and it is ordered by defining $(A, B) \leq (C, D)$ if, and only if, $A \subseteq C$. It was shown that this order always induces a complete lattice $\underline{\mathfrak{B}}(\mathbb{K}) := (\mathfrak{B}(\mathbb{K}), \leq, \wedge, \vee, \top, \bot)$, called the *concept lattice* of $\mathbb{K}$, cf. [4, 13], in which the infimum and the supremum operation satisfy the equations

$$\bigwedge_{t \in T} (A_t, B_t) = \left(\bigcap_{t \in T} A_t, \left(\bigcup_{t \in T} B_t \right)^{II} \right),$$

$$\text{and} \quad \bigvee_{t \in T} (A_t, B_t) = \left(\left(\bigcup_{t \in T} A_t \right)^{II}, \bigcap_{t \in T} B_t \right),$$

and where $\top = (\emptyset^I, \emptyset^{II})$ is the greatest element, and where $\bot = (\emptyset^{II}, \emptyset^I)$ is the smallest element, respectively. Furthermore, the concept lattice of $\mathbb{K}$ can be nicely represented as a *line diagram* as follows: Each formal concept is depicted as a vertex. Furthermore, there is an upward directed edge from each formal concept to its upper neighbors, i.e., to all those formal concepts which are greater with respect to $\leq$, but for which there is no other formal concept in between. The nodes are labeled as follows: an attribute $m \in M$ is an upper label of the *attribute concept* $(\{m\}^I, \{m\}^{II})$, and an object $g \in G$ is a lower label of the *object concept* $(\{g\}^{II}, \{g\}^I)$. Then, the extent of the formal concept represented by a vertex consists of all objects which label vertices reachable by a downward directed path, and dually the intent is obtained by gathering all attribute labels of vertices reachable by an upward directed path.

Let $\mathbb{K} \models \mathcal{L}$. A *pseudo-intent* of a formal context $\mathbb{K}$ relative to an implication set $\mathcal{L}$ is an attribute set $P \subseteq M$ which is no intent of $\mathbb{K}$, but is a model of $\mathcal{L}$, and satisfies $Q^{II} \subseteq P$ for all pseudo-intents $Q \subsetneq P$. The set of all those pseudo-intents is symbolized by $\mathsf{PsInt}(\mathbb{K}, \mathcal{L})$. Then the implication set

$$\mathsf{Can}(\mathbb{K}, \mathcal{L}) := \{ P \to P^{II} \mid P \in \mathsf{PsInt}(\mathbb{K}, \mathcal{L}) \}$$

constitutes an *implicational base* of $\mathbb{K}$ relative to $\mathcal{L}$, i.e., for each implication $X \to Y$ over M, the following equivalence is satisfied:

$$\mathbb{K} \models X \to Y \Leftrightarrow \mathsf{Can}(\mathbb{K}, \mathcal{L}) \cup \mathcal{L} \models X \to Y.$$

$\mathsf{Can}(\mathbb{K}, \mathcal{L})$ is called the *canonical base* of $\mathbb{K}$ relative to $\mathcal{L}$. It can be shown that it is a *minimal* implicational base of $\mathbb{K}$ relative to $\mathcal{L}$, i.e., there is no implicational base of $\mathbb{K}$ relative to $\mathcal{L}$ with smaller cardinality. Further information is given by [2,3,5,12]. The most prominent algorithm for computing the canonical base is certainly *NextClosure* developed by Bernhard Ganter [2,3]. A parallel algorithm called *NextClosures* is also available [7,10], and an implementation is provided in *Concept Explorer FX* [6]; its advantage is that its processing time scales almost inverse linear with respect to the number of available CPU cores.

Eventually, in case a given formal context is not complete in the sense that it does not contain enough objects to refute invalid implications, i.e., only contains some observed objects in the domain of interest, but one aims at exploring all valid implications over the given attribute set, a technique called *Attribute Exploration* can be utilized, which guides the user through the process of axiomatizing an implicational base for the underlying domain in a way the number of questions posed to the user is minimal. For a sophisticated introduction as well as for theoretical and technical details, the interested reader is rather referred to [1–3,8,12]. A parallel variant of the *Attribute Exploration* also exists, cf. [7,8], which is implemented in *Concept Explorer FX* [6].

3   Probabilistic Formal Concept Analysis

This section presents probabilistic extensions of the common notions of *Formal Concept Analysis*, which were first introduced in [9]. A *probability measure* $\mathbb{P}$ on a countable set W is a mapping $\mathbb{P}\colon \wp(W) \to [0,1]$ such that $\mathbb{P}(\emptyset) = 0$, $\mathbb{P}(W) = 1$, and $\mathbb{P}$ is σ-*additive*, i.e., for all countable families $(U_n)_{n\in\mathbb{N}}$ of pairwise disjoint sets $U_n \subseteq W$ it holds that $\mathbb{P}(\bigcup_{n\in\mathbb{N}} U_n) = \sum_{n\in\mathbb{N}} \mathbb{P}(U_n)$. A world $w \in W$ is *possible* if $\mathbb{P}\{w\} > 0$, and *impossible* otherwise. The set of all possible worlds is denoted by W_ε, and the set of all impossible worlds is denoted by W_0. Obviously, $W_\varepsilon \uplus W_0$ is a partition of W. Of course, such a probability measure can be completely characterized by the definition of the probabilities of the singleton subsets of W, since it holds true that $\mathbb{P}(U) = \mathbb{P}(\bigcup_{w\in U}\{w\}) = \sum_{w\in U} \mathbb{P}(\{w\})$.

Definition 1. *A probabilistic formal context* $\mathbb{K}$ *is a tuple* $(G, M, W, I, \mathbb{P})$ *that consists of a set* G *of objects, a set* M *of attributes, a countable set* W *of worlds, an* incidence relation $I \subseteq G \times M \times W$, *and a probability measure* $\mathbb{P}$ *on* W. *For a triple* $(g, m, w) \in I$ *we say that object* g *has attribute* m *in world* w. *Furthermore, we define the* derivations in world w *as operators* $\cdot^{I(w)}\colon \wp(G) \to \wp(M)$ *and* $\cdot^{I(w)}\colon \wp(M) \to \wp(G)$ *where*

$$A^{I(w)} := \{\, m \in M \mid \forall g \in A\colon (g, m, w) \in I \,\} \qquad \text{for object sets } A \subseteq G,$$

and $\quad B^{I(w)} := \{\, g \in G \mid \forall m \in B\colon (g, m, w) \in I \,\} \qquad$ *for attribute sets* $B \subseteq M$,

i.e., $A^{I(w)}$ *is the set of all common attributes of all objects in* A *in the world* w, *and* $B^{I(w)}$ *is the set of all objects that have all attributes in* B *in* w. *The formal context induced by a world* $w \in W$ *is defined as* $\mathbb{K}(w) := (G, M, I(w))$.

$\mathbb{K}_{\text{ex}}(w_1)$	m_1	m_2	m_3	$\mathbb{K}_{\text{ex}}(w_2)$	m_1	m_2	m_3	$\mathbb{K}_{\text{ex}}(w_3)$	m_1	m_2	m_3
g_1	$\times$	$\cdot$	$\times$	g_1	$\times$	$\cdot$	$\times$	g_1	$\times$	$\cdot$	$\times$
g_2	$\cdot$	$\times$	$\times$	g_2	$\cdot$	$\times$	$\cdot$	g_2	$\cdot$	$\times$	$\times$
g_3	$\cdot$	$\cdot$	$\times$	g_3	$\times$	$\cdot$	$\times$	g_3	$\cdot$	$\times$	$\cdot$

$$\mathbb{P}_{\text{ex}}(w_1) := \tfrac{1}{2} \qquad\qquad \mathbb{P}_{\text{ex}}(w_2) := \tfrac{1}{3} \qquad\qquad \mathbb{P}_{\text{ex}}(w_3) := \tfrac{1}{6}$$

Fig. 1. An exemplary probabilistic formal context $\mathbb{K}_{\text{ex}}$

As a running example for the current and the up-coming sections, we consider the probabilistic formal context $\mathbb{K}_{\text{ex}}$ presented in Fig. 1. It consists of three objects g_1, g_2, g_3, three attributes m_1, m_2, m_3, and three worlds w_1, w_2, w_3. In $\mathbb{K}_{\text{ex}}$ it holds true that the object g_1 has the attribute m_1 in all three worlds, and the object g_3 has the attribute m_3 in all worlds except in w_3.

Definition 2. *Let* $\mathbb{K}$ *be a probabilistic formal context. The almost certain scaling of* $\mathbb{K}$ *is the formal context* $\mathbb{K}_\varepsilon^\times := (G \times W_\varepsilon, M, I_\varepsilon^\times)$ *where* $((g, w), m) \in I^\times$ *if* $(g, m, w) \in I$.

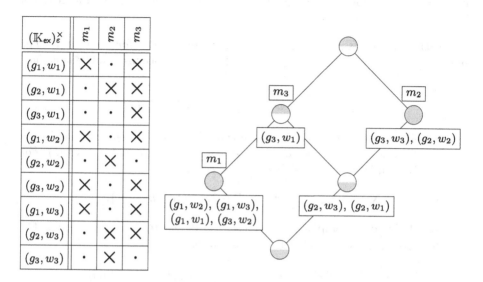

$(\mathbb{K}_{\mathbf{ex}})_\varepsilon^\times$	m_1	m_2	m_3
(g_1, w_1)	$\times$	$\cdot$	$\times$
(g_2, w_1)	$\cdot$	$\times$	$\times$
(g_3, w_1)	$\cdot$	$\cdot$	$\times$
(g_1, w_2)	$\times$	$\cdot$	$\times$
(g_2, w_2)	$\cdot$	$\times$	$\cdot$
(g_3, w_2)	$\times$	$\cdot$	$\times$
(g_1, w_3)	$\times$	$\cdot$	$\times$
(g_2, w_3)	$\cdot$	$\times$	$\times$
(g_3, w_3)	$\cdot$	$\times$	$\cdot$

Fig. 2. The certain scaling of $\mathbb{K}_{\mathbf{ex}}$ from Fig. 1 and its concept lattice

For our running example $\mathbb{K}_{\mathbf{ex}}$, the almost certain scaling is displayed in Fig. 2. As it can be read off it essentially consists of the subposition of the three formal contexts $\mathbb{K}_{\mathbf{ex}}(w_1)$, $\mathbb{K}_{\mathbf{ex}}(w_2)$, and $\mathbb{K}_{\mathbf{ex}}(w_3)$.

Lemma 3. *Let* $\mathbb{K} := (G, M, W, I, \mathbb{P})$ *be a probabilistic formal context. Then for all subsets* $B \subseteq M$ *and for all possible worlds* $w \in W_\varepsilon$, *it holds true that* $B^{I(w)} = B^{I_\varepsilon^\times I_\varepsilon^\times I(w)}$.

Proof. Let $X \subseteq M$. Then for all possible worlds $w \in W_\varepsilon$ it holds that

$$g \in X^{I(w)} \Leftrightarrow \forall\, m \in X \colon (g, m, w) \in I$$
$$\Leftrightarrow \forall\, m \in X \colon ((g, w), m) \in I_\varepsilon^\times \Leftrightarrow (g, w) \in X^{I_\varepsilon^\times},$$

and we conclude that $X^{I(w)} = \pi_1(X^{I_\varepsilon^\times} \cap (G \times \{w\}))$. Furthermore, we then infer $X^{I(w)} = X^{I_\varepsilon^\times I_\varepsilon^\times I(w)}$. $\qquad\square$

4 Implications over Probabilistic Attributes

In [9], the notion of probability of an implication in a probabilistic formal context has been defined. However, it was not possible to express implications between probabilistically quantified attributes, e.g., we could not state that having attribute m with probability $\frac{1}{3}$ implies having attribute n with probability $\frac{2}{3}$. In this section we will resolve this issue by defining the notion of probabilistic attributes, and considering implications over probabilistic attributes. Furthermore, a technique for the construction of bases of such probabilistic implications is proposed in the next Sect. 5.

Definition 4. *Let $\mathbb{K} := (G, M, W, I, \mathbb{P})$ be a probabilistic formal context. For object sets $A \subseteq G$ and attribute sets $B \subseteq M$, the incidence probability is defined as*

$$\mathbb{P}(A, B) := \mathbb{P}\{\, w \in W \mid A \times B \times \{w\} \subseteq I \,\}.$$

A probabilistic attribute over M is an expression $\mathsf{d} \geq p. B$ where $B \subseteq M$, and $p \in [0, 1]$. The set of all probabilistic attributes over M is denoted as $\mathsf{d}(M)$. For a subset $\mathbf{X} \subseteq \mathsf{d}(M)$, its extension in $\mathbb{K}$ is given by

$$\mathbf{X}^I := \{\, g \in G \mid \forall\, \mathsf{d} \geq p. B \in \mathbf{X} \colon \mathbb{P}(\{g\}, B) \geq p \,\}.$$

Considering our exemplary probabilistic formal context $\mathbb{K}_{\mathsf{ex}}$ from Fig. 1, the set $\{\mathsf{d} \geq \frac{1}{2}.\{m_1\}\}$ has the following extension in $\mathbb{K}_{\mathsf{ex}}$:

$$\{\mathsf{d} \geq \tfrac{1}{2}.\{m_1\}\}^I = \{\, g \in G \mid \mathbb{P}(\{g\}, \{m_1\}) \geq \tfrac{1}{2} \,\} = \{g_1\},$$

i.e., only the object g_1 has the attribute m_1 with a probability of at least $\frac{1}{2}$.

Lemma 5. *Let $\mathbb{K}$ be a probabilistic formal context, and $A \subseteq G$ as well as $B \subseteq M$. Then it holds true that $\mathbb{P}(A, B) = \mathbb{P}(A, B^{I^{\mathcal{E}} I^{\mathcal{E}}})$.*

Proof. In Lemma 3 we have shown that $B^{I(w)} = B^{I_{\mathcal{E}} I_{\mathcal{E}} I(w)}$ for all $B \subseteq M$ and all possible worlds $w \in W_{\varepsilon}$. Hence, we may conclude that

$$A \times B \times \{w\} \subseteq I \Leftrightarrow A \subseteq B^{I(w)} \Leftrightarrow A \subseteq B^{I_{\mathcal{E}} I_{\mathcal{E}} I(w)} \Leftrightarrow A \times B^{I^{\mathcal{E}} I^{\mathcal{E}}} \times \{w\} \subseteq I,$$

and so $\mathbb{P}(A, B) = \mathbb{P}(A, B^{I^{\mathcal{E}} I^{\mathcal{E}}})$. $\qquad\qquad\square$

Definition 6. *Let $\mathbb{K} := (G, M, W, I, \mathbb{P})$ be a probabilistic formal context.*

1. *A probabilistic implication over M is an implication over $\mathsf{d}(M)$, and the set of all probabilistic implications over M is denoted as $\mathsf{dImp}(M)$. A probabilistic implication $\mathbf{X} \to \mathbf{Y}$ is valid in $\mathbb{K}$ if $\mathbf{X}^I \subseteq \mathbf{Y}^I$ is satisfied, and we shall denote this by $\mathbb{K} \models \mathbf{X} \to \mathbf{Y}$.*
2. *An implication $P \to Q$ over $[0, 1]$ is valid in $\mathbb{K}$, denoted as $\mathbb{K} \models P \to Q$, if for all objects $g \in G$ and all attribute sets $B \subseteq M$, the following condition is satisfied:*

$$\mathbb{P}(\{g\}, B) \in P \text{ implies } \mathbb{P}(\{g\}, B) \in Q.$$

3. An implication $X \to Y$ over M is valid in $\mathbb{K}$, and we symbolize this by $\mathbb{K} \models X \to Y$, if for all objects $g \in G$ and all probability values $p \in [0,1]$, the following condition is satisfied:

$$\mathbb{P}(\{g\}, X) \geq p \text{ implies } \mathbb{P}(\{g\}, Y) \geq p.$$

An example for a probabilistic implication within the domain of the probabilistic formal context presented in Fig. 1 is

$$\{\mathsf{d} \geq \tfrac{1}{2}.\{m_1\}\} \to \{\mathsf{d} \geq \tfrac{2}{3}.\{m_2\}, \mathsf{d} \geq \tfrac{4}{5}.\{m_3\}\}.$$

However, it is not valid in $\mathbb{K}$ since the premise's extension $\{g_1\}$ of $\{\mathsf{d} \geq \tfrac{1}{2}.\{m_1\}\}$ is not a subset of the conclusion's extension

$$\{\mathsf{d} \geq \tfrac{2}{3}.\{m_2\}, \mathsf{d} \geq \tfrac{4}{5}.\{m_3\}\}^I = \{\mathsf{d} \geq \tfrac{2}{3}.\{m_2\}\}^I \cap \{\mathsf{d} \geq \tfrac{4}{5}.\{m_3\}\}^I$$
$$= \{g_2\} \cap \{g_1, g_3\} = \emptyset.$$

It can be easily verified that the probabilistic implication $\{\mathsf{d} \geq \tfrac{1}{2}.\{m_1\}\} \to \{\mathsf{d} \geq \tfrac{4}{5}.\{m_3\}\}$ is valid in $\mathbb{K}$.

Lemma 7. *Let $\mathbb{K} := (G, M, W, I, \mathbb{P})$ be a probabilistic formal context in which the implication $(p,q] \to \{q\}$ over $[0,1]$ is valid. Then, for each attribute set $X \subseteq M$ and each probability value $r \in (p,q]$, the probabilistic implication $\{\mathsf{d} \geq r.X\} \to \{\mathsf{d} \geq q.X\}$ is valid in $\mathbb{K}$, too.*

Proof. Assume that $\mathbb{K} \models (p,q] \to \{q\}$. It easily follows from Definition 6 that for each object $g \in G$, $\mathbb{P}(\{g\}, X) \in (p,q]$ implies $\mathbb{P}(\{g\}, X) = q$, and furthermore it is trivial that $\mathbb{P}(\{g\}, X) > q$ implies $\mathbb{P}(\{g\}, X) \geq q$. Hence, the probabilistic implication $\{\mathsf{d} \geq r.X\} \to \{\mathsf{d} \geq q.X\}$ is valid in $\mathbb{K}$ for each value $r \in (p,q]$. □

Lemma 8. *Let $\mathbb{K}$ be a probabilistic formal context, and assume that the implication $X \to Y \in \mathsf{Imp}(M)$ is valid in the almost certain scaling $\mathbb{K}_\varepsilon^\times$. Then $X \to Y$ is valid in $\mathbb{K}$, too, and in particular for each probability value $p \in [0,1]$, it holds true that $\mathbb{K} \models \{\mathsf{d} \geq p.X\} \to \{\mathsf{d} \geq p.Y\}$.*

Proof. Since $X \to Y$ is valid in $\mathbb{K}_\varepsilon^\times$, we conclude that $X^{I_\varepsilon^\times} \subseteq Y^{I_\varepsilon^\times}$, or equivalently $Y \subseteq X^{I_\varepsilon^\times I_\varepsilon^\times}$. By Lemma 5 we have that $\mathbb{P}(\{g\}, X) = \mathbb{P}(\{g\}, X^{I_\varepsilon^\times I_\varepsilon^\times})$. As a consequence we get that $\mathbb{P}(\{g\}, X) = \mathbb{P}(\{g\}, X^{I_\varepsilon^\times I_\varepsilon^\times}) \leq \mathbb{P}(\{g\}, Y)$, and thus $\mathbb{P}(\{g\}, X) \geq p$ implies $\mathbb{P}(\{g\}, Y) \geq p$. As an immediate corollary it then follows that the probabilistic implication $\{\mathsf{d} \geq p.X\} \to \{\mathsf{d} \geq p.Y\}$ is valid in $\mathbb{K}$. □

5 Probabilistic Implicational Knowledge Bases

In this main section, we define the notion of a probabilistic implicational knowledge base, and develop a method for the construction of a sound and complete knowledge base for a given probabilistic formal context. Furthermore, in case of finiteness of the input probabilistic formal context, the construction will yield a knowledge base which is finite as well.

Definition 9. *A probabilistic implicational knowledge base over an attribute set M is a triple $(\mathcal{V}, \mathcal{L}, \mathcal{P})$ where $\mathcal{V} \subseteq \mathsf{Imp}([0,1])$, $\mathcal{L} \subseteq \mathsf{Imp}(M)$, and $\mathcal{P} \subseteq \mathsf{dImp}(M)$. It is valid in a probabilistic formal context $\mathbb{K} := (G, M, W, I, \mathbb{P})$ if each implication in $\mathcal{V} \cup \mathcal{L} \cup \mathcal{P}$ is valid in $\mathbb{K}$. Then we shall denote this as $\mathbb{K} \models (\mathcal{V}, \mathcal{L}, \mathcal{P})$. The relation $\models$ is lifted as usual, i.e., for two probabilistic implicational knowledge bases $\mathcal{K}_1$ and $\mathcal{K}_2$, we say that $\mathcal{K}_1$ entails $\mathcal{K}_2$, symbolized as $\mathcal{K}_1 \models \mathcal{K}_2$, if $\mathcal{K}_2$ is valid in all probabilistic formal contexts in which $\mathcal{K}_1$ is valid, i.e., if for all probabilistic formal contexts $\mathbb{K}$, it holds true that $\mathbb{K} \models \mathcal{K}_1$ implies $\mathbb{K} \models \mathcal{K}_2$.*

Furthermore, a probabilistic implicational knowledge base for a probabilistic formal context $\mathbb{K}$ is a probabilistic implicational knowledge base $\mathcal{K}$ that satisfies the following condition for all probabilistic implications $\mathbf{X} \to \mathbf{Y}$ over M:

$$\mathbb{K} \models \mathbf{X} \to \mathbf{Y} \text{ if, and only if, } \mathcal{K} \models \mathbf{X} \to \mathbf{Y}.$$

5.1 Trivial Background Knowledge

Lemma 10. *Let $\mathbb{K} := (G, M, W, I, \mathbb{P})$ be a probabilistic formal context. Then the following probabilistic implications are valid in $\mathbb{K}$.*

1. $\{\mathsf{d} \geq p.\, X\} \to \{\mathsf{d} \geq q.\, Y\}$ *if $X \supseteq Y$ and $p \geq q$.*
2. $\{\mathsf{d} \geq p.\, X, \mathsf{d} \geq q.\, Y\} \to \{\mathsf{d} \geq p+q-1.\, X \cup Y\}$ *if $p + q - 1 > 0$.*

Proof. Clearly, $\mathbb{P}(\{g\}, X) \geq p$ implies that $\mathbb{P}(\{g\}, X) \geq q$. Furthermore, since $X \supseteq Y$ yields that $\{g\} \times X \times \{w\} \supseteq \{g\} \times Y \times \{w\}$, we conclude that $\mathbb{P}(\{g\}, Y) \geq q$. For the second implication, observe that in case $p+q-1 > 0$ the intersection of $\{w \in W \mid \forall x \in X \colon (g, x, w) \in I\}$ and $\{w \in W \mid \forall y \in Y \colon (g, y, w) \in I\}$ must be non-empty, and in particular must have a $\mathbb{P}$-measure of at least $p + q - 1$. $\square$

Define the background knowledge $\mathcal{T}(M)$ to consist of all trivial probabilistic implications over M, i.e.,

$$\mathcal{T}(M) := \{\mathbf{X} \to \mathbf{Y} \mid \mathbf{X} \to \mathbf{Y} \in \mathsf{dImp}(M) \text{ and } \emptyset \models \mathbf{X} \to \mathbf{Y}\}.$$

In particular, $\mathcal{T}(M)$ contains all implications from Lemma 10. Since our aim is to construct a knowledge base for a given probabilistic formal context, we do not have to explicitly compute these trivial implications, and we will hence utilize them as background knowledge, which is already present and known.

Lemma 11. *Let $\mathcal{P} \cup \{\mathbf{X} \to \mathbf{Y}\} \subseteq \mathsf{dImp}(M)$ be a set of probabilistic implications over M. If $\mathcal{P} \cup \mathcal{T}(M) \models \mathbf{X} \to \mathbf{Y}$ with respect to non-probabilistic entailment, then $\mathcal{P} \models \mathbf{X} \to \mathbf{Y}$ with respect to probabilistic entailment.*

Proof. Consider a probabilistic formal context $\mathbb{K}$ with $\mathbb{K} \models \mathcal{P}$. Then Lemma 14 implies that $\mathsf{d}(\mathbb{K}) \models \mathcal{P}$. Furthermore, it is trivial that $\mathsf{d}(\mathbb{K}) \models \mathcal{T}(M)$. Consequently, the probabilistic implication $\mathbf{X} \to \mathbf{Y}$ is valid in $\mathsf{d}(\mathbb{K})$, and another application of Lemma 14 yields that $\mathbb{K} \models \mathbf{X} \to \mathbf{Y}$. $\square$

5.2   Approximations of Probabilities

Assume that $\mathbb{K} := (G, M, W, I, \mathbb{P})$ is a probabilistic context. Then the set

$$V(\mathbb{K}) := \{\, \mathbb{P}(\{g\}, B) \mid g \in G \text{ and } B \subseteq M \,\}$$

contains all values that can occur when evaluating the validity of implications over probabilistic attributes in $\mathbb{K}$. Note that according to Lemma 5 it holds true that

$$V(\mathbb{K}) = \{\, \mathbb{P}(\{g\}, B) \mid g \in G \text{ and } B \in \mathsf{Int}(\mathbb{K}_\varepsilon^\times) \,\}.$$

Furthermore, we define an *upper approximation* of probability values as follows: For an arbitrary $p \in [0,1]$, let $\lceil p \rceil_{\mathbb{K}}$ be the smallest value above p which can occur when evaluating a probabilistic attribute in $\mathbb{K}$, i.e., we define

$$\lceil p \rceil_{\mathbb{K}} := 1 \wedge \bigwedge \{\, q \mid p \leq q \text{ and } q \in V(\mathbb{K}) \,\}.$$

Dually, we define the *lower approximation* of $p \in [0,1]$ in $\mathbb{K}$ as

$$\lfloor p \rfloor_{\mathbb{K}} := 0 \vee \bigvee \{\, q \mid q \leq p \text{ and } q \in V(\mathbb{K}) \,\}.$$

It is easy to verify that for all attribute sets $B \subseteq M$ and all probability values $p \in [0,1]$, the following entailment is valid:

$$\mathbb{K} \models \{\{\mathsf{d} \geq p.\, B\} \rightarrow \{\mathsf{d} \geq \lceil p \rceil_{\mathbb{K}}.\, B\}, \{\mathsf{d} \geq \lceil p \rceil_{\mathbb{K}}.\, B\} \rightarrow \{\mathsf{d} \geq p.\, B\}\}.$$

The second implication is in fact valid in arbitrary probabilistic contexts, since we can apply Lemma 10 with $p \leq \lceil p \rceil_{\mathbb{K}}$. For the first implication, observe that for all objects $g \in G$ and all attribute set $B \subseteq M$, it holds true that

$$\lfloor \mathbb{P}(\{g\}, B) \rfloor_{\mathbb{K}} = \mathbb{P}(\{g\}, B) = \lceil \mathbb{P}(\{g\}, B) \rceil_{\mathbb{K}},$$

and thus in particular $\mathbb{P}(\{g\}, B) \geq p$ if, and only if, $\mathbb{P}(\{g\}, B) \geq \lceil p \rceil_{\mathbb{K}}$. Analogously, $\mathbb{P}(\{g\}, B) \leq p$ is equivalent to $\mathbb{P}(\{g\}, B) \leq \lfloor p \rfloor_{\mathbb{K}}$.

Lemma 12. *Let $\mathbb{K}$ be a probabilistic formal context, and assume that $p \in (0,1)$ is a probability value. Then the implication $(\lfloor p \rfloor_{\mathbb{K}}, \lceil p \rceil_{\mathbb{K}}] \rightarrow \{\lceil p \rceil_{\mathbb{K}}\}$ is valid in $\mathbb{K}$. Furthermore, the implications $[0, \lceil 0 \rceil_{\mathbb{K}}] \rightarrow \{\lceil 0 \rceil_{\mathbb{K}}\}$ and $(\lfloor 1 \rfloor_{\mathbb{K}}, 1] \rightarrow \{1\}$ are valid in $\mathbb{K}$.*

Proof. Consider an arbitrary object $g \in G$ as well as an arbitrary attribute set $B \subseteq M$, and assume that $\lfloor p \rfloor_{\mathbb{K}} < \mathbb{P}(\{g\}, B) \leq \lceil p \rceil_{\mathbb{K}}$. Of course, it holds true that $(\lfloor p \rfloor_{\mathbb{K}}, \lceil p \rceil_{\mathbb{K}}] \cap V(\mathbb{K}) = \{\lceil p \rceil_{\mathbb{K}}\}$, and consequently $\mathbb{P}(\{g\}, B) = \lceil p \rceil_{\mathbb{K}}$.

Now consider the implication $[0, \lceil 0 \rceil_{\mathbb{K}}] \rightarrow \{\lceil 0 \rceil_{\mathbb{K}}\}$. In case $0 \in V(\mathbb{K})$ we have that $\lceil 0 \rceil_{\mathbb{K}} = 0$, and then the implication is trivial. Otherwise, it follows that $[0, \lceil 0 \rceil_{\mathbb{K}}] \cap V(\mathbb{K}) = \{\lceil 0 \rceil_{\mathbb{K}}\}$, and consequently $0 \leq \mathbb{P}(\{g\}, B) \leq \lceil 0 \rceil_{\mathbb{K}}$ implies $\mathbb{P}(\{g\}, B) = \lceil 0 \rceil_{\mathbb{K}}$.

Eventually, we prove the validity of the implication $(\lfloor 1 \rfloor_{\mathbb{K}}, 1] \rightarrow \{1\}$. If $1 \in V(\mathbb{K})$, then $\lfloor 1 \rfloor_{\mathbb{K}} = 1$, and hence the premise interval $(1, 1]$ is empty, i.e., the implication trivially holds in $\mathbb{K}$. Otherwise, $(\lfloor 1 \rfloor_{\mathbb{K}}, 1] \cap V(\mathbb{K}) = \emptyset$, and the implication is again trivial. $\qquad\square$

5.3   The Probabilistic Scaling

Definition 13. *Let* $\mathbb{K} := (G, M, W, I, \mathbb{P})$ *be a probabilistic formal context. The probabilistic scaling of* $\mathbb{K}$ *is defined as the formal context* $\mathsf{d}(\mathbb{K}) := (G, \mathsf{d}(M), I)$ *the incidence relation* I *of which is defined by*

$$(g, \mathsf{d} \geq p. B) \in I \text{ if } \mathbb{P}(\{g\}, B) \geq p,$$

and by $\mathsf{d}^*(\mathbb{K})$ *we denote the subcontext of* $\mathsf{d}(\mathbb{K})$ *with the attribute set*

$$\mathsf{d}^*(M) := \{\, \mathsf{d} \geq \lceil p \rceil_{\mathbb{K}}. B^{I^{\mathcal{E}} I^{\mathcal{E}}} \mid p \in [0,1],\ \lceil p \rceil_{\mathbb{K}} \neq 0,\ \text{and } B \subseteq M,\ B^{I^{\mathcal{E}} I^{\mathcal{E}}} \neq \emptyset \,\}.$$

Figure 3 shows the probabilistic scaling $\mathsf{d}^*(\mathbb{K}_{\mathsf{ex}})$ the attribute set of which is given by

$$\mathsf{d}^*(\{m_1, m_2, m_3\}) = \{\, \mathsf{d} \geq p. B \mid B \in \mathsf{Int}((\mathbb{K}_{\mathsf{ex}})_{\mathcal{E}}^{\times}) \setminus \{\emptyset\} \text{ and } p \in V(\mathbb{K}_{\mathsf{ex}}) \setminus \{0\} \,\}$$

$$= \left\{ \begin{array}{l} \mathsf{d} \geq p.\{m_2\}, \mathsf{d} \geq p.\{m_3\}, \mathsf{d} \geq p.\{m_1, m_3\}, \\ \mathsf{d} \geq p.\{m_2, m_3\}, \mathsf{d} \geq p.\{m_1, m_2, m_3\} \end{array} \middle| \, p \in \{\tfrac{1}{6}, \tfrac{1}{3}, \tfrac{2}{3}, \tfrac{5}{6}, 1\} \right\}.$$

Please note that the dual formal context is displayed, i.e., the incidence table is transposed. More formally, for a formal context $\mathbb{K} := (G, M, I)$ its *dual context* is $\mathbb{K}^{\partial} := (M, G, \{\, (m, g) \mid (g, m) \in I \,\})$.

Lemma 14. *Let* $\mathbb{K}$ *be a probabilistic formal context. Then for all probabilistic implications* $\{\, \mathsf{d} \geq p_t. X_t \mid t \in T \,\} \to \{\mathsf{d} \geq q. Y\}$, *the following statements are equivalent:*

1. $\mathbb{K} \models \{\, \mathsf{d} \geq p_t. X_t \mid t \in T \,\} \to \{\mathsf{d} \geq q. Y\}$
2. $\mathsf{d}(\mathbb{K}) \models \{\, \mathsf{d} \geq p_t. X_t \mid t \in T \,\} \to \{\mathsf{d} \geq q. Y\}$
3. $\mathsf{d}^*(\mathbb{K}) \models \{\, \mathsf{d} \geq \lceil p_t \rceil_{\mathbb{K}}. X_t^{I^{\mathcal{E}} I^{\mathcal{E}}} \mid t \in T \,\} \to \{\mathsf{d} \geq \lceil q \rceil_{\mathbb{K}}. Y^{I^{\mathcal{E}} I^{\mathcal{E}}}\}$

Proof. Statements 1 and 2 are equivalent by Definition 4. Furthermore, from Lemma 5 we conclude that $(\mathsf{d} \geq p. X)^I = (\mathsf{d} \geq p. X^{I^{\mathcal{E}} I^{\mathcal{E}}})^I$ for all probabilistic attributes $\mathsf{d} \geq p. X$. From this and the fact that $\mathbb{P}(\{g\}, B) \geq p$ is equivalent to $\mathbb{P}(\{g\}, B) \geq \lceil p \rceil_{\mathbb{K}}$, the equivalence of Statements 2 and 3 then follows easily. □

5.4   Construction of the Probabilistic Implicational Knowledge Base

Theorem 15. *Let* $\mathbb{K} := (G, M, W, I, \mathbb{P})$ *be a probabilistic context. Then* $\mathcal{K}(\mathbb{K}) := (V(\mathbb{K}), \mathsf{Can}(\mathbb{K}_{\mathcal{E}}^{\times}), \mathsf{Can}(\mathsf{d}^*(\mathbb{K}), \mathcal{T}(M)))$ *is a probabilistic implicational base for* $\mathbb{K}$, *where*

$$V(\mathbb{K}) := \{\, [0, \lceil 0 \rceil_{\mathbb{K}}] \to \{\lceil 0 \rceil_{\mathbb{K}}\} \mid 0 \neq \lceil 0 \rceil_{\mathbb{K}} \,\} \cup \{\, (\lfloor 1 \rfloor_{\mathbb{K}}, 1] \to \{1\} \mid \lfloor 1 \rfloor_{\mathbb{K}} \neq 1 \,\}$$
$$\cup \{\, (\lfloor p \rfloor_{\mathbb{K}}, \lceil p \rceil_{\mathbb{K}}] \to \{\lceil p \rceil_{\mathbb{K}}\} \mid p \in (\lceil 0 \rceil_{\mathbb{K}}, \lfloor 1 \rfloor_{\mathbb{K}}) \setminus V(\mathbb{K}) \,\}.$$

If $\mathbb{K}$ is finite, then $\mathcal{K}(\mathbb{K})$ is finite, too. Furthermore, finiteness of $\mathcal{K}(\mathbb{K})$ is also ensured in case of finiteness of both M and $V(\mathbb{K})$.

$(\mathrm{d}^*(\mathbb{K}_{\mathsf{ex}}))^\partial$	g_1	g_2	g_3
$\mathsf{d} \geq \frac{1}{6}.\{m_2\}$	·	X	X
$\mathsf{d} \geq \frac{1}{6}.\{m_3\}$	X	X	X
$\mathsf{d} \geq \frac{1}{6}.\{m_1, m_3\}$	X	·	X
$\mathsf{d} \geq \frac{1}{6}.\{m_2, m_3\}$	·	X	·
$\mathsf{d} \geq \frac{1}{6}.\{m_1, m_2, m_3\}$	·	·	·
$\mathsf{d} \geq \frac{1}{3}.\{m_2\}$	·	X	·
$\mathsf{d} \geq \frac{1}{3}.\{m_3\}$	X	X	X
$\mathsf{d} \geq \frac{1}{3}.\{m_1, m_3\}$	X	·	X
$\mathsf{d} \geq \frac{1}{3}.\{m_2, m_3\}$	·	X	·
$\mathsf{d} \geq \frac{1}{3}.\{m_1, m_2, m_3\}$	·	·	·
$\mathsf{d} \geq \frac{2}{3}.\{m_2\}$	·	X	·
$\mathsf{d} \geq \frac{2}{3}.\{m_3\}$	X	X	X
$\mathsf{d} \geq \frac{2}{3}.\{m_1, m_3\}$	X	·	·
$\mathsf{d} \geq \frac{2}{3}.\{m_2, m_3\}$	·	X	·
$\mathsf{d} \geq \frac{2}{3}.\{m_1, m_2, m_3\}$	·	·	·
$\mathsf{d} \geq \frac{5}{6}.\{m_2\}$	·	X	·
$\mathsf{d} \geq \frac{5}{6}.\{m_3\}$	X	·	X
$\mathsf{d} \geq \frac{5}{6}.\{m_1, m_3\}$	X	·	·
$\mathsf{d} \geq \frac{5}{6}.\{m_2, m_3\}$	·	·	·
$\mathsf{d} \geq \frac{5}{6}.\{m_1, m_2, m_3\}$	·	·	·
$\mathsf{d} \geq 1.\{m_2\}$	·	X	·
$\mathsf{d} \geq 1.\{m_3\}$	X	·	·
$\mathsf{d} \geq 1.\{m_1, m_3\}$	X	·	·
$\mathsf{d} \geq 1.\{m_2, m_3\}$	·	·	·
$\mathsf{d} \geq 1.\{m_1, m_2, m_3\}$	·	·	·

Fig. 3. The probabilistic scaling of the probabilistic formal context given in Fig. 1

Proof. We first prove soundness. Lemma 12 yields that $\mathcal{V}(\mathbb{K})$ is valid in $\mathbb{K}$, Lemma 8 proves that $\mathsf{Can}(\mathbb{K}_{\tilde{\varepsilon}}^{\times})$ is valid in $\mathbb{K}$, and Lemma 14 shows the validity of $\mathsf{Can}(\mathsf{d}^*(\mathbb{K}), \mathcal{T}(M))$ in $\mathbb{K}$.

We proceed with proving completeness. Assume that

$$\mathbb{K} \models \{\, \mathsf{d} \geq p_t . X_t \mid t \in T \,\} \to \{\, \mathsf{d} \geq q . Y \,\}.$$

Then by Lemma 14 the implication

$$\{\, \mathsf{d} \geq \lceil p_t \rceil_{\mathbb{K}} . X_t^{I_{\tilde{\varepsilon}}^{\times} I_{\tilde{\varepsilon}}^{\times}} \mid t \in T \,\} \to \{\, \mathsf{d} \geq \lceil q \rceil_{\mathbb{K}} . Y^{I_{\tilde{\varepsilon}}^{\times} I_{\tilde{\varepsilon}}^{\times}} \,\}$$

is valid in $\mathsf{d}^*(\mathbb{K})$, and must hence be entailed by $\mathsf{Can}(\mathsf{d}^*(\mathbb{K}), \mathcal{T}(M)) \cup \mathcal{T}(M)$ with respect to non-probabilistic entailment. Lemma 11 then yields that it must also be entailed by $\mathsf{Can}(\mathsf{d}^*(\mathbb{K}), \mathcal{T}(M))$ with respect to probabilistic entailment. Furthermore, the implications

$$(\lfloor p \rfloor_{\mathbb{K}}, \lceil p \rceil_{\mathbb{K}}] \to \{\lceil p \rceil_{\mathbb{K}}\}$$

for $p \in (\{\, p_t \mid t \in T \,\} \cup \{q\}) \setminus V(\mathbb{K})$ are contained in $\mathcal{V}(\mathbb{K})$, and the implications $X \to X^{I_{\tilde{\varepsilon}}^{\times} I_{\tilde{\varepsilon}}^{\times}}$ for $X \in \{\, X_t \mid t \in T \,\} \cup \{Y\}$ are entailed by $\mathsf{Can}(\mathbb{K}_{\tilde{\varepsilon}}^{\times})$. Utilizing Lemmas 7 and 8 and summing up, it holds true that

$$\begin{aligned}
\mathcal{K}(\mathbb{K}) \models\; & \{\, \{\mathsf{d} \geq p_t . X_t\} \to \{\mathsf{d} \geq \lceil p_t \rceil_{\mathbb{K}} . X_t\} \mid t \in T \,\} \\
& \cup \{\, \{\mathsf{d} \geq \lceil p_t \rceil_{\mathbb{K}} . X_t\} \to \{\mathsf{d} \geq \lceil p_t \rceil_{\mathbb{K}} . X_t^{I_{\tilde{\varepsilon}}^{\times} I_{\tilde{\varepsilon}}^{\times}}\} \mid t \in T \,\} \\
& \cup \{\, \{\mathsf{d} \geq \lceil p_t \rceil_{\mathbb{K}} . X_t^{I_{\tilde{\varepsilon}}^{\times} I_{\tilde{\varepsilon}}^{\times}} \mid t \in T \,\} \to \{\mathsf{d} \geq \lceil q \rceil_{\mathbb{K}} . Y^{I_{\tilde{\varepsilon}}^{\times} I_{\tilde{\varepsilon}}^{\times}}\} \,\},
\end{aligned}$$

and so $\mathcal{K}(\mathbb{K})$ also entails $\{\, \mathsf{d} \geq p_t . X_t \mid t \in T \,\} \to \{\mathsf{d} \geq \lceil q \rceil_{\mathbb{K}} . Y^{I_{\tilde{\varepsilon}}^{\times} I_{\tilde{\varepsilon}}^{\times}}\}$. Of course, the implications

$$\{\mathsf{d} \geq \lceil q \rceil_{\mathbb{K}} . Y^{I_{\tilde{\varepsilon}}^{\times} I_{\tilde{\varepsilon}}^{\times}}\} \to \{\mathsf{d} \geq q . Y^{I_{\tilde{\varepsilon}}^{\times} I_{\tilde{\varepsilon}}^{\times}}\} \text{ and } \{\mathsf{d} \geq q . Y^{I_{\tilde{\varepsilon}}^{\times} I_{\tilde{\varepsilon}}^{\times}}\} \to \{\mathsf{d} \geq q . Y\}$$

are trivial, i.e., are valid in all probabilistic formal contexts. Eventually, we have thus just shown that the considered probabilistic implication $\{\, \mathsf{d} \geq p_t . X_t \mid t \in T \,\} \to \{\mathsf{d} \geq q . Y\}$ follows from $\mathcal{K}(\mathbb{K})$.

As last step, we prove the claim on finiteness of $\mathcal{K}(\mathbb{K})$ if both M and $V(\mathbb{K})$ are finite. Clearly, then $\mathcal{V}(\mathbb{K})$ is finite. The almost certain scaling $\mathbb{K}_{\tilde{\varepsilon}}^{\times}$ then possess a finite attribute set, and thus its canonical base must be finite. It also follows that the attribute set $\mathsf{d}^*(M)$ of the probabilistic scaling $\mathsf{d}^*(\mathbb{K})$ is finite, and hence the canonical base of $\mathsf{d}^*(\mathbb{K})$ is finite as well. □

Returning back to our running example $\mathbb{K}_{\mathsf{ex}}$ from Fig. 1, we now construct its probabilistic implicational knowledge base $\mathcal{K}(\mathbb{K}_{\mathsf{ex}})$. The first component is the following set of implications between probability values:

$$\{(0, \tfrac{1}{6}] \to \{\tfrac{1}{6}\}, (\tfrac{1}{6}, \tfrac{1}{3}] \to \{\tfrac{1}{3}\}, (\tfrac{1}{3}, \tfrac{2}{3}] \to \{\tfrac{2}{3}\}, (\tfrac{2}{3}, \tfrac{5}{6}] \to \{\tfrac{5}{6}\}, (\tfrac{5}{6}, 1] \to \{1\}\}.$$

$$\text{Can}(\mathsf{d}^*(\mathbb{K}_{\text{ex}}),\mathcal{T}'(\{m_1,m_2,m_3\})) = \left\{ \begin{array}{l} \emptyset \\ \rightarrow \{\mathsf{d} \geq \frac{2}{3}.\{m_3\}\}, \\[4pt] \{\mathsf{d} \geq 1.\{m_1,m_3\}, \mathsf{d} \geq \frac{1}{6}.\{m_2\}\} \\ \rightarrow \{\mathsf{d} \geq 1.\{m_1,m_2,m_3\}\}, \\[4pt] \{\mathsf{d} \geq \frac{2}{3}.\{m_2,m_3\}, \mathsf{d} \geq \frac{5}{6}.\{m_3\}, \mathsf{d} \geq 1.\{m_2\}, \mathsf{d} \geq \frac{1}{3}.\{m_1,m_3\}\} \\ \rightarrow \{\mathsf{d} \geq 1.\{m_1,m_2,m_3\}\}, \\[4pt] \{\mathsf{d} \geq 1.\{m_3\}, \mathsf{d} \geq \frac{1}{3}.\{m_1,m_3\}\} \\ \rightarrow \{\mathsf{d} \geq 1.\{m_1,m_3\}\}, \\[4pt] \{\mathsf{d} \geq \frac{2}{3}.\{m_1,m_3\}, \mathsf{d} \geq \frac{5}{6}.\{m_3\}\} \\ \rightarrow \{\mathsf{d} \geq 1.\{m_1,m_3\}\}, \\[4pt] \{\mathsf{d} \geq \frac{2}{3}.\{m_3\}, \mathsf{d} \geq \frac{1}{6}.\{m_2,m_3\}\} \\ \rightarrow \{\mathsf{d} \geq \frac{2}{3}.\{m_2,m_3\}, \mathsf{d} \geq 1.\{m_2\}\}, \\[4pt] \{\mathsf{d} \geq \frac{2}{3}.\{m_3\}, \mathsf{d} \geq \frac{1}{3}.\{m_2\}\} \\ \rightarrow \{\mathsf{d} \geq \frac{2}{3}.\{m_2,m_3\}, \mathsf{d} \geq 1.\{m_2\}\}, \\[4pt] \{\mathsf{d} \geq \frac{5}{6}.\{m_3\}\} \\ \rightarrow \{\mathsf{d} \geq \frac{1}{3}.\{m_1,m_3\}\}, \\[4pt] \{\mathsf{d} \geq \frac{2}{3}.\{m_3\}, \mathsf{d} \geq \frac{1}{6}.\{m_1,m_3\}\} \\ \rightarrow \{\mathsf{d} \geq \frac{5}{6}.\{m_3\}, \mathsf{d} \geq \frac{1}{3}.\{m_1,m_3\}\} \end{array} \right\}$$

Fig. 4. The implicational base of $\mathsf{d}^*(\mathbb{K}_{\text{ex}})$ with respect to the background implications that are described in Lemma 10.

Note that we left out the trivial implication $\{0\} \rightarrow \{0\}$. The canonical base of the certain scaling $(\mathbb{K}_{\text{ex}})_\varepsilon^\times$ was computed as $\{\{m_1\} \rightarrow \{m_3\}\}$, which consequently is the second component $\mathcal{K}(\mathbb{K}_{\text{ex}})$. For the computation of the third component of $\mathcal{K}(\mathbb{K}_{\text{ex}})$, we consider the probabilistic scaling of $\mathbb{K}_{\text{ex}}$. In order to avoid the axiomatization of trivial implications, we construct the implicational base of $\mathsf{d}^*(\mathbb{K}_{\text{ex}})$ relative to the implication set containing all those probabilistic implications which are described in Lemma 10. This set $\mathcal{T}'(\{m_1,m_2,m_3\})$ of background knowledge contains, among others, the following implications:

$$\{\mathsf{d} \geq \tfrac{5}{6}.\{m_1,m_3\}\} \rightarrow \{\mathsf{d} \geq \tfrac{1}{3}.\{m_1\}\}$$
$$\{\mathsf{d} \geq \tfrac{5}{6}.\{m_1,m_3\}\} \rightarrow \{\mathsf{d} \geq \tfrac{2}{3}.\{m_3\}\}$$
$$\vdots$$
$$\{\mathsf{d} \geq \tfrac{2}{3}.\{m_2\}, \mathsf{d} \geq \tfrac{2}{3}.\{m_3\}\} \rightarrow \{\mathsf{d} \geq \tfrac{1}{3}.\{m_2,m_3\}\}$$
$$\{\mathsf{d} \geq \tfrac{5}{6}.\{m_1,m_3\}, \mathsf{d} \geq \tfrac{5}{6}.\{m_2,m_3\}\} \rightarrow \{\mathsf{d} \geq \tfrac{2}{3}.\{m_1,m_2,m_3\}\}$$
$$\vdots$$

The resulting probabilistic implication set is presented in Fig. 4. Please note that we possibly did not include all trivial probabilistic implications in the background knowledge, as we do not yet know an appropriate reasoning procedure.

6   Conclusion

In this document we have investigated a method for the axiomatization of rules that are valid in a given probabilistic data set, which is represented as a multi-world view over the same set of entities and vocabulary to describe the entities, and which is furthermore equipped with a probability measure on the set of worlds. We have developed such a method in the field of Formal Concept Analysis, where it is possible to assign properties to single objects. We have achieved the description of a technique for a sound and complete axiomatization of terminological knowledge which is valid in the input data set and expressible in the chosen description language. It is only natural to extend the results to a probabilistic version of the light-weight description logic $\mathcal{EL}^{\perp}$, which not only allows for assigning properties (*concept names*) to entities, but also allows for connecting pairs of entities by binary relations (*role names*). This will be subject of an upcoming publication.

It remains to apply the proposed method to concrete real-world data sets, e.g., in the medical domain, where worlds are represented by patients, or in natural sciences, where worlds are represented by repetitions of the same experiment, etc. From a theoretical perspective, it is interesting to investigate whether the constructed probabilistic implicational knowledge base is of a minimal size – as it holds true for the canonical base in the non-probabilistic case. Furthermore, so far we have only considered a model-theoretic semantics, and the induced semantic entailment between implication sets. For practical purposes, a syntactic entailment in the favor of the *Armstrong rules* in FCA is currently missing.

Acknowledgements. The author gratefully thanks the anonymous reviewers for their constructive hints and helpful remarks.

References

1. Ganter, B.: Attribute exploration with background knowledge. Theor. Comput. Sci. **217**(2), 215–233 (1999)
2. Ganter, B.: Two Basic Algorithms in Concept Analysis. FB4-Preprint 831. Technische Hochschule Darmstadt, Darmstadt, Germany (1984)
3. Ganter, B.: Two basic algorithms in concept analysis. In: Kwuida, L., Sertkaya, B. (eds.) ICFCA 2010. LNCS (LNAI), vol. 5986, pp. 312–340. Springer, Heidelberg (2010). doi:10.1007/978-3-642-11928-6_22
4. Ganter, B., Wille, R.: Formal Concept Analysis. Mathematical Foundations. Springer, Heidelberg (1999)
5. Guigues, J.-L., Duquenne, V.: Famille minimale d'implications informatives résultant d'un tableau de données binaires. Math. Sci. Humaines **95**, 5–18 (1986)

6. Kriegel, F.: Concept Explorer FX. Software for Formal Concept Analysis with Description Logic Extensions, 2010–2017. https://github.com/francesco-kriegel/conexp-fx

7. Kriegel, F.: NextClosures – Parallel Exploration of Constrained Closure Operators. LTCS-Report 15-01. Chair for Automata Theory, Technische Universität Dresden (2015)

8. Kriegel, F.: Parallel attribute exploration. In: Haemmerlé, O., Stapleton, G., Faron Zucker, C. (eds.) ICCS 2016. LNCS (LNAI), vol. 9717, pp. 91–106. Springer, Cham (2016). doi:10.1007/978-3-319-40985-6_8

9. Kriegel, F.: Probabilistic implicational bases in FCA and probabilistic bases of GCIs in $\mathcal{EL}^{\perp}$. In: Ben Yahia, S., Konecny, J. (eds.) Proceedings of the Twelfth International Conference on Concept Lattices and Their Applications. CEUR Workshop Proceedings, vol. 1466, Clermont-Ferrand, France, pp. 193–204. CEUR-WS.org, 13–16 October 2015

10. Kriegel, F., Borchmann, D.: NextClosures: parallel computation of the canonical base. In: Ben Yahia, S., Konecny, J. (eds.) Proceedings of the Twelfth International Conference on Concept Lattices and Their Applications. CEUR Workshop Proceedings, vol. 1466, Clermont-Ferrand, France, pp. 181–192. CEUR-WS.org, 13–16 October 2015

11. Maier, D.: The Theory of Relational Databases. Computer Science Press, Rockville (1983)

12. Stumme, G.: Attribute exploration with background implications and exceptions. In: Bock, H.-H., Polasek, W. (eds.) Data Analysis and Information Systems. Studies in Classification, Data Analysis, and Knowledge Organization, pp. 457–469. Springer, Heidelberg (1996)

13. Wille, R.: Restructuring lattice theory: an approach based on hierarchies of concepts. In: Rival, I. (ed.) Ordered Sets. LNCS, vol. 83, pp. 445–470. Springer, Netherlands (1982). doi:10.1007/978-94-009-7798-3_15

On Overfitting of Classifiers Making a Lattice

Tatiana Makhalova$^{(\boxtimes)}$ and Sergei O. Kuznetsov

National Research University Higher School of Economics,
Kochnovsky pr. 3, Moscow 125319, Russia
{tpmakhalova,skuznetsov}@hse.ru

Abstract. Obtaining accurate bounds of the probability of overfitting is a fundamental question in statistical learning theory. In this paper we propose exact combinatorial bounds for the family of classifiers making a lattice. We use some lattice properties to derive the probability of overfitting for a set of classifiers represented by concepts. The extent of a concept, in turn, matches the set of objects correctly classified by the corresponding classifier. Conducted experiments illustrate that the proposed bounds are consistent with the Monte Carlo bounds.

Keywords: Computational learning theory · Probability of overfitting · Lattice of classifiers

1 Introduction

Overfitting is usually defined as poor performance of classifier on test data in contrast to good performance on training data. Over the past few decades, formal definition of overfitting and obtaining good estimates of overfitting remains a very important problem in statistical learning theory. First bounds proposed by Vapnik and Chervonenkis [24] are unrealistically high. So far, the main reasons of this are realized, but there is still no general approach that would considerably improve the bounds.

Many attempts were made to obtain more accurate overfitting bounds. For example, there are bounds that use some measure concentration inequalities [5, 17,21,22]. Another approach to obtaining more accurate bounds is based on defining a new dimensionality on the space of classifiers induced by real-valued loss function (fat-shattering dimension [10], ε-covering number and ε-packing number [18], Rademacher complexity [11,12]).

Significant improvements have been made using PAC-Bayesian (Probably approximately correct) approach which estimate the expected true error by Kullback-Leibler divergence [15,16,19]. In several works the bounds were refined by taking into account similarity of classifiers [1,3,4,13,20].

In [25] it was proposed to estimate the probability of overfitting under a weak axiom, which allows one to deal with a finite instance space (a set of objects) rather then with an infinite one (below we will describe it formally). Within the simplified probabilistic framework some exact combinatorial bounds

© Springer International Publishing AG 2017
K. Bertet et al. (Eds.): ICFCA 2017, LNAI 10308, pp. 184–197, 2017.
DOI: 10.1007/978-3-319-59271-8_12

for several sets of classifiers have been obtained [2,23]. In [27] overfitting bounds are computed by a random walk through the set of classifiers.

In this paper we propose exact bounds for classifiers where vectors of correct answers comprise a lattice. This is a very typical case where e.g. classifiers are conjunctions of unary predicates, or in other terms, sets of binary attributes [6, 7,14]. An FCA-based approach to selecting the best performing classifier was proposed in [9]. The algorithm based on this approach has higher accuracy than that of other well-known methods, e.g., of AdaBoost and bagging with SVM. Our research is focused on obtaining theoretical estimates of overfitting.

In our approach we consider a formal context as an error matrix. Rows of the context correspond to a finite set of instances and its columns correspond to vectors of correct answers of classifiers. We assume that a learning algorithm observes only a subset of objects and – based on the information "classified correctly/misclassified" – selects the best classifiers, which have the minimal number of errors on an training set. For learning process we examine the worst case, where the learning algorithm selects the worst classifier among the best ones, i.e. with the maximal number of errors on the test set. We say that the learning algorithm overfits if the number of errors on the test set is much larger than that on the training set. Our purpose is to derive tight bounds of the probability of overfitting for the set of classifiers making a lattice.

The paper has the following structure. The next section briefly introduces the main notions of formal concept analysis and statistical learning theory. Section 4 contains inference of the exact combinatorial bounds of the probability of overfitting for the family of classifiers forming a lattice. Section 5 demonstrates empirically the difference between precise bounds and bounds computed by Monte Carlo method. In the last section we conclude and discuss directions of future work.

2   Formal Concept Analysis

Here we briefly recall the basic definitions of FCA [8] which we will use below. Given a set of objects (instances) G and a set of attributes M, we consider an incidence relation $I \subseteq G \times M$, i.e., $(g, m) \in I$ if object $g \in G$ has attribute $m \in M$. A formal context is a triple (G, M, I). The derivation operator $(\cdot)'$ is defined for $Y \subseteq G$ and $Z \subseteq M$ as follows:

$$Y' = \{m \in M \mid gIm \text{ for all } g \in Y\}$$
$$Z' = \{g \in G \mid gIm \text{ for all } m \in Z\}$$

Y' is the set of attributes common to all objects of Y and Z' is the set of objects sharing all attributes from Z. The double application of $(\cdot)'$ is a closure operator, i.e., $(\cdot)''$ is extensive, idempotent and monotone. Sets $Y \subseteq G$, $Z \subseteq M$, such that $Y = Y''$ and $Z = Z''$ are said to be closed. A (formal) concept is a pair (Y, Z), where $Y \subseteq G$, $Z \subseteq M$ and $Y' = Z$, $Z' = Y$. Y is called the (formal) extent and Z is called the (formal) intent of the concept (Y, Z). A concept lattice (or Galois lattice) is a partially ordered set of concepts, the order $\leq$ is defined as follows: $(Y, Z) \leq (C, D)$ iff $Y \subseteq C\,(D \subseteq Z)$, a pair (Y, Z) is a subconcept of (C, D) and (C, D) is a superconcept of (Y, Z). Each concept lattice has the largest element (G, G'), called the top element, and the lowest element (M', M), called the bottom element.

To make our notation closer to that of previous research on overfitting, in what follows by $\mathbb{X}$ we denote a finite set of instances and by A a finite set of binary attributes.

3   The Probability of Overfitting

Let $\mathbb{X} = \{x_1, x_2, ..., x_L\}$ be a finite object space (general sample) and $A = \{a_1, ..., a_N\}$ be a set of classifiers. Defined as $\mathbb{X}$, the general sample is assumed to be fixed and no other objects will arrive in future (closed world assumption). The classifiers are represented by the following function $I : \mathbb{X} \times A \rightarrow \{0, 1\}$, where indicator $I(x, a) = 1$ if classifier a classifies correctly object x. In our study we consider classifiers as black boxes (no information about their structure is given).

We assume that ℓ objects randomly selected as a training set and remaining $k = L - \ell$ objects are used as a test set to estimate the quality of learning. Let us denote by $\mathbb{X} = X \mid \overline{X}$ a random partition of $\mathbb{X}$ into training set X of size ℓ and test set $\overline{X}$ of size k, by $[\mathbb{X}]^\ell$ all subsets of the general sample of size ℓ. We use the week probabilistic axiom [25] to define the probability of selecting a particular subset of objects of size ℓ:

Axiom 1. *All partitions of general sample $\mathbb{X}$ into two subsets of a fixed size have equal chances to be realized.*

Axiom 1 can be considered from another point of view. Let us assume that all permutations of general sample $\mathbb{X}$ have equal chances to be realized. Taking the first ℓ objects of the permuted sample, we will observe that all such subsets arise with equal probability. The last is analogous to Axiom 1.

Under Axiom 1 we can state that all possible partitions of $\mathbb{X}$ into two subsets X of size ℓ and $\overline{X}$ of size $L - \ell$ may be realized with equal probability $1/\binom{L}{\ell}$, where $\binom{L}{\ell} = L! / (\ell!(L - \ell)!)$.

Let $p(a, X)$ denote the number of correct answers of classifier a on sample X. The precision of a on X (the rate of correct answers) and error rate are defined as $\alpha(a, X) = p(a, X)/|X|$ and $\nu(a, X) = (|X| - p(a, X))/|X|$, respectively.

We denote by $A(X) \subseteq A$ a subset of classifiers that make the minimal number of errors on the training set, i.e. classifiers with the maximal number of correctly classified objects. A learning algorithm $\mu : [\mathbb{X}]^\ell \rightarrow A$ minimizing empirical risk has the following form:

$$\mu X \in A(X) = \arg\max_{a \in A} p(a, X), \tag{1}$$

There are three main kinds of an empirical risk minimization (ERM) algorithm μ: pessimistic, optimistic and randomized. The pessimistic ERM chooses the worst classifiers among $A(X)$, i.e. it selects classifier $a \in A(X)$ with the maximal number of misclassified objects (i.e. with the minimal number of correctly classified objects) in the test set $\overline{X} = \mathbb{X} \setminus X$:

$$\mu X = \arg \min_{a \in A(X)} p(a, \overline{X}). \tag{2}$$

The optimistic ERM, on the contrary, takes out the best classifier from $A(X)$ based on the number of errors in $\overline{X}$:

$$\mu X = \arg \max_{a \in A(X)} p(a, \overline{X}). \tag{3}$$

The randomized ERM selects a classifier from $A(X)$ uniformly by chance: $P([\mu X = a]) = 1/|A(X)|$.

For simplicity of notation, we will write $\ell_1 = p(a, X)$ and $\ell_0 = \ell - \ell_1$ for the number of correctly classified and misclassified objects of a from training set X. Similarly, we will denote by $k_1 = p(a, \overline{X})$ and $k_0 = k - k_1$ the number of correctly classified and misclassified objects in test set $\overline{X}$.

Let us fix a number ε to denote the maximum allowable difference in the error rates on a test and training sets, i.e. error tolerance. A learning algorithm μ is said to overfit on X if classifier $\mu X = a$ makes much more errors on the test set than on the training set [26]:

$$\frac{k_0}{k} - \frac{\ell_0}{\ell} \geq \varepsilon.$$

Restricting the difference in the error rates, we guarantee that the accuracy of a classifier on the test set will be almost the same as on the training set.

The probability of overfitting under the weak probabilistic axiom (see Axiom 1) has the following form:

$$Q_\varepsilon = P\left(\left[\frac{k_0}{k} - \frac{\ell_0}{\ell} \geq \varepsilon\right]\right) \equiv \frac{1}{\binom{L}{\ell}} \sum_{\mathbb{X}=X|\overline{X}} \left[\frac{k_0}{k} - \frac{\ell_0}{\ell} \geq \varepsilon\right], \tag{4}$$

where $[\cdot]$ denotes the indicator function of an event.

The probability of obtaining ℓ_0 errors is described by a hypergeometric distribution. To verify this fact, it is sufficient to make the following observation. Each classifier is represented by L-dimensional binary vector $a^T = (v_1, v_2, ..., v_L)$, $v_i \in \{0, 1\}$ with m zeros and $L - m$ ones. According to the described procedure (see Formulas 1 and 2), we select ℓ components and observe ℓ_0 zeros, the remaining $k_0 = m - \ell_0$ zeros are in the test set. It is clear to see that ℓ_0 follows the hypergeometric distribution, which is a discrete probability distribution that describes the probability of ℓ_0 successes in ℓ draws, without replacement, from a finite population of size L that contains exactly m successes, wherein each draw is either a success or a failure. Thus, the probability of obtaining ℓ_0 errors has the following form:

$$h_L^{\ell,m}(\ell_0) = \frac{\binom{m}{\ell_0}\binom{L-m}{\ell_1}}{\binom{L}{\ell}}, \tag{5}$$

where $\ell_0 \in [\max(0, m-k), \min(m, \ell)]$.

The left tail of the distribution corresponds to the cases where we observe a small number of errors, while a larger number of errors remains unobservable. Thus, the probability of overfitting is the sum of values of the probability mass function given in Formula 5 on the left side of the distribution:

$$H_L^{\ell,m}(s_m) = \sum_{s=s_{\min}}^{s_m} h_L^{\ell,m}(s).$$

The rightmost value s_m is derived using the specified tolerance ε as follows:

$$\frac{k_0}{k} - \frac{\ell_0}{\ell} = \frac{m - \ell_0}{k} - \frac{\ell_0}{\ell} = \frac{\ell m - \ell_0 L}{\ell k} \geq \varepsilon,$$

$$\ell_0 \leq \frac{\ell}{L}(m - \varepsilon k).$$

By Formula 4 given above, the probability of overfitting takes the following form:

$$Q_\varepsilon = P\left(\left[\frac{k_0}{k} - \frac{\ell_0}{\ell} \geq \varepsilon\right]\right) = \sum_{s=\max(0, m-k)}^{\lfloor s_m \rfloor} h_L^{\ell,m}(s) = H_L^{\ell,m}\left(\frac{l}{L}(m - \varepsilon k)\right), \quad (6)$$

where $s_m \leq \ell/L(m - \varepsilon k)$ and $\lfloor \cdot \rfloor$ is the floor function, which rounds a number down to the nearest integer. More detailed information is given in [26].

It is worth noting that we can use another definition of overfitting and assume that a classifier overfits data if the error rate on test set exceeds the specified tolerance ε. The redefinition of overfitting only slightly changes Formula 6:

$$Q_\varepsilon = P\left(\left[\frac{k_0}{k} \geq \varepsilon\right]\right) = \frac{1}{\binom{L}{\ell}} \sum_{X=X|\overline{X}} \left[\frac{k_0}{k} \geq \varepsilon\right] = H_L^{\ell,m}(m - \varepsilon k).$$

4   The Overfitting Probability of the Lattice of Classifiers

To compute the probability of overfitting for a family of classifiers comprising a lattice we introduce Definition 1 which was suggested by Vorontsov as "Conjecture 1" [26].

Definition 1. *A learning algorithm μ on a sample space $\mathbb{X}$ is called **evident** if for each partition $\mathbb{X} = X \mid \overline{X}$ and each classifier $a \in A$, there exist disjoint sets $X_a, X_a^* \subseteq \mathbb{X}$, such that $[\mu X = a] = [X_a \subseteq X][X_a^* \subseteq \overline{X}]$.*

We consider a set of classifiers $A = \{a = (a_1, ..., a_L)\}$ which make a lattice. It can be derived from an "object-attribute" formal context as follows. Each concept $C = (Y, Z)$ we associate with classifier a that classify correctly only objects from Y, we will write a' to denote these objects, thus, the number of

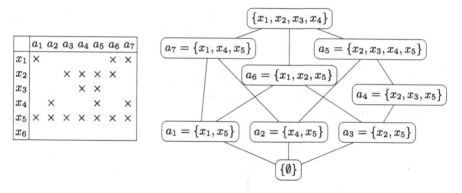

	a_1 a_2 a_3 a_4 a_5 a_6 a_7
x_1	× × ×
x_2	× × × ×
x_3	× ×
x_4	× × ×
x_5	× × × × × × ×
x_6	

Fig. 1. A set of classifiers represented by vectors of correct answers (on the left) and the corresponding lattice (on the right)

classifiers is equal to the number of concepts in the original lattice. From now, the attributes in a formal context correspond to classifiers and for any concept in a lattice that computed on this context there exists an attribute (classifier) in the given formal context. The table in Fig. 1 is an example of such kind of data.

The equivalence classes $\mathcal{S}$ of subsets of objects under closure is defined as follows: $\mathcal{S}(\mathcal{C}) = \{Y_{sub} \subseteq Y \mid Y'_{sub} = Z\}$, where $\mathcal{C} = (Y, Z)$ is a formal concept.

In this paper, we will consider equivalence classes in the level-wise manner, depending on size $|Y_{sub}|$. The equivalence class of the jth level ($1 \leq j \leq |Y|$) is defined as follows:

$$\mathcal{S}_j(\mathcal{C}) = \{Y_{sub} \subset Y \mid |Y_{sub}| = j, Y'_{sub} = Z\}.$$

Thus, an equivalence class can be represented as the union of equivalence subclasses: $\mathcal{S}(\mathcal{C}) = \mathcal{S}_1(\mathcal{C}) \cup \mathcal{S}_2(\mathcal{C}) \cup \ldots \cup \mathcal{S}_{|Y|}(\mathcal{C})$, elements in $\mathcal{S}_i(\mathcal{C})$ are generators of Y, and $\mathcal{S}_i(\mathcal{C})$ such that $i = \min_{j \in 1, \ldots, |Y|} \{\mathcal{S}_j(\mathcal{C}) \mid \mathcal{S}_j(\mathcal{C}) \neq \emptyset\}$ is a set of minimal generators for extent Y.

As it was mentioned before, each concept corresponds to a particular classifier a. Further, we will write $\mathcal{S}(a)$ for the concept $\mathcal{C} = (a', a'')$.

Example 1. The context in Fig. 1 (on the left) is comprised of 6 objects and 7 classifiers. For each classifier there exists a concept in the corresponding lattice (see Fig. 1, on the right). Classifier a_4, for instance, makes errors on objects x_1, x_4 and x_6.

Let us introduce the notation D, M_0 and M_1 for some subsets of objects from X. The sizes of these sets are d, m_0 and m_1, respectively. The set of objects D consists of those objects which are correctly classified by some classifiers, but not by all. Every classifier correctly classifies objects from M_1 and misclassified all objects from M_0 (see Fig. 2). Further, in our experiments, we will change the sample set by changing only the ratio of M_0 and M_1.

According to Definition 1 for any classifier $a \in A$ there exists at least two subsets X_a and X_a^*, such that their inclusion in training and test sets ensures that a will be selected by the learning algorithm μ. This assertion will be needed to prove Lemma 1.

In this way, we consider a random partition of $\mathbb{X}$ into two subsets, one of which is a training set X and another one is a test set $\overline{X}$. Both of subsets may include objects from D, M_1, M_0 as it is shown in Fig. 2.

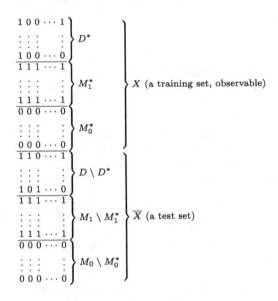

Fig. 2. A random partition $X \mid \overline{X}$ of the instance set $\mathbb{X} = D \cup M_0 \cup M_1$

Lemma 1. *If A makes a lattice, then each pessimistic ERM learning algorithm is evident. Moreover, the sets X_a, X_a^* can be chosen such that*

$$X_a = \arg \max_{X_i \in S(a), X_i \subseteq X} |X_i| \quad and \quad X_a^* = D \setminus a'.$$

In Lemma 1 it is sufficient to consider the equivalence classes $S(a)$ computed on D and not containing objects from M_1.

We divide the proof into two parts. The first one shows that

$$\mu X = a \Rightarrow X_a = \arg \max_{X_i \in S(a), X_i \subseteq X} |X_i| \text{ and } X_a^* = D \setminus a',$$

and the second one proves the opposite direction. The first part is also divided into two steps. Firstly, we show that $X_a^* = D \setminus a'$ and then $X_a = \arg \max_{X_i \in S(a), X_i \subseteq X} |X_i|$.

Proof. Let us suppose that $\mu X = a$. We need to show that $X_a^* = D \setminus a'$ is included in $\overline{X}$.

Suppose that $D \setminus a' \not\subseteq \overline{X}$. Thus, there is $x \in D \setminus a'$ such that $x \in X$. In this case μ does not minimize the number of error on the training set. We illustrate this by a counterexample given in Fig. 3. The training set is $X = \{x_2, x_3, x_4\}$, the test set is $\overline{X} = \{x_1, x_5, x_6\}$. Indeed, for classifier a_2 $X_a = \{x_3\} \in \mathcal{S}(a)$, but $D \setminus a' = \{x_3, x_4\} \setminus \{x_3\} \not\subseteq \overline{X}$. ERM μ chooses a_2 instead of a_1, a contradiction.

$$\begin{array}{c|ccc} & a_1 & a_2 & a_3 \\ x_1 & 1 & 1 & 1 \\ x_2 & 1 & 1 & 1 \\ x_3 & 1 & 1 & 0 \\ x_4 & 1 & 0 & 0 \\ x_5 & 0 & 0 & 0 \\ x_6 & 0 & 0 & 0 \end{array}$$

(a) $X = \{x_2, x_3, x_4\}$; $\overline{X} = \{x_1, x_5, x_6\}$

$$\begin{array}{c|ccc} & a_1 & a_2 & a_3 \\ x_2 & 1 & 1 & 1 \\ x_3 & 1 & 1 & 0 \\ x_4 & 1 & 0 & 0 \end{array}$$

(b) $A = \{a_1, a_2, a_3\}$
$A(X) = \arg\max_{a \in A} p(a, X) = \{a_1\}$

$$\begin{array}{c|ccc} & a_1 & a_2 & a_3 \\ x_1 & 1 & 1 & 1 \\ x_5 & 0 & 0 & 0 \\ x_6 & 0 & 0 & 0 \end{array}$$

(c) $A(X) = \{a_1\}$
$\arg\min_{a \in A} p(a, \overline{X}) = \{a_1\}$

Fig. 3. (a) The set of classifiers with $M_1 = \{x_1, x_2\}$, $D = \{x_3, x_4\}$, $M_0 = \{x_5, x_6\}$, (b) an ERM μ takes the classifiers with the minimal number of errors on X, (c) being pessimistic, μ chooses the classifier with the maximal number of errors on $\overline{X}$

Now we show that $\mu X = a$ implies $X_a = \arg\max_{X_i \in \mathcal{S}(a), X_i \subseteq X} |X_i|$. Suppose $X_a \notin \mathcal{S}(a)$. Here it is sufficient to show that at least one subset from $\mathcal{S}(a)$ is contained in the training set X (a stricter condition on the size of X_a makes sense only to ensure unambiguous choice of a classifier in Theorem 1 given below). From our assumption, it follows that there exists a classifier with the larger number of errors on the test set, i.e. μ is not the pessimistic ERM. We use a small counterexample in Fig. 4 to give a sketch of the general picture. The training set is $X = \{x_1, x_2, x_5\}$, the test set is $\overline{X} = \{x_3, x_4, x_6\}$. Indeed, for classifier a_2 the equivalence class is $\mathcal{S}(a_2) = \{\{2, 3\}, \{1, 2, 3\}\}$. Any subset $X_a \subseteq X$ is not

$$\begin{array}{c|cccc} & a_1 & a_2 & a_3 & a_4 \\ x_1 & 1 & 1 & 1 & 1 \\ x_2 & 0 & 1 & 1 & 0 \\ x_3 & 1 & 1 & 0 & 1 \\ x_4 & 1 & 0 & 0 & 0 \\ x_5 & 0 & 0 & 0 & 0 \\ x_6 & 0 & 0 & 0 & 0 \end{array}$$

(a) $X = \{x_1, x_2, x_5\}$; $\overline{X} = \{x_3, x_4, x_6\}$

$$\begin{array}{c|ccc} & a_1 & a_2 & a_3 \\ x_2 & 1 & 1 & 1 \\ x_3 & 0 & 1 & 1 \\ x_4 & 0 & 0 & 0 \end{array}$$

(b) $A = \{a_1, a_2, a_3\}$
$A(X) = \arg\max_{a \in A} p(a, X) = \{a_2, a_3\}$

$$\begin{array}{c|ccc} & a_1 & a_2 & a_3 \\ x_1 & 1 & 1 & 0 \\ x_5 & 0 & 0 & 0 \\ x_6 & 0 & 0 & 0 \end{array}$$

(c) $A(X) = \{a_2, a_3\}$
$\arg\min_{a \in A} p(a, \overline{X}) = \{a_3\}$

Fig. 4. (a) The set of classifiers with $M_1 = \{x_1\}$, $D = \{x_2, x_3, x_4\}$, $M_0 = \{x_5, x_6\}$, (b) an ERM μ takes the classifiers with the maximal number of correct answers on X, (c) being pessimistic, μ chooses the classifier with the maximal number of errors on $\overline{X}$

contained in $\mathcal{S}(a)$, but $X_a^* = D \setminus a' = \{x_4\} \subseteq \overline{X}$. Given X, μ considers a_2, a_3 as the empirical risk minimizers and consequently, being pessimistic, selects a_3. Thus, $\mu X = a_3$ contradicts our assumption that μ is pessimistic.

We proceed to show that

$$X_a = \arg\max_{X_i \in \mathcal{S}(a), X_i \subseteq X} |X_i| \text{ and } X_a^* = D \setminus a' \Rightarrow \mu X = a.$$

We will prove this statement even in more general settings:

$$X_a \in \mathcal{S}(a) \text{ and } X_a^* = D \setminus a' \Rightarrow \mu X = a. \tag{7}$$

For the sake of convenience, we will write $\{X \cap D \cap a'\}$ for X_a and $\{\overline{X} \cap D\}$ for X_a^*.

According to the introduced notation, we have

$$\arg\max_{a_i \in A} p(a_i, X) = \arg\max_{a_i \in A} |a_i' \cap X|,$$

as $p(a_i, X) = |a_i' \cap X|$. Some elements from M_1 and M_0 may be included in X, but their exclusion does not affect the set $A(X) = \{a \mid a = \arg\max_{a_i \in A} |a_i' \cap X|\}$, since all classifiers are not distinguished on M_1 and M_0. Consequently,

$$A(X) = \left\{a \mid a = \arg\max_{a_i \in A} |a_i' \cap X|\right\} = \left\{a \mid a = \arg\max_{a_i \in A} |a' \cap X \cap D|\right\},$$

and maximization of the cardinality of $|a' \cap X \cap D| = |X \cap D|$ yields $D \setminus a' \cap X = \emptyset$ for all $a \in A(X)$, otherwise another classifier will be selected as it has the minimal number of errors on training set (thus, $D \setminus a' \subset \overline{X}$ holds).

We may now use the set $A(X)$ to conclude that $X \cap D \cap a' \in \mathcal{S}(a)$ is sufficient to get $a \in A(X)$ as the result of the pessimistic ERM algorithm. It was shown above that all classifiers in $A(X)$ make no error on $X \cap D$, thus they differ from each other only on set $\overline{X} \cap D$. It remains to prove that the training set X must include $X_a \in \mathcal{S}(a)$, where a is the classifier with the minimal number of correctly classified objects in $\overline{X} \cap D$ (i.e. a is selected by μ). For such classifier a and any $a_i \in A(X) \setminus \{a\}$, $p(a, \overline{X}) < p(a_i, \overline{X})$ and $a \prec a_i$ in the corresponding lattice. The set $X \cap D$ cannot be included in the equivalence class of any $a_i \in A(X) \setminus \{a\}$, since a' is the minimal set that contains $X \cap D$ (we pessimistically minimize the number of correct answers on the test set), hence $X \cap D = a' \cap X \cap D \in \mathcal{S}(a)$.

Theorem 1. *If Definition 1 is valid, then the probability of overfitting for the family of classifiers comprising a lattice is given by the following formula:*

$$Q_\varepsilon(A) = \sum_{a \in A} \sum_{j=j_0}^{j_1} |S_j(a)| \frac{\binom{L-d}{\ell-j}}{\binom{L}{\ell}} H_{L-d}^{\ell-j, m_0}\left(\frac{\ell}{L}(\ell_0 + k_0 - \varepsilon k)\right),$$

where $j_0 = \max(0, \ell - m_0 - m_1)$ and $j_1 = \min(|a'|, \ell)$.

Proof. The proof falls naturally into two parts. In the first part we prove that the probability to select a classifier a given the random training set X of size ℓ is equal to $\sum_{j=\max(0,\ell-m_0-m_1)}^{\min(|a'|,\ell)} |S_j(a)| \frac{\binom{L-d}{\ell-j}}{\binom{L}{\ell}}$. The second part is quite involved: for the given value j we obtain the probability of overfitting by recalling Formula 6.

According to Lemma 1,

$$P([\mu X = a]) = \sum_{X_a \in S(a)} P([X_a \subseteq X][X_a^* \subseteq \overline{X}]).$$

The formula says that the probability of choosing a classifier a given a training set X is the sum of the probabilities of the following events: an element from the equivalence class is included in the training set and all misclassified objects (which can be classified correctly by other classifiers) are contained in the test set. By the definition of equivalence classes, if some sets $X_1, X_2 \in S(a)$, $X_1 \neq X_2$ then $X_1 \cup X_2 \in S(a)$. In the given formula X_a is a maximal set from $S(a)$ included in X, i.e. $X_a = \arg \max_{X_i \in S(a), X_i \subseteq X} |X_i|$. The minimal size of X_a cannot be smaller than $\ell - m_0 - m_1$, otherwise we do not have enough objects to be selected for training.

By Axiom 1 there are $\binom{L}{\ell}$ ways of choosing ℓ elements from L without repetition. We have only $\sum_{j=\max(0,\ell-m_0-m_1)}^{\min(|a'|,\ell)} |S_j(a)|$ possible generating sets X_a to be included in the training set. Since the maximal set from $S(a)$ cannot exceed the size of the training set X and the total number of objects correctly classified by a (ℓ and $|a'|$ respectively) an upper bound of $|X_a|$ is given by $\min(|a'|, \ell)$.

For each generating set of the jth level, i.e., $X_a \in S_j(a)$, there are $\binom{L-D}{\ell-j}/\binom{L}{\ell}$ ways to add the remaining elements of X. Since we take into account only the maximal set from $S(a)$ included in X, the other correctly classified object will be included in the test set, i.e. $(D \setminus a') \cup (a' \setminus X_a) \subseteq \overline{X}$.

We may now obtain the probability of overfitting for each value of j. Let us suppose that event $[X_a \subseteq X][X_a^* \subseteq \overline{X}]$ occurs for a fixed a and X_a of size j, where j is in the range from $\max(0, \ell - m_0 - m_1)$ to $\min(|a'|, \ell)$. Then it is known that $X_a \subseteq X$, $(a' \setminus X_a) \cup (D \setminus a') \subseteq \overline{X}$, the remaining elements from M_0 and M_1 are randomly selected to complete both X (to ℓ elements) and $\overline{X}$ (to $k = L - \ell$ elements). To complete X we need to select $\ell - j$ elements from $M_0 \cup M_1$, $|M_0 \cup M_1| = L - |D|$. The summary on the composition of the training and test set is given in Table 1.

Table 1. The information on objects from $\mathbb{X}$ given event $[X_a \subseteq X][X_a^* \subseteq \overline{X}]$ occurs

	Known		Unknown		Summary	
	# of errors	# of correct	# of errors	# of correct	# of errors	# of correct
$X, |X| = \ell$	–	$|X_a|$	w_0	w_1	$\ell_0 = w_0$	$\ell_1 = |X_a| + w_1$
$\overline{X}, |\overline{X}| = k$	$|D \setminus a'|$	$|a' \setminus X_a|$	z_0	z_1	$k_0 = |D \setminus a'| + z_0$	$k_1 = |a' \setminus X_a| + z_1$
Total	$d = |D|$		$m_0 = |M_0|$	$m_1 = |M_1|$	L	

We rewrite the condition of overfitting given in Formula 4 using information from Table 1:

$$\frac{k_0}{k} - \frac{\ell_0}{\ell} = \frac{|D \setminus a'| + z_0}{k} + \frac{w_0}{\ell} = \frac{|D \setminus a'| + m_0 - w_0}{k} + \frac{w_0}{\ell} =$$
$$= \frac{\ell(|D \setminus a'| + m_0) - w_0 L)}{k\ell} \geq \varepsilon;$$
$$w_0 \leq \frac{\ell}{L}(|D' \setminus a'| + m_0 - \varepsilon k).$$

Thus, the probability of overfitting for every generating set of the jth level is equal to $H_{L-D}^{\ell-j,m_0}\left(\frac{\ell}{L}(|D' \setminus a'| + m_0 - \varepsilon k)\right)$, the total number of objects misclassified by a is equal to $k_0 + \ell_0 = |D' \setminus a'| + m_0$.

Example 2. Let us consider the running example given in Fig. 1 to illustrate how Theorem 1 can be used. Let $\varepsilon = 0.05$. A classifier a_5 classifies correctly $\{2, 3, 4, 5\}$. The equivalence class computed on D is $\mathcal{S}(a) = \{\{2, 4\}, \{3, 4\}, \{2, 3, 4\}\}$. Let $\ell = 3$. Applying the formula from Theorem 1 to classifier a_5 we obtain

$$\underbrace{2\frac{\binom{6-4}{3-2}}{\binom{6}{3}}H_2^{3-2,1}\left(\frac{1}{2}(2-3\varepsilon)\right)}_{j=2} + \underbrace{\frac{\binom{6-4}{0}}{\binom{6}{3}}H_2^{0,1}\left(\frac{1}{2}(2-3\varepsilon)\right)}_{j=3} =$$
$$= \frac{4}{20}H_2^{1,1}(0) + \frac{1}{20}H_2^{0,1}(0) = \frac{1}{2} \cdot \frac{1}{4} = \frac{1}{8}.$$

The other summands of Q_ε can be calculated in the same way.

In [25] a similar approach was proposed to estimate the probability of overfitting under Axiom 1, which allows one to deal with a finite instance space (a set of objects) rather then the infinite one. Within the simplified probabilistic framework the exact combinatorial bounds for several sets of classifiers have been obtained. They include model sets such as a pair of classifiers, a layer of a Boolean cube, an interval of a Boolean cube, a monotonic chain, a unimodal chain, a unit neighborhood of the best classifier [26], the Hamming sphere and some subsets of it [23], a monotonic h-dimensional grid, a unimodal h-dimensional grid, and a pack of h monotonic chains [2]. In [27] overfitting bounds are computed by a random walk through the set of classifiers.

5   Experiments

In this section we consider the proposed overfitting bounds and compare them to the overfitting bounds computed by the Monte Carlo method with $N = 1000$. In our experiments we fix set D and change the sizes of M_1 and M_0, the general sample is $\mathbb{X} = D \cup M_0 \cup M_1$ to demonstrate how the precise bounds differ for "diverse" classifiers ($|D|/|\mathbb{X}|$ is close to 1) and for quite similar ones ($|D|/|\mathbb{X}|$ is close to 0). Thus, the structure of the set of classifiers (i.e. relations between vectors of correct answers) remains the same, but the error rates change by a particular value for all classifiers in the set.

To get the probability of overfitting by the Monte Carlo method on each iteration we randomly take a half of $\mathbb{X}$, which is considered as training set X, and compute $A(X)$ (see Formula 1), then, based on the remaining submatrix $(\mathbb{X} \setminus X) \times A$, we select a with minimal/maximal number of correct answers (see Formulas 2 and 3) or by chance. The average number of times when the error rate on the test set exceeds the error rate on the training set by ε is taken as the estimated probability of overfitting.

Figure 5 depicts the quality of classifiers on D. The blue bars of the histogram correspond to classifiers with a particular number of correctly classified objects, the height of the bars matches with the amount of such classifiers.

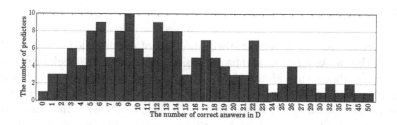

Fig. 5. A family of classifiers of size 151.

Figure 6 provides the dependencies of the probability of overfitting on the maximal permissible difference ε between error rates on test and training sets. As the ratio of objects from D is increasing (from 20% to 80%), the lengths of

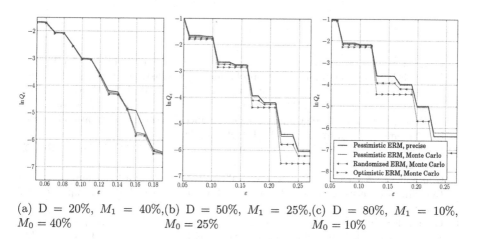

(a) D = 20%, M_1 = 40%, (b) D = 50%, M_1 = 25%, (c) D = 80%, M_1 = 10%, M_0 = 40% M_0 = 25% M_0 = 10%

Fig. 6. Dependence of the maximal permissible difference between error rates on test and training sets on the probability of overfitting

horizontal fragments of lines is increasing, the last allows us to conclude that quite small changes of ε do not affect the probability of overfitting, when the classifiers are similar. Moreover, the more homogenous a set of classifiers (when the rates of errors and correct answers are close to each other), the less the probability of overfitting. The optimistic estimates gives the lower bounds of the probability of overfitting. As can be seen from the figures, the theoretical bounds are consistent with the Monte Carlo pessimistic estimates. More than that, when the maximal permissible difference is close to zero (this case corresponds to a very strict condition on overfitting, i.e. the error rates on test and training sets must be almost the same) different types of estimates are close to each other, the last allows using the proposed pessimistic bounds instead of randomized or optimistic ones.

6   Conclusion

We have proposed exact combinatorial bounds of overfitting for the family of classifiers that make a lattice and have studied them under the weak probabilistic axiom. It was shown that Monte Carlo bounds agree with the theoretical bounds.

In spite of the fact that the proposed bounds may be used for a quite wide class of classifiers, this method operates like a black box. One of the most promising direction of future work is elaborating the overfitting model by taking into account the structure and generality of classifiers.

Acknowledgments. The authors thank Daniel Borchmann, Dmitry Ignatov and Konstantin Vorontsov for discussion and helpful comments. This paper was prepared within the framework of the Basic Research Program at the National Research University Higher School of Economics (HSE) and supported within the framework of a subsidy by the Russian Academic Excellence Project '5–100'.

References

1. Bax, E.: Similar classifiers and VC error bounds. CalTech-CS-TR-97-14 (1997)
2. Botov, P.V.: Exact estimates of the probability of overfitting for multidimensional modeling families of algorithms. Pattern Recogn. Image Anal. **21**(1), 52–65 (2011)
3. Bousquet, O., Elisseeff, A.: Algorithmic stability and generalization performance. In: Proceedings of the 2000 Conference on Advances in Neural Information Processing Systems, vol. 13, pp. 196–202. MIT Press (2001)
4. Bousquet, O., Elisseeff, A.: Stability and generalization. J. Mach. Learn. Res. **2**, 499–526 (2002)
5. Chernoff, H.: A measure of asymptotic efficiency for tests of a hypothesis based on the sum of observations. In: The Annals of Mathematical Statistics, pp. 493–507 (1952)
6. Ganter, B., Grigoriev, P.A., Kuznetsov, S.O., Samokhin, M.V.: Concept-based data mining with scaled labeled graphs. In: Wolff, K.E., Pfeiffer, H.D., Delugach, H.S. (eds.) ICCS-ConceptStruct 2004. LNCS, vol. 3127, pp. 94–108. Springer, Heidelberg (2004). doi:10.1007/978-3-540-27769-9_6

7. Ganter, B., Kuznetsov, S.O.: Hypotheses and version spaces. In: Ganter, B., Moor, A., Lex, W. (eds.) ICCS-ConceptStruct 2003. LNCS, vol. 2746, pp. 83–95. Springer, Heidelberg (2003). doi:10.1007/978-3-540-45091-7_6

8. Ganter, B., Wille, R.: Formal Concept Analysis: Mathematical Foundations. Springer, Heidelberg (1999)

9. Kashnitsky, Y., Ignatov, D.I.: Can FCA-based recommender system suggest a proper classifier? In: Proceedings of the International Workshop "What can FCA do for Artificial Intelligence?" (FCA4AI at ECAI 2014), CEUR Workshop Proceedings, pp. 17–26 (2014)

10. Kearns, M.J., Schapire, R.E.: Efficient distribution-free learning of probabilistic concepts. In: Proceedings of 31st Annual Symposium on Foundations of Computer Science, pp. 382–391. IEEE (1990)

11. Koltchinskii, V.: Rademacher penalties and structural risk minimization. IEEE Trans. Inf. Theor. **47**(5), 1902–1914 (2001)

12. Koltchinskii, V., Panchenko, D.: Rademacher processes and bounding the risk of function learning. In: Giné, E., Mason, D.M., Wellner, J.A. (eds.) High Dimensional Probability II, vol. 47, pp. 443–457. Springer, Heidelberg (2000)

13. Koltchinskii, V., Panchenko, D.: Empirical margin distributions and bounding the generalization error of combined classifiers. Ann. Stat. **30**(1), 1–50 (2002)

14. Kuznetsov, S.O., Poelmans, J.: Knowledge representation and processing with formal concept analysis. Wiley Interdisc. Rev.: Data Min. Knowl. Discov. **3**(3), 200–215 (2013)

15. Langford, J.: Quantitatively tight sample complexity bounds. Ph.D. thesis/Carnegie Mellon thesis (2002)

16. McAllester, D.A.: Pac-Bayesian model averaging. In: Proceedings of the Twelfth Annual Conference on Computational Learning Theory, pp. 164–170. ACM (1999)

17. McDiarmid, C.: On the method of bounded differences. Surv. Comb. **141**(1), 148–188 (1989)

18. Philips, P., et al.: Data-dependent analysis of learning algorithms (2005)

19. Seeger, M.: Pac-Bayesian generalisation error bounds for Gaussian process classification. J. Mach. Learn. Res. **3**, 233–269 (2002)

20. Sill, J.: Monotonicity and connectedness in learning systems. Ph.D. thesis, California Institute of Technology (1998)

21. Talagrand, M.: Sharper bounds for Gaussian and empirical processes. Ann. Probab. **22**(1), 28–76 (1994)

22. Talagrand, M.: Concentration of measure and isoperimetric inequalities in product spaces. Publications Mathématiques de l'Institut des Hautes Etudes Scientifiques **81**(1), 73–205 (1995)

23. Tolstikhin, I.O.: The probability of overfitting for the compact and sparse sets of predictors. In: Intelligent Data Processing: Theory and Applications: IDP-8, pp. 83–86 (2010)

24. Vapnik, V.N., Chervonenkis, A.Y.: On the uniform convergence of relative frequencies of events to their probabilities. Theory Probab. Appl. **16**(2), 264–280 (1971)

25. Vorontsov, K.V.: Combinatorial probability and the tightness of generalization bounds. Pattern Recogn. Image Anal. **18**(2), 243–259 (2008)

26. Vorontsov, K.V.: Exact combinatorial bounds on the probability of overfitting for empirical risk minimization. Pattern Recogn. Image Anal. **20**(3), 269–285 (2010)

27. Vorontsov, K.V., Frey, A.I., Sokolov, E.A.: Computable combinatorial overfitting bounds. Mach. Learn. **1**, 6 (2013)

Learning Thresholds in Formal Concept Analysis

Uta Priss ⓘ

Zentrum für erfolgreiches Lehren und Lernen,
Ostfalia University of Applied Sciences, Wolfenbüttel, Germany
http://www.upriss.org.uk

Abstract. This paper views Formal Concept Analysis (FCA) from an educational perspective. Novice users of FCA who are not mathematicians might find diagrams of concept lattices counter-intuitive and challenging to read. According to educational theory, learning thresholds are concepts that are difficult to learn and easy to be misunderstood. Experts of a domain are often not aware of such learning thresholds. This paper explores learning thresholds occurring in FCA teaching material drawing on examples from a discrete structures class taught to first year computer science students.

1 Introduction

Formal Concept Analysis (FCA[1]) has over the years grown from a small research area in the 1980s to a reasonably-sized research community with thousands of published papers. Rudolf Wille originally perceived FCA as an example of 'restructuring mathematics' because it provides a tool for data analysis with applications in non-mathematical disciplines (Wille 1982). Early applications of FCA in psychology were aimed at establishing FCA as an alternative to statistical analyses (Spangenberg and Wolff 1993). Because FCA formalises conceptual hierarchies it should be a natural tool for investigating and modelling hierarchical structures in dictionaries, lexical databases, biological taxonomies, library classification systems and so on. Thus FCA has the potential of being used as a tool in a wide variety of non-mathematical fields. But so far it is still mostly used by researchers with mathematical or computational training. A few papers were published in other areas since the 1990s. But these papers were mostly written by mathematicians or computer scientists and not by researchers who belong to these fields and are not mathematicians. This is in contrast to other mathematical techniques such as statistics which appear to have a much wider application domain.

As an example, we manually examined the first 100 documents that were retrieved by a query for 'formal concept analysis and linguistics' in a bibliographic search engine. It seems that the majority of the retrieved papers were

---

[1] Because FCA is the topic of this conference, this paper does not provide an introduction to FCA. Further information about FCA can be found, for example, on-line (http://www.upriss.org.uk/fca/) and in the main FCA textbook by Ganter and Wille (1999).

© Springer International Publishing AG 2017
K. Bertet et al. (Eds.): ICFCA 2017, LNAI 10308, pp. 198–210, 2017.
DOI: 10.1007/978-3-319-59271-8_13

either written by a computer scientist or mathematician or submitted to an FCA or a computer science conference. The topics of these papers centred on ontologies, data mining/processing/retrieval and logic or artificial intelligence in the widest sense. A query for 'formal concept analysis and psychology' yields a similar result. On the other hand, searching for 'statistics and linguistics' retrieves mostly papers in the area of computational linguistics and searching for 'statistics and psychology' retrieves papers on research methods, textbooks for psychologists and papers written for a psychological audience. Thus it appears that FCA researchers are still writing papers that attempt to demonstrate that and how FCA can be used in linguistics and psychology whereas statistics is an established tool in these disciplines. The question arises as to why it seems to be so difficult for FCA to establish itself as a commonly-used tool in non-mathematical or non-computational fields even though there are so many papers showing that and how FCA can be used in such fields. Clearly a large number of factors could be causing this, such as availability and usability of FCA software or a general resistance to change and to adopting new paradigms. One possible factor which is examined in this paper, however, is that the mathematics underlying FCA might be surprisingly difficult for non-mathematicians to understand. We are not arguing that FCA is more difficult to learn than, for example, statistics but simply that the difficulties faced by FCA novices should not be underestimated.

From a mathematical viewpoint, the core definitions of FCA are short and simple and can easily be explained to fellow mathematicians. But it may be that non-mathematicians need a significant amount of time and motivation in order to learn FCA. This is not contradicted by Eklund et al.'s (2004) observation that novices can read the line diagrams in their FCA software because their experiment was more focussed on the usability of their software than establishing whether users have a mathematically correct understanding of lattices. Again, there can be many reasons why learning FCA is difficult but it is at least a possibility that FCA contains a few 'threshold concepts' which are challenging to learn. According to Meyer and Land (2003) threshold concepts are difficult concepts which function like a portal into a domain of knowledge. Because their notion of 'concept' is not the same as the one in FCA we use the term 'learning threshold' instead of 'threshold concept' in the remainder of this paper. Meyer & Land state that experts are people who have overcome the learning thresholds of a domain. Learning thresholds have several characteristic properties: they are transformative because they invoke a shift in understanding, values, feeling or attitudes. They are irreversible because once one has mastered a learning threshold one cannot later unlearn it again. They are integrative because they establish connections with other concepts. But they are also troublesome because they might contradict prior beliefs or intuitions. Unfortunately because learning thresholds are both integrative and irreversible, experts tend to forget what their thinking used to be like before they acquired the concepts. That means that experts often lack understanding for the exact difficulties that novices are experiencing. Thus the notion of 'learning threshold' can be a learning threshold itself for teachers in training.

The purpose of this paper is to start a discussion within the FCA community about encountering learning thresholds within the teaching material of FCA. Once teachers have identified learning thresholds and possible misconceptions in a domain they can devote more time, materials and exercises to the teaching of difficult concepts. One possible outcome of such a discussion would be to establish a 'concept inventory' for FCA. Concept inventories are lists of learning thresholds for a domain. For example Almstrum et al. (2006) develop a concept inventory for discrete mathematics and at the same time describe the process of establishing such an inventory. Presumably the list of learning thresholds for the basic notions of FCA would not be long. But the questions are: what concepts belong to this list and why are they difficult to learn? This paper examines some concepts that might belong to the list of learning thresholds of FCA. Deciding on a list of learning thresholds is usually a community effort. Thus it is hoped that this paper will stimulate a discussion amongst FCA teachers about this topic.

The background for this paper was teaching a class on discrete structures to 70 first year computer science students (abbreviated by DS in this paper). The class was taught using a just-in-time teaching method[2]. The lecturer took notes about any concepts that appeared difficult because many students asked about them, many students had problems with exercises relating to them or there was a particularly lengthy class discussion about them. The class covered the usual discrete structures topics (logic, sets, functions, relations, groups, graphs) and concluded with the topic of partially ordered sets and lattices discussed in the last two 1.5 h class sessions. In addition to compiling a list of learning threshold candidates, it was attempted to investigate why the concepts appeared to be difficult to learn.

In analysing the list of difficult concepts gathered from the DS class it appeared that students have a general misunderstanding of how mathematical concepts are to be used. This analysis which employs semiotic-conceptual analysis (SCA) according to Priss (2016) is described in the next section. Section 3 then uses a further semiotic analysis to identify learning thresholds related to line diagrams of partially ordered sets and concept lattices. The paper finishes with a concluding section.

2   The Notion of 'Formal Concept' is a Learning Threshold

This section introduces a slightly more general notion of 'formal concept' which is seen as the core building block of mathematics and the main reason for why mathematics can be difficult to learn. The definitions in this section are not

---

[2] In just-in-time teaching, students read textbook pages and submit exercises, comments and questions before each class session. The lecturer then prepares each class session so that it addresses exactly the questions and problems the students are having. This method provides a wealth of continuous feedback both for learners and for teachers about the learning process.

strictly formal because they attempt to build a bridge between the normal FCA definitions and less formal notions of 'concept'. A more formal description of SCA is provided by Priss (2016, 2017). The following notion of 'open set' is adapted from linguistics (Lyons 1968, p. 436).

Definition 1: An *open set* is a set for which there is a precise method for determining whether an element belongs to it or not but which is too large to be explicitly listed and which does not have an algorithmic construction rule.

The sets of even numbers or of all finite strings are not open because they have algorithmic construction rules. Examples of open sets are the sets of prime numbers, the words of the English language and currently existing species of primates. The last two examples are finite open sets at any point in time even though there may be more elements added to them in the future. The reference to an 'abstract idea' in the next definition is non-formal but is added in order to avoid having any pair of sets being called 'concept'. Whether something is a concept depends on the formal condition and on the informal property of someone considering it to be an abstract idea.

Definition 2

(a) A *concept* is an abstract idea corresponding to a pair of two sets (extension and intension) which can be crisp, rough, fuzzy or open.
(b) A *formal instance concept* is a concept whose extension is finite and whose attributes are clearly determined by the extension.
(c) A *mathematical concept* is a concept for which a necessary and sufficient set of attributes in its intension can be identified which determine exactly whether an item is in the extension of the concept.
(d) A *formal concept* is a formal instance concept and/or a mathematical concept.
(e) An *associative concept* is a concept that is not formal.

Instance concepts arise, for example, when someone perceives an object or a set of objects. Depending on whether the attributes of that object are clear an instance concept can be formal or associative. For example, a cat perceiving a mouse will most likely form an associative concept in its mind whereas a person thinking about number 5 may be thinking a formal concept. Values of variables in programming languages are formal instance concepts. Their extension is their value and their intension is the properties of the value such as its datatype. Any mathematical definition establishes a mathematical concept. Extensions and intensions of mathematical concepts can be open sets. For example the set of zeros of Riemann's zeta function is currently open. For any existing number it can be determined whether it belongs to the set but it is currently not possible to list the complete set. Although a mathematical concept has a set of necessary and sufficient attributes, the set of all of its attributes tends to be open. Mathematical concepts also occur in other scientific disciplines. For example, the concept of the plant genus 'bellis' (to which the common daisy belongs) has a necessary and sufficient definition. With modern genetic methods it can be clearly distinguished which plants belong to this genus. But it is not possible to list all species that belong to bellis because some may not yet have been discovered. It is also not possible to provide a definitive list of all attributes of bellis.

The notion of formal concepts in Definition 2 extends the normal FCA definition. Normal FCA formal concepts are both formal instance concepts and mathematical concepts. Intensions of FCA concepts are always sufficient. The subset of necessary attributes can be determined by the FCA methods of clarifying and reducing (Ganter and Wille 1999). For formal concepts a clear ordering can be defined either extensionally for formal instance concepts or intensionally for mathematical concepts. For mathematical concepts this corresponds to evaluating attribute implications.

Concepts formed when interpreting natural language tend to be associative and not formal. For example, it is impossible to provide a set of necessary and sufficient conditions for a concept such as 'democracy'. Because cats have an associative concept of 'mouse' they might chase anything that is mouse-like but not a mouse. Even everyday concepts are difficult to exactly define and delimit from other concepts. For example 'chair' is difficult to precisely distinguish from 'armchair', 'recliner' and 'bench'. Cognitive theories such as Rosch's (1973) prototype theory explain that the extensions and intensions of such concepts tend to be prototypical and fuzzy. While concepts are the core units of thought, signs are the core units of communication. The next definition is taken from Priss (2017). Again the reference to 'unit of communication' is non-formal but important in order to avoid having any kind of triple which fulfils the formal condition automatically being a sign.

Definition 3: A *sign* is a unit of communication corresponding to a triple (i, r, d) consisting of an interpretation i, a representamen r and a denotation d with the condition that i and r together uniquely determine d.

A sign always involves at least two layers of interpretation. One interpretation i is an explicit component of the sign, but there is also a second interpretation by someone who observes or participates in a communicative act. This observer decides what i, r and d are. This second level is not explicitly mentioned because it usually happens in the mind of a person. The same applies to FCA: there is always someone who creates a formal context but this is not explicitly mentioned. To provide an example of signs, the set of even numbers can be represented verbally as 'set of even numbers', as $\{2, 4, 6, ..\}$ or as '$\{n \in \mathcal{N} \mid n/2 = 0(\bmod 2)\}$'. These are three different representamens r_1, r_2, r_3 which are interpreted by mathematicians i to have the same denotation d. This results in three signs (i, r_1, d), (i, r_2, d) and (i, r_3, d). Signs with different representamens and equal denotations are called strong synonyms. Thus the three signs in this example are strong synonyms. Students, however, might interpret some or all of these representamens incorrectly. Thus if sign use by students is not synonymous to sign use by teachers this indicates that students have misunderstood something.

It might be possible to connect Definitions 2 and 3 because denotations can be considered to be concepts. But representamens and interpretations can also be modelled as concepts. Furthermore, interpretations, representamens and denotations can also be signs and if a concept is represented in some form then it must be a sign itself. Thus the relationship between Definitions 2 and 3 is complex.

For some signs, the three components r, i and d are not totally independent of each other. For example, icons are signs where the representamen and the denotation are similar to each other in some respect such as a traffic sign for 'bike path' showing the picture of a bike. The signs for abstract associative concepts (such as 'democracy') tend to have three independent components. Because it is not possible to provide an exact definition of the denotation of democracy, the representamen is always needed when communicating this sign. It is also not possible to talk about democracy without explaining what interpretation is used. This is contrary to formal concepts. The set of even numbers can be defined as '$\{n \in \mathcal{N} \mid n/2 = 0(\mod 2)\}$' and discussed without calling it 'set of even numbers'. Thus the definition of even numbers can be a representamen itself. The same holds for the formal instance concept with the extension $\{1\}$. Both are examples of anonymous signs as defined below.

Definition 4: An *i-anonymous sign* is a sign (i, r, d) with $r = d$. If i is clear, it can also be referred to as an *anonymous sign*.

Thus (i, d, d) is a sign where the representamen equals the denotation. This is an extreme form of an icon. In mathematics it is often sufficient to assume a single general interpretation which consists of understanding the notation. For example, a whole textbook on mathematics might use a single interpretation. Mathematical variables tend to be just placeholders for their values and thus anonymous signs. Programming values are also anonymous. Variables in programming languages, however, are not anonymous signs because they change their values at run-time depending on the state (which can be considered an interpretation). The following conjecture is based on the idea that formal concepts are fully described by their definition. It may be convenient to give a name to a formal concept but it is not necessary to do so. Associative concepts do not have a precise definition. The relationship between the representamen that is used for an associative concept, its interpretation and its fuzzy definition provides further information. If an associative concept is to be communicated it requires a triadic sign. The representamen need not be a word. It could be a cat meowing in the vicinity of a fridge and looking at its owner in order to express 'I am hungry'. But for an associative concept this triad cannot be reduced.

Conjecture 1

(a) A formal concept can be used as an anonymous sign.
(b) The denotation of an anonymous sign corresponds to a formal concept.

Nevertheless because values of variables in programming languages are formal concepts one cannot conclude that signs with formal concepts as denotations are always anonymous signs. A claim of Priss (2016) is that the concepts underlying mathematical definitions are always formal but students often interpret them in an associative manner. Until students have a grasp of the nature of formal concepts, every mathematical concept is a learning threshold for them because they are using an inappropriate cognitive approach. This claim cannot be formally proven but there is evidence for it. For example, when asking students in the DS class what a graph is (according to graph theory), their first answer was that it is something that is graphically represented. But that is a prototypical attribute of graphs that is neither necessary nor sufficient. Thus such an answer is incorrect. A correct answer is easy to produce by stating the definition. Since the students were allowed to use the textbook for this answer, they could have just looked it up. An incorrect answer for such a simple question shows that the students have not yet grasped the nature of mathematical concepts and what it means to understand a mathematical concept.

As another example, when teaching FCA to linguists I have many times experienced at least one linguist objecting: 'what you call a concept is not a concept'. Such a statement is mathematically non-sensical because a formal concept is exactly what it is defined to be. Whether it is called 'concept' or something else is just a convention. But for a (non-mathematically trained) linguist the notion 'concept' is associatively defined with some reference to abstract ideas. The rest of Definition 2 and in particular the notion that a concept is a pair of sets is meaningless for a linguist. From a mathematical viewpoint an appropriate argument that could be discussed with linguists is whether the mathematical model suggested in the formal definition approximates the associative concepts that linguists have about concepts. But this discussion is only possible if the distinction between formal and associative concepts is clear.

Further evidence for the importance of formal concepts in mathematics comes from Moore (1994) and Edwards and Ward (2004) who highlight the role of formal definitions in university mathematics. This is in contrast to primary and secondary school where mathematics is often taught in an associative manner using practical examples and introductory exercises. First year university students are having difficulties with mathematics because they are relying too much on associations instead of concept definitions. Another source of evidence for a difference between associative and formal concepts is the work by Amalric and Dehaene (2016) who argue that expert mathematicians use different parts of their brains when they are listening to mathematical and non-mathematical statements. Unfortunately, the difference between associative and formal concepts also raises questions about whether concrete examples help or hinder the teaching of abstract mathematical ideas. Kaminski et al. (2008) started a debate on this topic which is still ongoing but there seems to be a consensus that transfer from concrete to abstract is not easy.

3   A Semiotic Analysis of Further FCA Learning Thresholds

The last section argues that representamens and denotations coincide for mathematical concepts. Therefore analysing mathematical representamens coincides with analysing mathematical meanings. Mathematical concepts usually have many synonymous representamens. For example, Fig. 1 shows two different representamens of a set operation. From a semiotic-conceptual perspective it is of interest to analyse how structures amongst representamens relate to each other and to denotational structures. For example, in Fig. 1 the circles on the left correspond to the curly brackets on the right. On the left the intersection appears static, on the right it is more obvious that intersection is an operation. It is probably more apparent on the left that the sequence of elements in a set is not important. The right representamen might be misinterpreted by students as imposing a fixed sequence on set elements. The left representamen might be more difficult to observe for students with mild dyslexia because there is no clear reading direction for an image. The signs on the left and right are synonyms. Although visualisations might seem more intuitive they tend to be less precise than formulas and include irrelevant information. For example the size of the circles on the left is irrelevant. They do not coincide with the sizes of the sets. Mathematicians tend to frequently use visualisations and for example scribble on paper or blackboards while they are thinking. Amalric and Dehaene (2016) show that the part of the brain that is used by mathematicians while thinking about mathematics is the one responsible for spatial tasks, numbers and formulas in non-mathematicians.

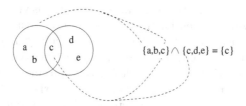

Fig. 1. Two different representamens for the same denotation

From a semiotic viewpoint there are two major research questions about mathematical representamens: (a) how the formal language of mathematics functions and (b) how denotational structures are preserved, highlighted or hidden in different kinds of representamens or translations between representamens. The idea that a semiotic analysis should focus on how structures are represented and translated is similar to Goguen's (1999) algebraic semiotics. We suspect that one major difficulty in teaching mathematics is that teachers are used to the different types of representamens and know what to look for and what to ignore whereas students might misinterpret them. The next two sections provide a semiotic analysis of line diagrams of partially ordered sets and lattices, respectively, based on our experiences with the DS class.

3.1 Reading Line Diagrams of Partially Ordered Sets

In the 90s Rudolf Wille's research group organised workshops for non-mathematicians to learn FCA. As far as I remember, the workshops lasted for at least 3 hours and started with letting the workshop participants manually construct concept lattices from formal contexts by first generating a list of extensions (or intensions) and then creating the concept hierarchy from that list. It is quite possible that Wille's teaching method would avoid some of the conceptual difficulties described in this paper. But most people probably encounter FCA first via line diagrams which were also the starting point for FCA in the DS class. Furthermore manual construction of lattices is time consuming and requires learners to be highly motivated to learn FCA. Teaching by constructing examples also has limits. For example, because concept lattices are always complete, one cannot teach what it means for a lattice to be complete or incomplete using concept lattices.

As mentioned in the introduction, Eklund et al. (2004) observe that users can interact with the line diagrams in their FCA software in order to conduct queries. Our experience with the DS class showed, however, that without some detailed instruction students employ incorrect interpretations when they first encounter line diagrams. There exists an overwhelming amount of research about the use of visualisation in learning and teaching of mathematics (Presmeg 2006). Nevertheless we have not been able to find any existing literature on the specifics of learning to use line diagrams. Before introducing lattices as a topic in the DS class, the students had already seen the visualisations shown in Fig. 2. The lattices on the left were introduced earlier in the semester as visualisations of power sets and divisors without mentioning lattices. At that point the students were asked to construct such lattices for the relations of subset and divides based on some examples. Thus the students had already constructed examples of lattices. On the right of Fig. 2 there are examples of a graph and a visualisation of an equivalence relation. These were visualisations used earlier in the semester which could potentially be a source of misconceptions about line diagrams because, for example, reflexivity and transitivity are explicitly represented in such diagrams but omitted in line diagrams.

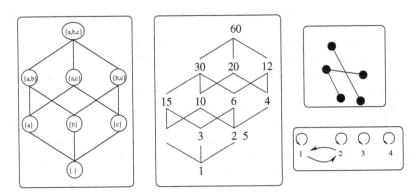

Fig. 2. Different types of graphs: partially ordered sets, graph and equivalence relation

The main source of misconceptions about line diagrams in the DS class, how-ever, appeared to be vector addition (cf. Fig. 3). Students were commenting that a coordinate system seems to be missing from line diagrams. Because linear alge-bra is taught in German secondary schools but graph theory is not, vector spaces appear to be the primary visualisation model for the students. A major difference between line diagrams and vector addition visualisations is that line diagrams are discrete and represent exactly the elements that exist whereas the vector addition visualisation imposes a few lines on a continuous space. Thus it was at first difficult for the students to realise that whether an element is smaller than another element depends on the existence of lines in the line diagram whereas it depends on distances and coordinates for vectors. For example, when asked about the ordering of nodes in a line diagram one student used a ruler in order to measure distances. Other examples of prior knowledge which the students mentioned were class inheritance in Java and levels. Because class inheritance in Java forms a partially ordered set, this is probably a case of supportive prior knowledge. The notion of 'levels', however, is probably more distracting than helpful because levels in a partially ordered set change depending on whether or not they are counted from above and below. Most likely asking about 'levels' is further evidence of an assumed underlying coordinate system and a miscon-ception that the lengths of the edges or the absolute height of nodes in a line diagram carry meaning.

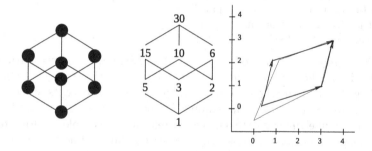

Fig. 3. Conflicting interpretations: line diagrams and vector addition

The example of divisor lattices (in the centre of Fig. 3) appeared to be most intuitive for the students. They were familiar with the notions of greatest com-mon divisor (gcd) and least common multiple (lcm). The students were able to read these and the division relation from the diagram and to create similar dia-grams themselves. Nevertheless it is not so easy for first year students to then perform the abstraction to lattices in general. Accepting that the two lattices on the left in Fig. 2 are both examples of a shared abstract idea is difficult. This involves understanding that the lines in line diagrams represent an ordering rela-tion which in some examples corresponds to subset in others to division. The transfer from finding gcds in the middle example in Fig. 3 to finding infimas in the left example in Fig. 4 is another hurdle. The gcds can be found by either following lines or calculating the numbers and then searching for the number in the diagram. It appears that some students (who may have forms of dyslexia) have difficulty reading line diagrams and finding infima without some aid such as tracing the lines with their fingers.

In summary, it appears that line diagrams are not instantly intuitive to many users because they conflict with other types of graphical representations which the users already know. The main learning thresholds in this context appear to be the challenge of overcoming the misconception of line diagrams as embedded into vector spaces and the abstraction from the specific examples of divisor and powerset lattices to the underlying shared algebraic structure. Students with some form of dyslexia might find visualisations difficult to read in general.

3.2   Understanding Concept Lattices

After introducing line diagrams, the next steps in the DS class were to introduce lattices and then concept lattices. Priss (2016) collects a list of typical questions about and problems with lattices that users tend to have when they first encounter concept lattices. This list was also confirmed in the DS class:

- What is the purpose of the top and bottom node?
- Why are there unlabelled nodes?
- How can the extensions and intensions be read from the diagram?
- What is the relationship between nodes that do not have an edge between them but can be reached via a path?
- What is a supremum or an infimum?
- How can one tell whether it is a lattice?

The SCA analysis conducted by Priss (2016) resulted in the two lattices presented in Fig. 4. The left hand side presents a lattice for the structures contained in line diagrams and the right hand side a lattice for the concepts of lattice theory. The figure shows that a fair number of concepts is involved and that the mapping from line diagrams to lattices is reasonably complex. Ultimately, in order to answer questions such as 'why are there unlabelled nodes' one needs to know what nodes and edges are, how they can be traversed and extensions and intensions be formed. One also needs to know what concepts, joins, meets, operators and sets are. In order to understand what a lattice is, one needs knowledge of all of the concepts in Fig. 4, an understanding of lattices as an abstraction and examples of partially ordered sets which are not lattices.

To conclude this section it should be stressed that a few students in the DS class appeared to have a very good grasp of lattices towards the end of the class sessions on this topic. Most of the students who passed the class achieved at least 75% of the points related to the exam question about partially ordered sets and lattices. Thus the misconceptions that students were initially having can be overcome. The students can learn to interpret line diagrams correctly after some time spent with reading, instruction and exercises. But some students are having difficulties with the required abstraction and with the visualisations. FCA contains several learning thresholds and is not instantly intuitive.

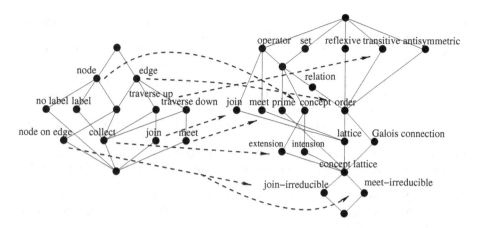

Fig. 4. Structures of line diagrams and concept lattices (Priss 2016)

4 Conclusion

This paper discusses learning thresholds in the teaching materials of FCA. A major learning threshold that applies to all of mathematics is the realisation that mathematical concepts are very different from associative concepts underlying natural language. A different cognitive strategy must be used for both types of concepts. Visualisations such as line diagrams are helpful for users who understand them but cannot be assumed to be instantly intuitive for people who have never seen them before. It would be of interest to conduct a more comprehensive analysis of mathematical notation and visualisations from a semiotic-conceptual perspective. Most pedagogical studies tend to focus on particular examples – we are not aware of an existing larger scale analysis of the semiotic structures of mathematical notation and visualisation.

References

Almstrum, V.L., Henderson, P.B., Harvey, V., Heeren, C., Marion, W., Riedesel, C., Soh, L.K., Tew, A.E.: Concept inventories in computer science for the topic discrete mathematics. ACM SIGCSE Bull. **38**(4), 132–145 (2006)

Amalric, M., Dehaene, S.: Origins of the brain networks for advanced mathematics in expert mathematicians. Proc. Natl. Acad. Sci. **113**(18), 4909–4917 (2016)

Ganter, B., Wille, R.: Formal Concept Analysis. Mathematical Foundations. Springer, Heidelberg (1999)

Edwards, B.S., Ward, M.B.: Surprises from mathematics education research: student (mis)use of mathematical definitions. Am. Math. Mon. **111**(5), 411–424 (2004)

Eklund, P., Ducrou, J., Brawn, P.: Concept lattices for information visualization: can novices read line-diagrams? In: Eklund, P. (ed.) ICFCA 2004. LNCS, vol. 2961, pp. 57–73. Springer, Heidelberg (2004). doi:10.1007/978-3-540-24651-0_7

Goguen, J.: An introduction to algebraic semiotics, with application to user interface design. In: Nehaniv, C.L. (ed.) CMAA 1998. LNCS, vol. 1562, pp. 242–291. Springer, Heidelberg (1999). doi:10.1007/3-540-48834-0_15

Kaminski, J.A., Sloutsky, V.M., Heckler, A.F.: The advantage of abstract examples in learning math. Science **320**(5875), 454–455 (2008)

Lyons, J.: Introduction to Theoretical Linguistics. Cambridge University Press, Cambridge (1968)

Meyer, J.H.F., Land, R.: Threshold concepts and troublesome knowledge 1 - linkages to ways of thinking and practising. In: Rust, C. (ed.) Improving Student Learning - Ten Years On. OCSLD, Oxford (2003)

Moore, R.C.: Making the transition to formal proof. Educ. Stud. Math. **27**(3), 249–266 (1994)

Presmeg, N.C.: Research on visualization in learning and teaching mathematics. In: Handbook of Research on the Psychology of Mathematics Education, pp. 205–235 (2006)

Priss, U.: Associative and formal concepts. In: Priss, U., Corbett, D., Angelova, G. (eds.) ICCS-ConceptStruct 2002. LNCS, vol. 2393, pp. 354–368. Springer, Heidelberg (2002). doi:10.1007/3-540-45483-7_27

Priss, U.: A semiotic-conceptual analysis of conceptual learning. In: Haemmerlé, O., Stapleton, G., Faron Zucker, C. (eds.) ICCS 2016. LNCS, vol. 9717, pp. 122–136. Springer, Cham (2016). doi:10.1007/978-3-319-40985-6_10

Priss, U.: Semiotic-conceptual analysis: a proposal. Int. J. Gen. Syst. (2017, to appear)

Rosch, E.: Natural categories. Cogn. Psychol. **4**, 328–350 (1973)

Spangenberg, N., Wolff, K.E.: Datenreduktion durch die Formale Begriffsanalyse von Repertory Grids. In: Scheer, J.W., Catina, A. (eds.) Einführung in die Repertory Grid Technik. Klinische Forschung und Praxis 2, pp. 38–54. Verlag Hans Huber (1993)

Wille, R.: Restructuring lattice theory: an approach based on hierarchies of concepts. In: Rival, I. (ed.) Ordered Sets, pp. 445–470. Reidel, Dordrecht-Boston (1982)

The Linear Algebra in Extended Formal Concept Analysis Over Idempotent Semifields

Francisco José Valverde-Albacete$^{(\boxtimes)}$ (iD) and Carmen Peláez-Moreno (iD)

Departamento de Teoría de la Señal y de las Comunicaciones,
Universidad Carlos III de Madrid, 28911 Leganés, Spain
{fva,carmen}@tsc.uc3m.es

Abstract. We report on progress relating $\mathcal{K}$-valued FCA to $\mathcal{K}$-Linear Algebra where $\mathcal{K}$ is an idempotent semifield. We first find that the standard machinery of linear algebra points to Galois adjunctions as the preferred construction, which generates either Neighbourhood Lattices of attributes or objects. For the Neighbourhood of objects we provide the adjoints, their respective closure and interior operators and the general structure of the lattices, both of objects and attributes. Next, these results and those previous on Galois connections are set against the backdrop of Extended Formal Concept Analysis. Our results show that for a $\mathcal{K}$-valued formal context (G, M, R)—where $|G| = g$, $|M| = m$ and $R \in K^{g \times m}$—there are only two different "shapes" of lattices each of which comes in four different "colours", suggesting a notion of a 4-concept associated to a formal concept. Finally, we draw some conclusions as to the use of these as data exploration constructs, allowing many different "readings" on the contextualized data.

1 Introduction

In [1] a generalization of Formal Concept Analysis was presented where incidences have values in a complete idempotent semifield $\mathcal{K}$. This is a complete idempotent semiring with a multiplicative semiring structure where the unit is distinct from top of the semiring, unlike in e.g. inclines. This setting was later extended to the other four types of Galois connections or adjunctions arising from a single $\mathcal{K}$-valued incidence and named Extended Formal Concept Analysis [2,3].

In later work [4,5], for the case of a Galois connection the concept lattices in this extension were more extensively investigated in the framework of linear algebra over complete idempotent semifields. It was made evident that extended concept lattices are the idempotent algebra analogues of vector spaces, that is complete idempotent semimodules, and they have also ties to the eigenspaces of some matrices related to the multivalued incidence [4]. Indeed in a more

---

C. Peláez-Moreno—FVA and CPM have been partially supported by the Spanish Government-MinECo projects TEC2014-53390-P and TEC2014-61729-EXP for this work.

K. Bertet et al. (Eds.): ICFCA 2017, LNAI 10308, pp. 211–227, 2017.
DOI: 10.1007/978-3-319-59271-8_14

extensive investigation of these issues it was found that standard FCA can be easily embedded into $\mathcal{K}$-FCA when properly treating the values of the binary incidence and a particular choice of the φ free-parameter [5, Sect. 3.8].

However, these strands of research walk away from the standard presentation of the semimodules associated to a linear form in standard algebra, as summarized, for instance in [6]. With this parallel in mind, Cuninghame-Green investigated the semimodules associated to a max-plus linear form and found a number of results that parallel the standard ones [7]. In the years elapsing a number of results have also been found, possibly best summarized in [8,9].

In this paper we clarify the role of Galois Adjunctions in the "standard way" of conceiving linear forms over complete idempotent semimodules. This paves the way for re-visiting Extended FCA as induced by a $\overline{\mathcal{K}}$-valued formal context. We believe that the clarification of these issues has a potential impact on a number of applications of Formal Concept Analysis at large.

For that purpose we first recall in Sect. 2 the algebra of semimodules over idempotent semifields, including a summary of the relation of the $\overline{\mathcal{K}}$-concept lattices to $\overline{\mathcal{K}}$-semimodules. In Sect. 3 we present our results on the definition of the 4-fold FCA of formal contexts and we finish with a discussion of the prospect for data mining such step affords us.

2   Galois Connections Over Idempotent Semifields

2.1   Idempotent Semirings, Semifields and Semimodules

Semirings. A *semiring* is an algebra $\mathcal{S} = \langle S, \oplus, \otimes, \epsilon, e \rangle$ whose additive structure, $\langle S, \oplus, \epsilon \rangle$, is a commutative monoid and whose multiplicative structure, $\langle S \backslash \{\epsilon\}, \otimes, e \rangle$, is a monoid with multiplication distributing over addition from right and left and with additive neutral element absorbing for $\otimes$, i.e. $\forall a \in S,\ \epsilon \otimes a = \epsilon$.

Every commutative semiring accepts a canonical preorder, $a \leq b$ if and only if there exists $c \in D$ with $a \oplus c = b$. A *dioid* is a semiring $\mathcal{D}$ where this relation is actually an order. Dioids are zerosumfree and entire, that is they have no non-null additive or multiplicative factors of zero. Commutative complete dioids are already complete residuated lattices.

An *idempotent semiring* is a dioid whose addition is idempotent, and a *selective semiring* one where the arguments attaining the value of the additive operation can be identified.

Example 1. Examples of idempotent dioids are

1. The *Boolean lattice* $\mathbb{B} = \langle \{0,1\}, \vee, \wedge, 0, 1 \rangle$
2. All fuzzy semirings, e.g. $\langle [0,1], \max, \min, 0, 1 \rangle$
3. The *min-plus algebra* $\mathbb{R}_{\min,+} = \langle \mathbb{R} \cup \{\infty\}, \min, +, \infty, 0 \rangle$
4. The *max-plus algebra* $\mathbb{R}_{\max,+} = \langle \mathbb{R} \cup \{-\infty\}, \max, +, -\infty, 0 \rangle$ □

Of the semirings above, only the boolean lattice and the fuzzy semirings are complete dioids, since the rest lack the *top* element $\top$ as an adequate inverse for the bottom in the order.

A semiring is a *semifield* if there exists a multiplicative inverse for every element $a \in S$, except the null, notated as a^{-1}, and *radicable* if the equation $a^b = c$ can be solved for a. As exemplified above, idempotent semifields are incomplete in their natural order, but there are procedures for *completing* such structures [3] and we will not differentiate between *complete or completed* structures. Note, first, that in complete semifields $e \neq \top$ which distinguishes them from inclines, and also that the inverse for the null is prescribed as $\perp^{-1} = \top$.

Semimodules. Let $S = \langle S, +, \times, \epsilon_S, e_S \rangle$ be a commutative semiring. A S-*semimodule* $\mathcal{X} = \langle X, \oplus, \odot, \epsilon_X \rangle$ is a commutative monoid $\langle X, \oplus, \epsilon_X \rangle$ endowed with a scalar action $(\lambda, x) \mapsto \lambda \odot x$ satisfying the following conditions for all $\lambda, \mu \in S, \ x, x' \in X$:

$$(\lambda \times \mu) \odot x = \lambda \odot (\mu \odot x) \qquad \lambda \odot (x \oplus x') = \lambda \odot x \oplus \lambda \odot x' \qquad (1)$$
$$(\lambda + \mu) \odot x = \lambda \odot x \oplus \mu \odot x \qquad \lambda \odot \epsilon_X = \epsilon_X = \epsilon_D \otimes x$$
$$e_D \odot x = x$$

If S is commutative, idempotent or complete, then $\mathcal{X}$ is also commutative, idempotent or complete.

Consider a set of vectors $\mathcal{Z} \subseteq \mathcal{X}$ where $\mathcal{X}$ is an S-semiring then the *span* of $\mathcal{Z}$, $\langle Z \rangle_S \subseteq X$ is the subsemimodule of $\mathcal{X}$ generated by linear combinations of finitely many such vectors. Note that a semimodule $\mathcal{V}$ is *finitely generated* if there exists a finite set of vectors $\mathcal{Z} \subseteq \mathcal{X}$ such that $\mathcal{V} = \langle Z \rangle_S$.

Matrices form a S-semimodule $S^{g \times m}$ for given g, m. In this paper, we only use finite-dimensional semimodules where we can identify *right S-semimodules* with column vectors, e.g. $\mathcal{X} \equiv S^{g \times 1}$ and *left S-semimodules* with row vectors.

Inverses and Conjugates. If $\mathcal{X} \subseteq \overline{\mathcal{K}}^{n \times 1}$ is a right semimodule over a complete semifield $\overline{\mathcal{K}}$, the *(pointwise) inverse* [7], $\mathcal{X}^{-1} \subseteq (\overline{\mathcal{K}}^{-1})^{n \times 1}$ is a right subsemimodule such that if $x \in X$, then $(x^{-1})_i = (x_i)^{-1}$, and the *conjugate* $\mathcal{X}^* \subseteq (\mathcal{K}^{-1})^{1 \times n}$ is a left *left*-semimodule, by $\mathcal{X}^* = (\mathcal{X}^{-1})^{\mathrm{T}} = (\mathcal{X}^{\mathrm{T}})^{-1}$. When $\mathcal{X}$ is reduced to a naturally-ordered semifield $\mathcal{X} \equiv \mathcal{K}$ we simply speak of the inverse or dual $\mathcal{K}^* = \mathcal{K}^{-1}$, since this duality is that of partial orders on the natural order.

In fact, the interaction of duality, the natural order and the abstract notation of the semifields $\overline{\mathcal{K}}$ and $\overline{\mathcal{K}}^*$ is awkward, so we use a special notation for them: complete, naturally-ordered semifields appear as enriched structures by combining their signatures as suggested by Fig. 1. In these paired structures, meets can be expressed by means of joins and inversion as $a \mathbin{\dot{\oplus}} b = (a^{-1} \oplus b^{-1})^{-1}$, and vice-versa and the dotted notation is used to suggest how multiplication works for extremal elements $\perp \otimes \top = \perp$ and $\perp \mathbin{\dot{\otimes}} \top = \top$.

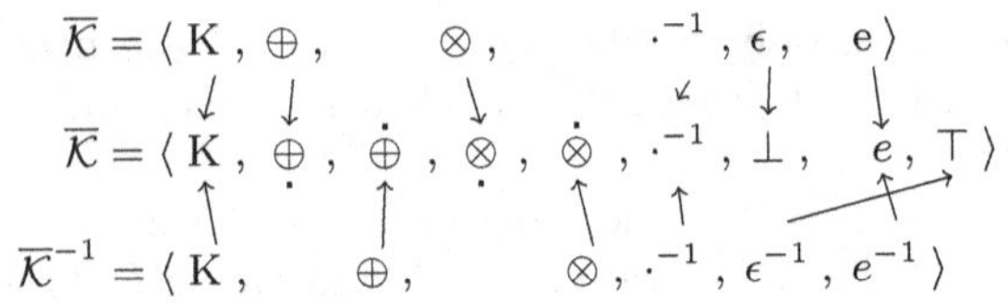

$$\overline{\mathcal{K}} = \langle \mathrm{K}, \oplus, \qquad \otimes, \qquad \cdot^{-1}, \epsilon, \quad e \rangle$$

$$\overline{\mathcal{K}} = \langle \mathrm{K}, \oplus, \dot{\oplus}, \otimes, \dot{\otimes}, \cdot^{-1}, \bot, \quad e, \top \rangle$$

$$\overline{\mathcal{K}}^{-1} = \langle \mathrm{K}, \qquad \oplus, \qquad \otimes, \cdot^{-1}, \epsilon^{-1}, e^{-1} \rangle$$

Fig. 1. Construction of the enriched structure of a completed naturally-ordered semifield with its inverse. The name remains that of the original basis for the construction, e.g. $\overline{\mathcal{K}}$ since they share the natural order.

One further advantage is that residuation can be expressed in terms of inverses, and this extends to semimodules:

$$(x \odot \lambda)^{-1} = x^{-1} \dot{\odot} \lambda^{-1} \qquad\qquad (x_1 \oplus x_2)^{-1} = x_1^{-1} \dot{\oplus} x_2^{-1}$$

In fact, for complete idempotent semifields, the following matrix algebra equations are proven in [7, Chap. 8].

Proposition 1. *Let $\mathcal{K}$ be an idempotent semifield, and $A \in \mathcal{K}^{m \times n}$. Then:*

1. $A \otimes (A^* \dot{\otimes} A) = A \dot{\otimes} (A^* \otimes A) = (A \dot{\otimes} A^*) \otimes A = (A \otimes A^*) \dot{\otimes} A = A$ *and*
   $A^* \otimes (A \dot{\otimes} A^*) = A^* \dot{\otimes} (A \otimes A^*) = (A^* \dot{\otimes} A) \otimes A^* = (A^* \otimes A) \dot{\otimes} A^* = A^*$.
2. *Alternating $A - A^*$ products of 4 matrices can be shortened as in:*

$$A^* \dot{\otimes} (A \otimes (A^* \dot{\otimes} A)) = A^* \dot{\otimes} A = (A^* \dot{\otimes} A) \otimes (A^* \dot{\otimes} A)$$

3. *Alternating $A - A^*$ products of 3 matrices and another terminal, arbitrary matrix can be shortened as in:*

$$A^* \dot{\otimes} (A \otimes (A^* \dot{\otimes} M)) = A^* \dot{\otimes} M = (A^* \dot{\otimes} A) \otimes (A^* \dot{\otimes} M)$$

4. *The following inequalities apply:*

$$A^* \dot{\otimes} (A \otimes M) \geq M \qquad\qquad A^* \otimes (A \dot{\otimes} M) \leq M$$

2.2 $\mathcal{K}$-Formal Concept Analysis Basics

Consider the formal K-context $(G, M, R)_{\overline{\mathcal{K}}}$ where K is a set that carries an idempotent semifield structure $\overline{\mathcal{K}}$, G is a set of $|G| = g$ objects, M a set of $|M| = m$ attributes, and $R \in K^{g \times m}$ is a matrix with entries in K. Consider the spaces $\mathcal{X} = \overline{\mathcal{K}}^g$ and $\mathcal{Y} = \overline{\mathcal{K}}^m$ and a bracket $\langle x \mid R \mid y \rangle_{\text{oi}} = x^* \dot{\otimes} R \dot{\otimes} y^{-1}$ between them. Then the following results can be harvested from [4,5][1].

---

[1] The notation follows the more developed one in [5].

Proposition 2. *Given the formal K-context $(G, M, R)_{\overline{K}}$ and an invertible element $\varphi = \gamma \dot{\otimes} \mu \in K$, then:*

1. *The polars of the formal context $(G, M, R)_{\overline{K}}$*

$$x_R^{\uparrow} = R^{\mathrm{T}} \dot{\otimes} x^{-1} \qquad\qquad y_R^{\downarrow} = R \dot{\otimes} y^{-1} \qquad (2)$$

form a Galois connection $(\cdot_R^{\uparrow}, \cdot_R^{\downarrow}) : \tilde{\mathcal{X}}^{\gamma} \times \tilde{\mathcal{Y}}^{\mu}$ between the scaled spaces $\tilde{\mathcal{X}}^{\gamma} = \gamma \otimes \mathcal{X}$ and $\tilde{\mathcal{Y}}^{\mu} = \mu \otimes \mathcal{Y}$.

2. *The polars acting on the ambient spaces define two sets, the system of extents $\mathfrak{B}_G^{\gamma}(G, M, R)_{\overline{K}}$ and the system of intents $\mathfrak{B}_M^{\mu}(G, M, R)_{\overline{K}}$ of the Galois connection.*

$$\mathfrak{B}_G^{\gamma}(G, M, R)_{\overline{K}} = (\tilde{Y}^{\mu})_R^{\downarrow} \qquad\qquad \mathfrak{B}_M^{\mu}(G, M, R)_{\overline{K}} = (\tilde{X}^{\gamma})_R^{\uparrow} \qquad (3)$$

which are bijective through the polars, for which reason, we call $(\cdot)_R^{\uparrow} : \tilde{X}^{\gamma} \to \mathfrak{B}_M^{\mu}(G, M, R)$ the polar of (or generating) intents and $(\cdot)_R^{\downarrow} : \tilde{Y}^{\mu} \to \mathfrak{B}_G^{\gamma}(G, M, R)$ the polar of (or generating) extents. Furthermore the composition of the polars generate closure operators:

$$\pi_R(x) = (x_R^{\uparrow})_R^{\downarrow} = R \dot{\otimes} (R^* \otimes x) \qquad \pi_{R^{\mathrm{T}}}(y) = (y_R^{\downarrow})_R^{\uparrow} = R^{\mathrm{T}} \dot{\otimes} (R^{-1} \otimes y). \qquad (4)$$

which are the identities on the system of extents and intents, respectively.

3. *The system of extents and intents are $\overline{K}^{\mathrm{d}}$-subsemimodules of the spaces generated by the rows (resp, columns) or the incidence R. Furthermore, they are the closures of the $\overline{K}$-semimodules generated by the projection matrices $R \dot{\otimes} R^*$ and $R^{\mathrm{T}} \dot{\otimes} R^{-1}$:*

$$\underline{\mathfrak{B}}_G^{\gamma}(G, M, R)_{\overline{K}} = \langle R \rangle_{\overline{K}^{\mathrm{d}}} = \pi_R(\langle R \dot{\otimes} R^* \rangle_{\overline{K}})$$

$$\underline{\mathfrak{B}}_M^{\mu}(G, M, R)_{\overline{K}} = \langle R^{\mathrm{T}} \rangle_{\overline{K}^{\mathrm{d}}} = \pi_{R^{\mathrm{T}}}(\langle R^{\mathrm{T}} \dot{\otimes} R^{-1} \rangle_{\overline{K}})$$

4. *The bijection between systems of extents and intents is really a dual lattice isomorphism with φ-formal concepts $(a, b) \in \underline{\mathfrak{B}}_{\mathrm{OI}}^{\gamma \dot{\otimes} \mu}(G, M, R)_{\overline{K}}$ so that*

$$(a)_R^{\uparrow} = b \iff (b)_R^{\downarrow} = a.$$

For the last result we have used the positional notation introduced in [3] for extended concept lattices in which the subindices refer to the existence (I) or not (O) of an inversion in the lattice for $\mathcal{X}$ or $\mathcal{Y}$. However, for the summary of the Galois connection in Fig. 2, where the individual lattices of extents and intents appear, we use a more traditional notation for them. Note that we have glossed over the process of scaling the spaces [4, Sect. 3.1] and a number of other results.

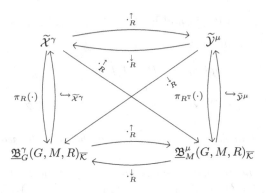

Fig. 2. The Galois connection $(\cdot_R^\uparrow, \cdot_R^\downarrow) : \tilde{\mathcal{X}}^\gamma \times \tilde{\mathcal{Y}}^\mu$ for $(G, M, R)_{\overline{\mathcal{K}}}$ and $\varphi = \gamma \dot{\otimes} \mu$. (From [4]).

3   Extended FCA in Complete Idempotent Semifields

In the following we present an exploration of similar ideas to those expressed in Sect. 2.2 but along the lines suggested in [3] to make different lattices appear.

3.1   The Neighbourhood of Objects

Consider the bracket $\langle x \mid R \mid y \rangle_{OO} = x^* \dot{\otimes} R \dot{\otimes} y$ over the $\overline{\mathcal{K}}$-semimodules $\mathcal{X} \equiv \overline{\mathcal{K}}^g$ and $\mathcal{Y} \equiv \overline{\mathcal{K}}^m$. The adjunct-generating inequality $x^* \dot{\otimes} R \dot{\otimes} y \geq \varphi = \gamma \dot{\otimes} \mu$ may be scaled and residuated [4] to obtain:

$$x_R^{\dot{\exists}} = (x^* \dot{\otimes} R)\backslash e = R^* \dot{\otimes} x \qquad\qquad y_R^{\dot{\forall}} = (y^* \dot{\otimes} R^*)\backslash e = R \dot{\otimes} y \qquad (5)$$

over normalized spaces $\tilde{\mathcal{X}}^\gamma$ and $\tilde{\mathcal{Y}}^\mu$.

Proposition 3. $(\cdot_R^{\dot{\exists}}, \cdot_R^{\dot{\forall}}) : \tilde{\mathcal{X}}^\gamma \leftrightharpoons \tilde{\mathcal{Y}}^\mu$ is a Galois adjunction on the left between the semimodules $\tilde{\mathcal{X}}^\gamma \cong \widetilde{(\overline{\mathcal{K}}^g)}^\gamma$ and $\tilde{\mathcal{Y}}^\mu \cong \widetilde{(\overline{\mathcal{K}}^m)}^\mu$.

Proof. For $x \in \tilde{\mathcal{X}}^\gamma$, $y \in \tilde{\mathcal{Y}}^\mu$, we need to prove $x_R^{\dot{\exists}} \leq y \Leftrightarrow x \leq y_R^{\dot{\forall}}$. If $x_R^{\dot{\exists}} = R^* \dot{\otimes} x \leq y$, then by residuation $x \leq R^* \dot{\otimes} y = y_R^{\dot{\forall}}$.   □

This immediately puts at our disposal a number of results.

Proposition 4 (Properties of the left adjunction). *Consider the left adjunction* $(\cdot_R^{\dot{\exists}}, \cdot_R^{\dot{\forall}}) : \tilde{\mathcal{X}}^\gamma \leftrightharpoons \tilde{\mathcal{Y}}^\mu$. *Then,*

1. *The adjuncts are monotone functions.* $(\cdot)_R^{\dot{\exists}}$ *is join preserving and* $(\cdot)_R^{\dot{\forall}}$ *is meet preserving:*

$$(x_1 \dot{\oplus} x_2)_R^{\dot{\exists}} = x_1{}_R^{\dot{\exists}} \dot{\oplus} x_2{}_R^{\dot{\exists}} \qquad\qquad (y_1 \dot{\oplus} y_2)_R^{\overset{\dot{\forall}}{}} = y_1{}_R^{\dot{\forall}} \dot{\oplus} y_2{}_R^{\dot{\forall}}. \qquad (6)$$

2. *The compositions of the adjuncts:* $\pi_R : X \to X, \kappa_R : Y \to Y$ *are*

$$\pi_R(x) = (x_R^{\exists})_R^{\forall} = R \,\dot{\otimes}\, (R^* \otimes x) \qquad \kappa_{R^*}(y) = (y_R^{\forall})_R^{\exists} = R^* \,\dot{\otimes}\, (R \,\dot{\otimes}\, y) \qquad (7)$$

with $\pi_R(x)$ *a closure, that is, an extensive and idempotent operator, and* $\kappa_R(y)$ *a kernel or interior operator, a contractive, idempotent operator.*

$$\pi_R(x) \geq x \qquad\qquad\qquad \kappa_{R^*}(y) \leq y$$
$$\pi_R(\pi_R(x)) = \pi_R(x) \qquad\qquad \kappa_{R^*}(\kappa_{R^*}(y)) = \kappa_{R^*}(y)$$

3. *The adjuncts are mutual pseudo-inverses:*

$$(\cdot)_R^{\exists} \circ (\cdot)_R^{\forall} \circ (\cdot)_R^{\exists} = (\cdot)_R^{\exists} \qquad\qquad (\cdot)_R^{\forall} \circ (\cdot)_R^{\exists} \circ (\cdot)_R^{\forall} = (\cdot)_R^{\forall}$$

Proof. We prove a stronger statement than property 1 for sets of attributes:

$$(y_1 \,\dot{\otimes}\, \lambda_1 \,\dot{\oplus}\, y_2 \,\dot{\otimes}\, \lambda_2)_R^{\forall} = R \,\dot{\otimes}\, (y_1 \,\dot{\otimes}\, \lambda_1 \,\dot{\oplus}\, y_2 \,\dot{\otimes}\, \lambda_2) = R \,\dot{\otimes}\, (y_1 \,\dot{\otimes}\, \lambda_1) \,\dot{\oplus}\, R \,\dot{\otimes}\, (y_2 \,\dot{\otimes}\, \lambda_2)$$
$$= y_1{}_R^{\forall} \otimes \lambda_1 \,\dot{\oplus}\, y_2{}_R^{\forall} \otimes \lambda_2 \qquad\qquad (8)$$

When $\lambda_1 = \lambda_2 = e$ we get the statement. For property 2, the closed form of the operators is obtained from the definitions. From Proposition 1.4 we find that $\kappa_{R^*}(y) = R^* \otimes (R \,\dot{\otimes}\, y) \leq y$, that is, it is contractive. Next, from Proposition 1.3 we know that $R^* \otimes (R \,\dot{\otimes}\, (R^* \otimes M)) = R^* \otimes M$, whence $\kappa_{R^*}(\kappa_{R^*}(y)) = R^* \otimes (R \,\dot{\otimes}\, (R^* \otimes (R \,\dot{\otimes}\, y))) = R^* \otimes (R \,\dot{\otimes}\, y) = \kappa_{R^*}(y)$. Finally, using Proposition 1.3,

$$((x_R^{\exists})_R^{\forall})_R^{\exists} = R^* \otimes (R \,\dot{\otimes}\, (R^* \otimes x)) = R^* \otimes x = x_R^{\exists}.$$

The proofs for the closure of sets of objects can be found in [4]. □

The adjuncts are in this case idempotent semimodule homomorphisms:

Proposition 5. *Let* $(\cdot_R^{\exists}, \cdot_R^{\forall}) : \tilde{\mathcal{X}}^{\gamma} \leftrightarrows \tilde{\mathcal{Y}}^{\mu}$ *be a left adjunction. Then* $(\cdot)_R^{\exists}$ *is a homomorphism of* $\mathcal{K}$-*semimodules and* $(\cdot)_R^{\forall}$ *is a homomorphism of* $\mathcal{K}^{-1}$-*semimodules.*

Proof. For the adjunct of intents, consider $x_1{}_R^{\exists} = R^* \otimes x_1$ and $x_2{}_R^{\exists} = R^* \otimes x_2$.

$$(x_1 \otimes \lambda_1 \,\dot{\oplus}\, x_2 \otimes \lambda_2)_R^{\exists} = R^* \otimes (x_1 \otimes \lambda_1 \,\dot{\oplus}\, x_2 \otimes \lambda_2)$$
$$= R^* \otimes (x_1 \otimes \lambda_1) \,\dot{\oplus}\, R^* \otimes (x_2 \otimes \lambda_2)$$
$$= x_1{}_R^{\exists} \otimes \lambda_1 \,\dot{\oplus}\, x_2{}_R^{\exists} \otimes \lambda_2$$

The proof for $(\cdot)_R^{\forall}$ is that of (8). □

Applying the adjuncts to the whole of the ambient spaces, we define two sets, the *neighbourhood system of extents* $(\widetilde{Y}^\mu)_R^\vee$ and the *neighbourhood system of intents* $(\widetilde{X}^\gamma)_R^\exists$ of the adjunction[2]. Next, call the fixpoints of the closure and interior operators, respectively:

$$\text{fix}(\pi_R(\cdot)) = \{a \in \widetilde{X}^\gamma \mid \pi_R(a) = a\} \quad \text{fix}(\kappa_{R^*}(\cdot)) = \{b \in \widetilde{Y}^\mu \mid \kappa_{R^*}(b) = b\}.$$

These are also called *the set of closed, respectively open, elements* of the Galois connection that generates $\pi_R(\cdot)$ and $\kappa_{R^*}(\cdot)$.

However, they are not new elements:

Lemma 1. *The sets of extents and intents are the sets of fixpoints of the closure operators of the adjunction.*

$$\text{fix}(\pi_R(\cdot)) = (\widetilde{Y}^\mu)_R^\vee \qquad\qquad \text{fix}(\kappa_{R^*}(\cdot)) = (\widetilde{X}^\gamma)_R^\exists \qquad (9)$$

Proof. For all $a \in (\widetilde{Y}^\mu)_R^\vee \subseteq \widetilde{X}^\gamma$ we have $(a)_R^\exists \in (\widetilde{X}^\gamma)_R^\exists$ whence $((\widetilde{Y}^\mu)_R^\vee)_R^\exists \in \widetilde{X}^\gamma$.

On the other hand, consider a $d \in (\widetilde{X}^\gamma)_R^\exists$ then there must be $x \in \widetilde{X}^\gamma$ such that $d = (x)_R^\exists$, and call $a = (d)_R^\vee$. Then $(a)_R^\exists = ((d)_R^\vee)_R^\exists = ((x_R^\exists)_R^\vee)_R^\exists$ and since the adjuncts are pseudoinverses $((d)_R^\vee)_R^\exists = x_R^\exists = d$. $\qquad\qquad\square$

In the previous proof, the name for the fixpoints of $d \in \text{fix}(\kappa_{R^*}(\cdot))$ is intentional and carries a meaning to be explained later.

Notice that from (4) and (7) it is immediately evident that the closure of extents for the Galois connection and the left adjunction are the same. Respecting prior notation we give FCA-meaningful names to the sets of fixpoints:

$$\mathfrak{B}_G^\gamma(G, M, R)_{\overline{K}} = (\widetilde{Y}^\mu)_R^\vee \qquad\qquad \mathfrak{N}_M^\mu(G, M, R)_{\overline{K}} = (\widetilde{X}^\gamma)_R^\exists \qquad (10)$$

For this reason, we call $(\cdot)_R^\exists : \widetilde{X}^\gamma \to \mathfrak{N}_M^\mu(G, M, R)_{\overline{K}}$ the *(object left) adjunct of (or generating) intents* and $(\cdot)_R^\vee : \widetilde{Y}^\mu \to \mathfrak{B}_G^\gamma(G, M, R)_{\overline{K}}$ the *(object right) adjunct of (or generating) extents*.

Fixpoints of either operator are very easily generable for an adjunction:

Proposition 6. $\mathfrak{B}_G^\gamma(G, M, R)_{\overline{K}}$ *and* $\mathfrak{N}_M^\mu(G, M, R)_{\overline{K}}$ *are complete* $\overline{K}^{-1}$- *and* $\overline{K}$-*subsemimodules of* $\widetilde{X}^\gamma$ *and* $\widetilde{Y}^\mu$, *generated by the columns of R and R^*, respectively.*

Proof. Given a generic vector $z \in \widetilde{Y}^\mu$, then by Lemma 1.2 we have

$$\kappa_{R^*}(R^* \otimes z) = R^* \otimes (R \dot{\otimes} (R^* \otimes z)) = R^* \otimes z$$

This means that any $\overline{K}$-combination of columns of R^* is a fixpoint of $\kappa_{R^*}(\cdot)$, that is $\langle R^* \rangle_{\overline{K}} \subseteq \text{fix}(\kappa_{R^*}(\cdot))$. Now, consider $d \in \text{fix}(\kappa_{R^*}(\cdot))$. Then $R^* \otimes (R \dot{\otimes} d) = d$ whence $R^* \otimes z = d$, so $\text{fix}(\kappa_{R^*}(\cdot)) \subseteq \langle R^* \rangle_{\overline{K}}$. Using the same technique we find $\text{fix}(\pi_R(\cdot)) = \dot{\langle} R \rangle_{\overline{K}^{-1}}$. $\qquad\qquad\square$

---

[2] With a little abuse of the terms "extent" and "intent".

Figure 3 shows a commutative diagram of the adjunction defined by $(G, M, R)_{\overline{\mathcal{K}}}$ and $\varphi = \gamma \dot{\otimes} \mu$. But in fact a stronger result follows.

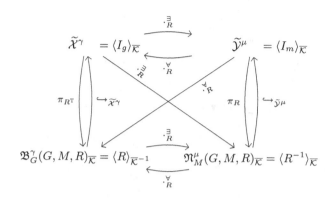

Fig. 3. The adjunction $(\cdot\frac{\exists}{R}, \cdot\frac{\forall}{R}) : \widetilde{\mathcal{X}}^{\gamma} \leftrightarrows \widetilde{\mathcal{Y}}^{\mu}$ for $(G, M, R)_{\overline{\mathcal{K}}}$ and $\varphi = \gamma \dot{\otimes} \mu$.

Proposition 7. $\mathfrak{B}_G^{\gamma}(G, M, R)_{\overline{\mathcal{K}}}$ and $\mathfrak{M}_M^{\mu}(G, M, R)_{\overline{\mathcal{K}}}$ are complete lattices included in $\widetilde{X}^{\gamma}$ and $\widetilde{Y}^{\mu}$.

Proof. As complete semimodules they are complete meet (respectively join) sub-semilattices. Since they include a top (respectively a bottom) element, they are complete lattices.   □

This last fact allows us to speak of a "neighbourhood-based" FCA:

Corollary 1. *The bijection between systems of extents and intents is really a dual lattice isomorphism with φ-formal pairs* $(a, d) \in \mathfrak{B}_{oo}^{\gamma \dot{\otimes} \mu}(G, M, R)_{\overline{\mathcal{K}}}$ *so that*

$$(a)_R^{\exists} = d \iff (d)_R^{\forall} = a.$$

The schematic diagram of Fig. 4 makes these mechanisms evident, but adds some unproven facts to these lattices, to be explained in full elsewhere. Note the difference in orientation with the $\widetilde{Y}^{\mu}$ in the Galois connection of [4], and that the same caveat as in the Galois connections hold here: although $\mathfrak{B}_G^{\gamma}(G, M, R)_{\overline{\mathcal{K}}}$ and $\mathfrak{M}_M^{\mu}(G, M, R)_{\overline{\mathcal{K}}}$ are lattices, they are not *sublattices* of $\widetilde{X}^{\gamma}$ and $\widetilde{Y}^{\mu}$ since they do not agree on the joins, respectively on the meets.

From this point onwards, we could follow the same line as pursued in [4], by investigating the two lattice structures associated to the fixpoints, the relation of the adjunction to the eigenspaces, etc. But we prefer to follow a different track: we will consider the other adjunctions and connections generated by the formal context $(G, M, R)_{\overline{\mathcal{K}}}$.

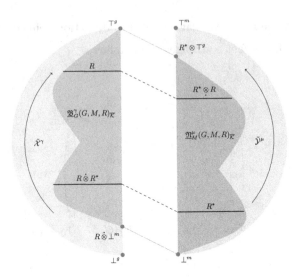

Fig. 4. Extended schematics of the adjunction $\mathfrak{B}_{oo}^{\gamma\,\dot{\otimes}\,\mu}(G,M,R)_{\overline{\mathcal{K}}}$ between $\tilde{X}^\gamma$ and $\tilde{Y}^\mu$ (outer clouds). Extents $\mathfrak{B}_G^\gamma(G,M,R)_{\overline{\mathcal{K}}}$ (left inner cloud) and intents $\mathfrak{M}_M^\mu(G,M,R)_{\overline{\mathcal{K}}}$ (right inner cloud) are dually isomorphic semimodules $\overline{\mathcal{K}}^{-1}$-generated by R and $\overline{\mathcal{K}}$-generated by R^* respectively.

3.2   The Neighbourhood of Attributes

For the case of the right adjunct, we will provide the results in a similar way to those of Sect. 2.2, but considering the bracket $\langle x \mid R \mid y \rangle_{II} = x^T \dot{\otimes} R \otimes y^{-1}$ over the $\overline{\mathcal{K}}$-semimodules $\mathcal{X} \equiv \overline{\mathcal{K}}^g$ and $\mathcal{Y} \equiv \overline{\mathcal{K}}^m$.

Proposition 8. *Given the formal K-context $(G,M,R)_{\overline{\mathcal{K}}}$ and an invertible element $\varphi = \gamma \dot{\otimes} \mu \in K$, then:*

1. *The mappings $(\cdot_{R^T}^\forall, \cdot_{R^T}^\exists) : \tilde{X}^\gamma \rightleftharpoons \tilde{Y}^\mu$ of the formal context $(G,M,R)_{\overline{\mathcal{K}}}$*

$$x_{R^T}^\forall = R^T \dot{\otimes} x \qquad\qquad y_{R^T}^\exists = R^{-1} \otimes y \qquad (11)$$

form an adjunction on the right between the scaled semimodules $\tilde{X}^\gamma \cong \widetilde{(\mathcal{K}^g)}^\gamma$ and $\tilde{Y}^\mu \cong \widetilde{(\mathcal{K}^m)}^\mu$, such that the adjuncts are monotone, mutually pseudo-inverse functions, $(\cdot)_{R^T}^\forall$ is meet preserving and $(\cdot)_{R^T}^\exists$ is join preserving:

$$(x_1 \dot{\oplus} x_2)_{R^T}^\forall = x_{1\,R^T}^\forall \dot{\oplus} x_{2\,R^T}^\forall \qquad (y_1 \oplus y_2)_{R^T}^\exists = y_{1\,R^T}^\exists \oplus y_{2\,R^T}^\exists. \qquad (12)$$

2. *The adjuncts acting on the ambient spaces define two sets, the system of extents $\mathfrak{N}_G^\gamma(G,M,R)_{\overline{\mathcal{K}}}$ and the system of intents $\mathfrak{B}_M^\mu(G,M,R)_{\overline{\mathcal{K}}}$ of the Galois connection.*

$$\mathfrak{N}_G^\gamma(G,M,R)_{\overline{\mathcal{K}}} = (\tilde{Y}^\mu)_{R^T}^\exists \qquad\qquad \mathfrak{B}_M^\mu(G,M,R)_{\overline{\mathcal{K}}} = (\tilde{X}^\gamma)_{R^T}^\forall \qquad (13)$$

for which the adjuncts form a bijection, hence $(\cdot)_{R^\mathrm{T}}^{\vee} : \widetilde{X}^\gamma \to \mathfrak{B}_M^\mu(G, M, R)$ *is the adjunct of (or generating) intents and* $(\cdot)_{R^\mathrm{T}}^{\exists} : \widetilde{Y}^\mu \to \mathfrak{N}_G^\gamma(G, M, R)$ *is the adjunct of (or generating) extents. Furthermore, their compositions are*

$$\kappa_{R^{-1}}(x) = R^{-1} \dot{\otimes} (R^\mathrm{T} \otimes x) \qquad \pi_{R^\mathrm{T}}(y) = R^\mathrm{T} \otimes (R^{-1} \dot{\otimes} y) \qquad (14)$$

with $\kappa_{R^{-1}}(x)$ *an interior operator, that is, a contractive and idempotent operator, and* $\pi_{R^\mathrm{T}}(y)$ *a closure operator, an expansive, idempotent operator which are the identities on the system of extents and intents, respectively.*

3. *The system of extents and intents are complete* $\overline{K}$- *and* $\overline{K}^{-1}$-*subsemimodules of* $\widetilde{X}^\gamma$ *and* $\widetilde{Y}^\mu$, *generated by the columns of* R^{-1} *and* R^T, *respectively.*

$$\mathfrak{N}_G^\gamma(G, M, R)_{\overline{K}} = \langle R^{-1} \rangle_{\overline{K}} \qquad \mathfrak{B}_M^\mu(G, M, R)_{\overline{K}} = \langle R^\mathrm{T} \rangle_{\overline{K}^{-1}}$$

4. *The bijection between systems of extents and intents is really a lattice isomorphism with* φ-*formal pairs* $(b, c) \in \underline{\mathfrak{B}}_{\Pi}^{\gamma \dot{\otimes} \mu}(G, M, R)_{\overline{K}}$ *so that*

$$(c)_{R^\mathrm{T}}^{\vee} = b \iff (b)_{R^\mathrm{T}}^{\exists} = c.$$

Proof (Sketch). *The adjunct-generating inequality* $x^\mathrm{T} \dot{\otimes} R \dot{\otimes} y^{-1} \geq \varphi = \gamma \dot{\otimes} \mu$ *may be scaled and residuated* [4] *to solve for* x^{-1} *and* y^{-1}

$$y^{-1} \leq (x^\mathrm{T} \dot{\otimes} R) \backslash e = R^* \underset{\cdot}{\otimes} x^{-1} \qquad x^{-1} \geq (y^\mathrm{T} \otimes R^*) \backslash e = R \dot{\otimes} y^{-1}$$

Inverting and comparing this to (5) suggests the notation for the adjuncts. To prove that it is an adjunction on the right we notice that $(\cdot_{R^\mathrm{T}}^{\exists}, \cdot_{R^\mathrm{T}}^{\vee}) : \widetilde{Y}^\mu \leftrightarrows \widetilde{X}^\gamma$ is an adjunction on the left and there is an interchange in the roles of the semimodules reflected in transposition of the incidence matrix, with respect to Sect. 3.1. Therefore the proof of these results would proceed as in Proposition 3, etc. □

An analogue of Fig. 4 will be introduced in the wider context of Sect. 3.4.

3.3 The Lattice of Alterity

The last connection between $\mathcal{X}$ and $\mathcal{Y}$ is induced by bracket $\langle x \mid R \mid y \rangle_{IO} = x^\mathrm{T} \dot{\otimes} R \dot{\otimes} y$ between them. This generates, for lack of a better name, the lattice of "alterity", also called "of alienness" or "unrelatedness" in [3]. We collect most results related to them in a single proposition.

Proposition 9. *Given the formal K-context $(G, M, R)_{\overline{K}}$ and an invertible element $\varphi = \gamma \dot{\otimes} \mu \in K$, then:*

1. *The dual polars of the formal context* $(G, M, R)_{\overline{\mathcal{K}}}$

$$x^{\uparrow}_{R^{-1}} = R^* \otimes x^{-1} \qquad\qquad y^{\downarrow}_{R^{-1}} = R^{-1} \otimes y^{-1} \qquad (15)$$

form a co-Galois connection $(\cdot^{\uparrow}_{R^{-1}}, \cdot^{\downarrow}_{R^{-1}}) : \widetilde{\mathcal{X}}^{\gamma} \times \widetilde{\mathcal{Y}}^{\mu}$ *between the scaled spaces* $\widetilde{\mathcal{X}}^{\gamma} = \gamma \otimes \mathcal{X}$ *and* $\widetilde{\mathcal{Y}}^{\mu} = \mu \otimes \mathcal{Y}$.

2. *The dual polars acting on the ambient spaces define two sets, the system of extents* $\mathfrak{N}^{\gamma}_{G}(G, M, R)_{\overline{\mathcal{K}}}$ *and the system of intents* $\mathfrak{N}^{\mu}_{M}(G, M, R)_{\overline{\mathcal{K}}}$ *of the co-Galois connection.*

$$\mathfrak{N}^{\gamma}_{G}(G, M, R) = (\widetilde{Y}^{\mu})^{\downarrow}_{R^{-1}} \qquad\qquad \mathfrak{N}^{\mu}_{M}(G, M, R) = (\widetilde{X}^{\gamma})^{\uparrow}_{R^{-1}} \qquad (16)$$

which are bijective through the dual polars, for which reason, we call $\cdot^{\uparrow}_{R^{-1}}$: $\widetilde{X}^{\gamma} \to \mathfrak{N}^{\mu}_{M}(G, M, R)_{\overline{\mathcal{K}}}$ *the dual polar of (or generating) intents and* $\cdot^{\downarrow}_{R^{-1}}$: $\widetilde{Y}^{\mu} \to \mathfrak{N}^{\gamma}_{G}(G, M, R)_{\overline{\mathcal{K}}}$ *the dual polar of (or generating) extents. Furthermore the composition of the dual polars generate interior operators:*

$$\kappa_{R^{-1}}(x) = R^{-1} \otimes (R^{\mathrm{T}} \dot{\otimes} x) \qquad \kappa_{R^*}(y) = R^* \otimes (R^{-1} \dot{\otimes} y). \qquad (17)$$

which are the identities on the system of extents and intents, respectively.

3. *The system of extents and intents are* $\overline{\mathcal{K}}$-*subsemimodules of the spaces generated by the rows (resp, columns) or the incidence* R.

$$\mathfrak{N}^{\gamma}_{G}(G, M, R)_{\overline{\mathcal{K}}} = \langle R^{-1} \rangle_{\overline{\mathcal{K}}} \qquad\qquad \mathfrak{N}^{\mu}_{M}(G, M, R)_{\overline{\mathcal{K}}} = \langle R^* \rangle_{\overline{\mathcal{K}}}$$

4. *Furthermore, the bijection between systems of extents and intents is really a dual lattice isomorphism with dual* φ-*formal concepts* $(d, c) \in \underline{\mathfrak{B}}^{\gamma \dot{\otimes} \mu}_{10}(G, M, R)_{\overline{\mathcal{K}}}$.

$$(c)^{\uparrow}_{R^{-1}} = d \iff (d)^{\downarrow}_{R^{-1}} = c$$

Proof (Sketch). The bracket generates the polar by means of the following inequalities:

$$x^{\mathrm{T}} \dot{\otimes} R \dot{\otimes} y \geq e \iff y^{\mathrm{T}} \dot{\otimes} R^{\mathrm{T}} \dot{\otimes} x \geq e \qquad (18)$$

The rest of the results follow by techniques similar to those applied to Galois connections and adjunctions. □

3.4   The Four-Fold Connection

The preceding material sheds a new light over FCA as a whole.

Lemma 2. *The closure and interior systems of the connections of* $(G, M, R)_{\overline{\mathcal{K}}}$ *are identical two by two.*

Proof. Compare (4), (7), (14) and (17) and collect all the systems of extents and intents two by two.   □

Actually we can collect as a corollary similar equivalences on the lattices.

Corollary 2. *There are only four lattices of the closure or interior operators of* $(G, M, R)_{\overline{\mathcal{K}}}$ *given* $\varphi = \gamma \,\dot{\otimes}\, \mu$. *These are all isomorphic or dually isomorphic.*

$$\pi_R(\widetilde{\mathcal{X}}^\gamma) = \underline{\mathfrak{B}}_G^\gamma(G, M, R)_{\overline{\mathcal{K}}} = \langle R \rangle_{\overline{\mathcal{K}}^{-1}} \qquad \pi_{R^{\mathsf{T}}}(\widetilde{\mathcal{Y}}^\mu) = \underline{\mathfrak{B}}_M^\mu(G, M, R)_{\overline{\mathcal{K}}} = \langle R^{\mathsf{T}} \rangle_{\overline{\mathcal{K}}^{-1}}$$

$$\kappa_{R^*}(\widetilde{\mathcal{Y}}^\mu) = \underline{\mathfrak{N}}_M^\mu(G, M, R)_{\overline{\mathcal{K}}} = \langle R^* \rangle_{\overline{\mathcal{K}}} \qquad \kappa_{R^{-1}}(\widetilde{\mathcal{X}}^\gamma) = \underline{\mathfrak{N}}_G^\gamma(G, M, R)_{\overline{\mathcal{K}}} = \langle R^{-1} \rangle_{\overline{\mathcal{K}}}$$

In this sense, Fig. 5, is a more complete picture than Figs. 2 or 3: it collects all of the polars, dual polars, left and right adjuncts and their actions on the normalized spaces and their closure/interior systems.

Therefore, it makes sense to define the following (meta) concept:

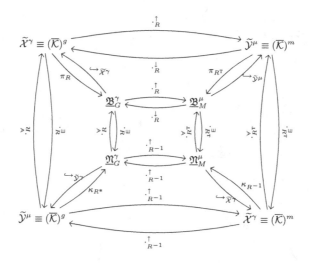

Fig. 5. A domain diagram of the four-fold connection between the spaces implicit in a $\overline{\mathcal{K}}$-valued formal context $(G, M, R)_{\overline{\mathcal{K}}}$. Some mappings missing for clarity.

Definition 1. *The 4-formal concept* (a, b, c, d) *is a 4-tuple such that* $a \in \underline{\mathfrak{B}}_G^\gamma(G, M, R)$, $b \in \underline{\mathfrak{B}}_M^\mu(G, M, R)$, $c \in \underline{\mathfrak{N}}_G^\gamma(G, M, R)$, *and* $d \in \underline{\mathfrak{N}}_M^\mu(G, M, R)$ *and all the following relations hold:*

$$a = (b)_R^\downarrow = (d)_R^\vee \qquad\qquad b = (a)_R^\uparrow = (c)_{R^{\mathsf{T}}}^\vee \qquad (19)$$

$$d = (c)_{R^{-1}}^\uparrow = (a)_R^\exists \qquad\qquad c = (d)_{R^{-1}}^\downarrow = (b)_{R^{\mathsf{T}}}^\exists \qquad (20)$$

So, for instance, if $x \in \widetilde{\mathcal{X}}^\gamma$, *the 4-tuple* $(((x)_R^\uparrow)_R^\downarrow, (x)_R^\uparrow, ((x)_R^\uparrow)_{R^{\mathsf{T}}}^\exists, (((x)_R^\uparrow)_{R^{\mathsf{T}}}^\exists)_{R^{-1}}^\uparrow)$ *is a 4-concept.*

Cuninghame-Green already developed a construction similar to that of Fig. 5 [7, Chap. 22]. Although the connections can be seen there, the notion that the images of the polars and adjuncts in the connections are actually complete lattices is missing, as is the notion that formal 4-concepts are a unit of the construction. We draw in Fig. 6 these four connections in the manner of Sect. 3 [7, Chap. 22] filling a little bit more of the details.

Actually, he went beyond the four-fold connection shown above to display a complementary one where the *bias* is taken in the inverse semifield $\overline{\mathcal{K}}^{-1}$. This suggests another lattice related to $\underline{\underline{\mathfrak{B}}}^{\varphi}(G, M, R)_{\overline{\mathcal{K}}}$, that of $\underline{\underline{\mathfrak{B}}}^{\varphi}(G, M, R)_{\overline{\mathcal{K}}^{-1}}$. However, a quick investigation shows that we may relate this to the original semifield.

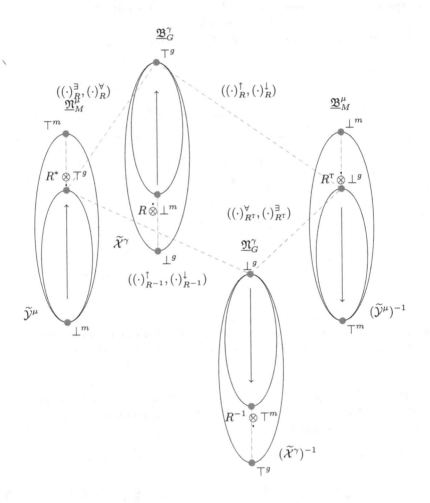

Fig. 6. Schematics of the four-fold connection of the lattices around $(G, M, R)_{\overline{\mathcal{K}}}$. The increasing sense of the orders is schematised with arrows. The dashed slanted lines connect the points of the 4-concept $(\mathsf{T}^g,\ R^{\mathsf{T}} \dot{\otimes} \perp^g,\ \perp^g,\ R^* \otimes \mathsf{T}^g)$.

Proposition 10. *Let (G, M, R) be a formal context and take the point of view of $\overline{\mathcal{K}}^{-1}$. Then $\mathfrak{B}_{01}^{\varphi}(G, M, R)_{\overline{\mathcal{K}}^{-1}} = \mathfrak{B}_{10}^{\varphi}(G, M, R^{-1})_{\overline{\mathcal{K}}}$.*

Proof. Taking the point of view of $\overline{\mathcal{K}}^{-1}$, consider the bracket $\langle x \mid R \mid y \rangle = x \backslash R / y^{\mathsf{T}} = x^* \otimes R \otimes y^{-1}$. Then, following the proof in [4] we arrive at the new polars for the Galois connection between $\mathcal{X} \equiv (\overline{\mathcal{K}}^{-1})^g$ and $\mathcal{Y} \equiv (\overline{\mathcal{K}}^{-1})^m$:

$$x_R^{\uparrow} = R^{\mathsf{T}} \mathbin{\dot{\otimes}} x^{-1} \qquad\qquad y_R^{\downarrow} = R \mathbin{\dot{\otimes}} y^{-1} \qquad (21)$$

whence the closures (in $\overline{\mathcal{K}}^{-1}$) are:

$$\kappa_R(x) = R \otimes (R^* \mathbin{\dot{\otimes}} x) \leq x \qquad \kappa_R(y) = R^{\mathsf{T}} \mathbin{\dot{\otimes}} (R^{-1} \mathbin{\dot{\otimes}} y) \leq y \qquad (22)$$

Note that these are interior operators for $\overline{\mathcal{K}}$, but they are none of those found in (7), (14) or (17). However if we substitute R for R^{-1} we realise that these are the closure operators of $\mathfrak{B}_{10}^{\varphi}(G, M, R^{-1})_{\overline{\mathcal{K}}}$, whence the result. $\qquad\square$

This suggests that the qualities of the connections might be swapped two by two: the Galois connection under the point of view of $\overline{\mathcal{K}}^{-1}$ is actually a co-Galois connection, under the point of view of $\overline{\mathcal{K}}$, and so on. The details of this assertion are left for future work on this issue.

4 Discussion

This discussion will be centered around the data science applications of $\mathcal{K}$-FCA. In the data practice of FCA *contextualization* represents a modelling bottleneck where the practitioner asserts its domain expertise in delimiting formal object and attributes and the relation that mediates between them as the formal context (G, M, R). In $\mathcal{K}$-FCA this also includes a later, user-informed *scaling or thresholding* of the data to access standard FCA results for *post-hoc* binary data.

The present investigation shows that the choice of semifield to interpret (G, M, R) represents a *modelling bias*:

1. Considering the relevant algebra $\overline{\mathcal{K}}$ to be a semifield opens up for interpretation the 4-fold Galois connections and the isomorphic lattices in Sect. 3.
2. But considering the relevant algebra to be its inverse $\overline{\mathcal{K}}^{-1}$ essentially obtains the Galois connections of (G, M, R^{-1}).

This does not mean that the information in the context as seen through the *prisms* of any of the 4-fold connections is the same. Each of the connections and the lattices in there has to be interpreted in their own particular light, e.g. concept lattices in their peculiar *colour* and neighbourhood lattices in theirs [10]. For instance, a case of analysis in [3] is shown where the underlying data are confusion matrices from human perceptual studies and machine classifiers where these differences are made explicit.

We suggest this conscious metaphor to deal with $\mathcal{K}$-FCA as an exploratory technique based in the following steps:

1. CONTEXTUALIZATION: coalescing the initial (G, M, R).
2. BIAS INTRODUCTION: choosing whether to analyze in $\overline{\mathcal{K}}$ or $\overline{\mathcal{K}}^{-1}$.
3. SCALING OR THRESHOLDING: choosing $\varphi = \gamma \overset{.}{\otimes} \mu$, and
4. COLOURING: choosing which type of connection or lattice to observe.

We may also wonder if this is the only lattice-theoretic information we can extract from the context. The work of [11] suggests this is not so: the context lattice of the contrary context $(M, G, (R^{-1})^{\mathrm{T}})$ is an important theoretical notion and we may expect it to be equally important from the application point of view. However, the universal representation capabilities of Concept Lattices has already suggested an interpretation for such lattices [12,13] when the incidence is an order relation.

In an interesting "final remarks" of the seminals [14] Wille renounces any attempt at exhaustiveness of the lattice restructuring program and recommends: "Besides the interpretation by hierarchies of concepts, other basic interpretations of lattices should be introduced; ... " In this light, we should wonder what other alternative conceptualizations of the information carried by a context might be, what other "matter" can we see in a formal context.

References

1. Valverde-Albacete, F.J., Peláez-Moreno, C.: Towards a generalisation of formal concept analysis for data mining purposes. In: Missaoui, R., Schmidt, J. (eds.) ICFCA 2006. LNCS, vol. 3874, pp. 161–176. Springer, Heidelberg (2006). doi:10.1007/11671404_11
2. Valverde-Albacete, F.J., Peláez-Moreno, C.: Further Galois connections between semimodules over idempotent semirings. In: Diatta, J., Eklund, P. (eds.) Proceedings of the 4th Conference on Concept Lattices and Applications (CLA 2007), Montpellier, France, pp. 199–212 (2007)
3. Valverde-Albacete, F.J., Peláez-Moreno, C.: Extending conceptualisation modes for generalised formal concept analysis. Inf. Sci. **181**, 1888–1909 (2011)
4. Valverde-Albacete, F.J., Peláez-Moreno, C.: The linear algebra in formal concept analysis over idempotent semifields. In: Baixeries, J., Sacarea, C., Ojeda-Aciego, M. (eds.) ICFCA 2015. LNCS, vol. 9113, pp. 97–113. Springer, Cham (2015). doi:10.1007/978-3-319-19545-2_6
5. Valverde-Albacete, F.J., Peláez-Moreno, C.: K-formal concept analysis as linear algebra over idempotent semifields (2017). Submitted
6. Strang, G.: The fundamental theorem of linear algebra. Am. Math. Mon. **100**, 848–855 (1993)
7. Cuninghame-Green, R.: Minimax Algebra. LNEMS, vol. 166. Springer, Heidelberg (1979)

8. Gaubert, S.: Two lectures on max-plus algebra. Support de cours de la 26-iéme École de Printemps d'Informatique Théorique (1998). http://amadeus.inria.fr/ gaubert/papers.html

9. Butkovič, P.: Max-Linear Systems: Theory and Algorithms. Springer Monographs in Mathematics. Springer, Heidelberg (2010)

10. Düntsch, I., Gediga, G.: Modal-style operators in qualitative data analysis. In: Proceedings of the 2002 IEEE International Conference on Data Mining, ICDM 2002, pp. 155–162 (2002)

11. Deiters, K., Erné, M.: Negations and contrapositions of complete lattices. Discrete Math. **181**, 91–111 (1998)

12. Wille, R.: Finite distributive lattices as concept lattices. Atti Inc. Logica Math. **2**, 635–648 (1985)

13. Reuter, K.: The jump number and the lattice of maximal antichains. Discrete Math. **88**, 289–307 (1991)

14. Wille, R.: Restructuring lattice theory: an approach based on hierarchies of concepts. In: Ordered Sets (Banff, Alta 1981), Reidel, pp. 445–470 (1982)

Distributed and Parallel Computation of the Canonical Direct Basis

Jean-François Viaud[1]([✉]), Karell Bertet[1], Rokia Missaoui[2],
and Christophe Demko[1]

[1] Laboratory L3i, University of La Rochelle, La Rochelle, France
{jviaud,kbertet,cdemko}@univ-lr.fr
[2] University of Québec in Outaouais, Gatineau, Canada
rokia.missaoui@uqo.ca

Abstract. Mining association rules, including implications, is an important topic in Knowledge Discovery research area and in Formal Concept Analysis (FCA). In this paper, we present a novel algorithm that computes in a parallel way the canonical direct unit basis of a formal context in FCA. To that end, the algorithm first performs a horizontal split of the initial context into subcontexts and then exploits the notion of minimal dual transversal to merge the canonical direct unit bases generated from subcontexts.

Keywords: Formal context · Implication · Canonical direct unit basis · Distributed and parallel computing · Dual transversal

1 Introduction

With the increasing size of data and the need for data analytics in frequently distributed environments, parallel and distributed processing becomes a necessity to efficiently gain deeper insight into data and identify relevant patterns [28]. In the Knowledge Discovery area, there is still a great interest and many significant contributions in association rule mining. For instance, a novel parallel algorithm for the new multi-core shared memory platforms has been proposed in [26]. Each parallel process independently mines association rules and takes into account the table density to select the most appropriate mining strategy.

In Formal Concept Analysis (FCA) [10], computing implications from binary data is also widely studied. Many efficient algorithms have been designed to compute implication bases, *i.e.*, sets of implications from which all others can be inferred. The most known implication basis is the canonical (or Guigues-Duquenne or stem) basis due to its minimal number of implications. Sequential algorithms for the canonical basis computation have been implemented [18] as well as parallel ones [9,12,14]. However, distributed computation in the FCA studies remains scarce [13,23,27]. As an illustration of parallel computing, the work in [19] tackles the problem of generating an implication basis using the join of closure systems expressed by their corresponding implicational bases and

© Springer International Publishing AG 2017
K. Bertet et al. (Eds.): ICFCA 2017, LNAI 10308, pp. 228–241, 2017.
DOI: 10.1007/978-3-319-59271-8_15

concludes that the proposed algorithm is polynomial when the implicational bases are direct. The recent work by Kriegel and Borchmann [14] investigates a parallel computing of the stem basis by generating all concept intents and pseudo-intents through an exploration of the concept lattice in a bottom-up manner and level-wise order.

Two other bases have also been studied, namely the canonical direct basis (CDB) discovered independently by a couple of authors [4] and the D-basis [2]. The canonical direct basis is direct, which means that closures can be linearly computed. The D-basis is direct when its implications are properly ordered.

In this paper, we focus on a distributed and parallel computation of the canonical direct basis. To the best of our knowledge, this kind of computation has not been defined so far.

The paper is organized as follows. The next section presents the background of Formal Concept Analysis and implication bases. Then, the main contribution of our paper is given in Sect. 3 and relies on the theoretical notion of minimal dual transversal and its computation. Finally, we conclude the paper.

2 Structural Framework

We assume in the sequel that all sets (and thus lattices) are finite.

2.1 Lattices and Formal Concept Analysis

Algebraic Lattice. Let us first recall that a *lattice* $(L, \leq)$ is an ordered set in which every pair (x, y) of elements has a least upper bound, called *join* $x \vee y$, and a greatest lower bound, called *meet* $x \wedge y$. As we are only considering finite structures, every subset $\mathcal{A} \subset L$ has a join and meet (e.g. finite lattices are complete).

Concept or Galois Lattice. A (formal) *context* (O, A, R) (see Table 1) is defined by a set O of objects, a set A of attributes, and a binary relation $R \subset O \times A$, between O and A. Two operators are derived:

Table 1. A binary context

	a	b	c	d	e
1	x	x			
2		x	x	x	
3				x	x
4		x			
5	x		x		
6	x	x	x	x	x

– for each subset $X \subset O$, we define $X' = \{m \in A, o\ R\ m\ \forall o \in X\}$ and dually,
– for each subset $Y \subset A$, we define $Y' = \{o \in O, o\ R\ m\ \forall m \in Y\}$.

A (formal) *concept* (X,Y) represents a maximal objects-attributes correspondence such that $X' = Y$ and $Y' = X$. The sets X and Y are respectively called *extent* and *intent* of the concept. The generated concepts derived from a context are ordered as follows:

$$(X_1, Y_1) \le (X_2, Y_2) \iff X_1 \subseteq X_2 \iff Y_2 \subseteq Y_1$$

The whole set of formal concepts together with this order relation form a complete lattice, called the *concept lattice* of the context (O, A, R).

Special Elements. An element $j \in L$ is *join-irreducible* if it is not a least upper bound of any subset of elements not containing it. The set of join irreducible elements is noted J_L. *Meet-irreducible* elements are defined dually and their set is M_L. As a direct consequence, an element $j \in L$ is join-irreducible if and only if it has a unique immediate predecessor denoted j^- (i.e. $j^- < j$ and there is no x such that $j^- < x < j$). Moreover, if L is a concept lattice, for each j there exists an $o \in O$ such that $j = (o'', o')$. Dually, an element $m \in L$ is meet-irreducible if and only if it has a unique immediate successor denoted m^+. Moreover, if L is a concept lattice, for each m there exists an $a \in A$ such that $m = (a', a'')$.

In Fig. 1, join-irreducible nodes are labelled with a number and meet-irreducible nodes are labelled with a letter.

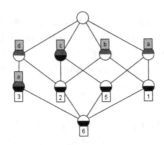

Fig. 1. The concept lattice of the context given in Table 1.

2.2   Association Rules and Bases

Association Rules. An association rule is a relationship between Y and Z as disjoint subsets of A, usually written $Y \rightarrow Z$. The set Y is called rule premise while Z is its conclusion. The support of an association rule $Y \rightarrow Z$ is the ratio of the cardinality of $(Y \cup Z)'$ over the cardinality of O. The confidence of an association rule $Y \rightarrow Z$ is the ratio of the cardinality of $(Y \cup Z)'$ over the cardinality of Y'. An association rule with a confidence equal to one is an implication.

For example, using the context given in Table 1, $bd \rightarrow c$ has a support of $\frac{1}{3}$ and a confidence equal to 1.

Bases of Implications. Given a context K, the set Γ_K of all implications is usually considerably large and contains redundant implications. For instance, when $B \to C, D$ holds, the implications $B \to C$ and $B \to D$ become redundant.

Implications can then be deduced from other ones using an inference system such as the set of Armstrong's axioms [3].

- Reflexivity: if $Z \subseteq Y$ then $Y \to Z$.
- Augmentation: if $Y \to Z$ then $Y \cup T \to Z \cup T$
- Transitivity: if $Y \to Z$ and $Z \to T$ then $Y \to T$

An alternate set of axioms, called Simplication Logic for Functional Dependencies [6–8,16,17,20,21] can also be used.

In this framework, the challenge is to find concise implication bases from which all implications can be deduced [15].

Definition 1. *Given a context $K = (O, A, R)$ and its set of implications Γ_K. A subset $\mathcal{B} \subseteq \Gamma_K$ is an implication basis if every implication of Γ_K can be deduced from $\mathcal{B}$ using inference axioms.*

Three different bases have been widely studied:

- The canonical basis, also known as Guigues-Duquenne basis or stem basis [11,18,24,25]
- The canonical direct basis, which has been defined many times with different properties and names [4,22], including the generic basis.
- The ordered direct implicational basis [1,2].

Notice that *direct* means that the operator $\bullet \to \bullet''$ can be computed from a direct basis using its rules at most once.

Canonical Direct Basis. In the following, we give more details about the canonical direct basis with a focus on the implication premises as shown below:

Definition 2. *Let $K = (O, A, R)$ be a context.*

- *Given a closed set of attributes $T = T''$, a subset $Y \subseteq A$ is a generator of T, if $Y'' = T$.*
- *Among all generators of a given set, some are minimal for the inclusion and are called minimal generators.*
- *Premises of the implications in the canonical direct basis are minimal generators.*

Considering our previous example whose context is given in Table 1, the canonical direct basis is given in Table 2.

Table 2. The canonical direct basis of the context given in Table 1.

$bd \rightarrow c$	$bc \rightarrow d$
$cd \rightarrow b$	$e \rightarrow d$
$abc \rightarrow e$	$ad \rightarrow bce$
$ae \rightarrow bc$	$be \rightarrow acd$
$ce \rightarrow abd$	

3   Main Contribution

Our approach towards parallel and distributed computation of the canonical direct basis (CDB) consists to first split the initial context horizontally (i.e., splitting the set of objects into subsets), compute the corresponding CDB of each subcontext and finally merge all of the generated CDBs.

Of course, the merge is not straightforward since minimal generators in the subcontexts can be smaller than minimal generators in the initial context.

3.1   Dual Transversal

The merge process is based on the following definition:

Definition 3. *Let $(A_i^j)_{1 \leq j \leq m, 1 \leq i \leq n_j}$ be a finite family of sets corresponding to the premises of implications (with the same unary-attribute conclusion) related to the m generated subcontexts, where n_j is the cardinality of A_i^j.*

- *The set τ is a dual transversal if for all $1 \leq j \leq m$, there exists $1 \leq i_j \leq n_j$ such that:*

$$\tau = \cup_{1 \leq j \leq m} A_{i_j}^j$$

  This means that τ contains at least one member from each family $(A_i^j)_{1 \leq i \leq n_j}$, j being given.
- *A dual transversal τ is minimal if any dual transversal $\tau' \subseteq \tau$ is such that $\tau' = \tau$.*

Consider the following example with two set families (i.e., $m = 2$). Take $A_1 = \{b, e\}$ and $A_2 = \{c, e, ac, ae, be, ce\}$. Then e, ae, abc and ace are dual transversals. However, ace is not minimal since ae is also a dual transversal.

3.2   Computing the CDB with Minimal Dual Transversals

Steps. The main steps of our procedure are as follows:

- First, split the initially given context $K = (O, A, R)$ into sub-contexts $K_j = (O_j, A, R \cap O_j \times A)$, such that $O = \cup O_j$ and K is the subposition of the K_j for $1 \leq j \leq m$.

- Compute the canonical direct unit basis of each sub-context and make sure that each implication whose conclusion contains p attributes is decomposed into p implications with unary-attribute conclusion.
- For each unary-attribute conclusion, merge implications from canonical unit direct bases using minimal dual transversals. This last merge step is the main solution of our paper and is detailed below.

Example. Consider the context in Table 1 and let us split it vertically into the two subcontexts shown in Tables 3 and 4. The generated bases from the two subcontexts are given in Table 5.

Table 3. Upper part of the initial context.

	a	b	c	d	e
1	x	x			
2		x	x	x	
3				x	x

Table 4. Lower part of the initial context.

	a	b	c	d	e
4			x		
5	x		x		
6	x	x	x	x	x

Theorem 1. *Let K be the context split into m sub-contexts K_j with $1 \leq j \leq m$. Let a be a unary-attribute conclusion, and $A_i^j \to a$, $i \in \{1, \ldots, n_j\}$ implications of the canonical direct unit basis computed for each K_j. Let $\{\tau_k \mid 1 \leq k \leq p\}$ be the set of minimal dual transversals of $(A_i^j)_{1 \leq j \leq m, 1 \leq i \leq n_j}$. Then $\tau_k \to a$ is an implication of the canonical direct unit basis of K with a in its conclusion.*

Proof. By definition of minimal generators, A_i^j is a set of minimal attributes such that $A_i^j \to a$ is an implication in K_j.

Since τ_k is a dual transversal, $\tau_k \to a$ is an implication in each K_j so is in K.

Moreover, τ_k is minimal and $\tau_k \to a$ is an implication. Then τ_k is a minimal generator.

Reciprocally, each minimal generator can be considered as a minimal dual transversal.

In order to compute the canonical direct basis of K, this theorem states that we only need to compute the canonical direct unit basis of all the K_j and then merge them by means of the minimal dual transversals.

Table 5. The canonical direct unit basis of each one of the sub-contexts of the initial context given in Table 1.

Upper context	Lower context
$be \rightarrow a$	$b \rightarrow a$
$ce \rightarrow a$	$d \rightarrow a$
	$e \rightarrow a$
$a \rightarrow b$	$d \rightarrow b$
$c \rightarrow b$	$e \rightarrow b$
$ac \rightarrow b$	
$ad \rightarrow b$	
$ae \rightarrow b$	
$ce \rightarrow b$	
$bd \rightarrow c$	$a \rightarrow c$
$ad \rightarrow c$	$b \rightarrow c$
$ae \rightarrow c$	$d \rightarrow c$
$be \rightarrow c$	$e \rightarrow c$
$c \rightarrow d$	$b \rightarrow d$
$e \rightarrow d$	$b \rightarrow d$
$ac \rightarrow d$	
$ae \rightarrow d$	
$be \rightarrow d$	
$ce \rightarrow d$	
$ac \rightarrow e$	$b \rightarrow e$
$ad \rightarrow e$	$d \rightarrow e$

Algorithm. In the following we propose an algorithm to compute minimal dual transversals. As stated before, we decompose implications with more than one attribute in their conclusion into implications with unary-attribute conclusion. To get the whole canonical direct basis, the following computation has to be done for each attribute in the conclusion.

Notice that the computation of the canonical direct basis by means of minimal tranversals is NP-complete [5].

As in the proof of the theorem, consider A_i^j the premises of the canonical direct unit basis of the K_j. Minimal dual transversals are then computed iteratively, considering successively each j. Algorithm 1 and its description are given below.

When $j = 1$, we consider that the unique context is K_1, so by definition, minimal dual transversals are in the set A_i^1. Let L_1 be the set of these minimal dual transversals.

Suppose now that the merge process has been done up to Step $j - 1$. So we have the set L_{j-1} of minimal dual transversals covering contexts K_1 to K_{j-1}.

Algorithm 1. Merging algorithm to compute minimal unit transversals having a as a conclusion.

Data: $K \leftarrow$ Initial formal context split into m subcontexts K_j with $1 \leq j \leq m$;
$A_i^j \leftarrow$ premises of the CDB of K_j.
Output: L_m containing the premises of the canoninal basis of K.
```
/* Special treatment for j = 1                                    */
```
1   $L_1 \leftarrow$ Set of all the A_i^1;
2 **for** $j \leftarrow 2$ **to** m **do**
3   **forall** $\alpha \in L_{j-1}$ and $\beta \in (A_i^j)_{1 \leq i \leq n_j}$ **do**
4   **if** L_j contains a set included in $\alpha \cup \beta$ **then**
5 | Remove $\alpha \cup \beta$
6   **else**
7 | Add $\alpha \cup \beta$ to L_j
8   Remove every set of L_j strictly included in $\alpha \cup \beta$
9 **Return** L_m

In other words, L_{j-1} contains all the premises of the canonical direct unit basis of the subposition of $K_1, \ldots, K_{j-1}$ having a as a conclusion.

At step j, we consider successively all the A_i^j and all the $\alpha \in L_{j-1}$. At the beginning, L_j is supposed to be empty. Now consider $\alpha \cup A_i^j$:

- If L_j contains a set included in $\alpha \cup A_i^j$, then $\alpha \cup A_i^j$ is removed. If not, $\alpha \cup A_i^j$ is added to L_j.
- Every set of L_j **strictly** included in $\alpha \cup A_i^j$ is removed. Notice that, if $\alpha \cup A_i^j \in L_j$, it is not removed at this step.

At the end of the process, we get all the minimal dual transversals since the minimal property is maintained at each step.

To illustrate our algorithm, we consider the context given in Table 1 split into the two subcontexts given in Tables 3 and 4.

Let us focus on implications with conclusion e. At the first step, we immediately get $L_1 = \{ac, ad\}$. Iterations of the second step are given in Table 6.

At the end of the four iterations of Table 6, $L_2 = \{abc, ad\}$ and these sets are minimal generators for the implications $abc \rightarrow e$ and $ad \rightarrow e$ respectively.

Table 6. Second step of the computation of minimal dual transversals with the conclusion e.

Iteration	Element of L_1	A_i^2	Dual transversal	Removed at iteration
1	ac	b	abc	
2	ac	d	acd	4
3	ad	b	abd	4
4	ad	d	ad	

We can then repeat this process for the rest of attributes in A and thus get the canonical direct unit basis.

3.3 Example

In this subsection, we consider the classical context [10], named "Live In Water". The context is given in Table 7 and its canonical direct basis is given in Table 8.

Table 7. Context "Live In Water". a: lives in water, b: lives on land, c: needs chlorophyll, d: dicotyledon, e: monocotyledon, f: can move, g: has limbs, h: breast feeds.

	a	b	c	d	e	f	g	h
Bean		x	x	x				
Bream	x					x	x	
Corn		x	x		x			
Dog		x				x	x	x
Fish leech	x					x		
Frog	x	x				x	x	
Reed	x	x	x		x			
Water weeds	x		x		x			

Table 8. The canonical direct basis of the context "Live In Water".

$h \rightarrow bfg$	$g \rightarrow f$	$e \rightarrow c$
$eh \rightarrow ad$	$eg \rightarrow abdh$	$ef \rightarrow abdgh$
$d \rightarrow bc$	$dh \rightarrow ae$	$dg \rightarrow aeh$
$df \rightarrow ae.g.h$	$de \rightarrow afgh$	$ch \rightarrow ade$
$cg \rightarrow abdeh$	$cf \rightarrow abde.g.h$	$bf \rightarrow g$
$ah \rightarrow cde$	$ad \rightarrow efgh$	$ac \rightarrow e$

This context is split into three pieces, containing respectively the following groups: {bean, bream, corn}, {dog, fishleech, frog} and {reed, waterweeds}. Canonical Direct Bases of these three contexts are given in Table 9.

With these three bases, we can use our merging algorithm to generate the canonical direct basis of the whole context. In the following, we consider the attribute-conclusion a, so Algorithm 1 receives as input the following premises: $\{f, g, h, de\}$ from the first subcontext, $\{c, d, e\}$ from the second one, and $\{b, c, d, e, f, g, h\}$ from the third one.

The first subcontext ($j = 1$) has a special treatment, and:

$$L_1 = \{f, g, h, de\}$$

Table 9. The canonical direct bases of the three subcontexts.

First subcontext: bean, bream, corn	Second subcontext: dog, fishleech, frog
$h \rightarrow abcdefg$	$h \rightarrow bfg$
$g \rightarrow af$	$g \rightarrow bf$
$f \rightarrow ag$	$e \rightarrow abcdfgh$
$e \rightarrow bc$	$d \rightarrow abcefgh$
$eg \rightarrow dh$	$c \rightarrow abdefgh$
$ef \rightarrow dh$	$b \rightarrow fg$
$d \rightarrow bc$	$a \rightarrow f$
$dg \rightarrow eh$	$ah \rightarrow cde$
$df \rightarrow eh$	Third subcontext: reed, waterweeds
$de \rightarrow afgh$	$h \rightarrow abcdefg$
$c \rightarrow b$	$g \rightarrow abcdefh$
$cg \rightarrow deh$	$f \rightarrow abcde.g.h$
$cf \rightarrow deh$	$e \rightarrow ac$
$b \rightarrow c$	$d \rightarrow abcefgh$
$bg \rightarrow deh$	$c \rightarrow ae$
$bf \rightarrow deh$	$b \rightarrow ace$
$a \rightarrow fg$	$a \rightarrow ce$
$ae \rightarrow dh$	
$ad \rightarrow eh$	
$ac \rightarrow deh$	
$ab \rightarrow deh$	

When $j = 2$, we consider each union of an element in $L_1 = \{f, g, h, de\}$ with an element in $\{c, d, e\}$ so that we get:

$$\{cf, df, ef, cg, dg, eg, ch, dh, eh, cde, de\}$$

Table 10. Steps of the computation of L_2.

x	L_2
	$\{\}$
cf	$\{cf\}$
df	$\{cf, df\}$
ef	$\{cf, df, ef\}$
cg	$\{cf, df, ef, cg\}$
dg	$\{cf, df, ef, cg, dg\}$
eg	$\{cf, df, ef, cg, dg, eg\}$
ch	$\{cf, df, ef, cg, dg, eg, ch\}$
dh	$\{cf, df, ef, cg, dg, eg, ch, dh\}$
eh	$\{cf, df, ef, cg, dg, eg, ch, dh, eh\}$
cde	$\{cf, df, ef, cg, dg, eg, ch, dh, eh, cde\}$
de	$\{cf, df, ef, cg, dg, eg, ch, dh, eh, de\}$

Each element x of this set is considered successively. Then x is added to L_2 except if L_2 already contains a smaller one. Moreover, every element of L_2 strictly containing x is removed. All the steps to compute L_2 are given in Table 10.

We find $L_2 = \{cf, df, ef, cg, dg, eg, ch, dh, eh, de\}$, and go further with $j = 3$. Again, each union of an element in L_2 with an element in $\{b, c, d, e, f, g, h\}$ is considered, so that we get:

Table 11. Steps of the computation of L_3.

y	L_3
	$\{\}$
bcf	$\{bcf\}$
cf	$\{cf\}$
cdf, cef, cfg, cfh	$\{cf\}$
bdf	$\{cf, bdf\}$
df	$\{cf, df\}$
def, dfg, dfh	$\{cf, df\}$
bef	$\{cf, df, bef\}$
ef	$\{cf, df, ef\}$
efg, efh	$\{cf, df, ef\}$
bcg	$\{cf, df, ef, bcg\}$
cg	$\{cf, df, ef, cg\}$
cdg, ceg, cgh	$\{cf, df, ef, cg\}$
bdg	$\{cf, df, ef, cg, bdg\}$
dg	$\{cf, df, ef, cg, dg\}$
deg, dgh	$\{cf, df, ef, cg, dg\}$
beg	$\{cf, df, ef, cg, dg, beg\}$
eg	$\{cf, df, ef, cg, dg, e.g.\}$
egh	$\{cf, df, ef, cg, dg, e.g.\}$
bch	$\{cf, df, ef, cg, dg, e.g., bch\}$
ch	$\{cf, df, ef, cg, dg, e.g., ch\}$
cdh, ceh	$\{cf, df, ef, cg, dg, e.g., ch\}$
bdh	$\{cf, df, ef, cg, dg, e.g., ch, bdh\}$
dh	$\{cf, df, ef, cg, dg, e.g., ch, dh\}$
deh	$\{cf, df, ef, cg, dg, e.g., ch, dh\}$
beh	$\{cf, df, ef, cg, dg, e.g., ch, dh, beh\}$
eh	$\{cf, df, ef, cg, dg, e.g., ch, dh, eh\}$
bde	$\{cf, df, ef, cg, dg, e.g., ch, dh, eh, bde\}$
cde	$\{cf, df, ef, cg, dg, e.g., ch, dh, eh, bde, cde\}$
de	$\{cf, df, ef, cg, dg, e.g., ch, dh, eh, de\}$

$$\{bcf, cf, cdf, cef, cfg, cfh, bdf, df, def, dfg, dfh, bef,$$
$$ef, efg, efh, bcg, cg, cdg, ceg, cgh, bdg, dg, deg, dgh, beg,$$
$$eg, egh, bch, ch, cdh, ceh, bdh, dh, deh, beh, eh, bde, cde, de\}$$

Each element y of this set is used to compute L_3 as shown in Table 11.

We find $L_3 = \{cf, df, ef, cg, dg, e.g., ch, dh, eh, de\}$, which is the set of premises of the canonical direct basis of the initial context, having a in their conclusion.

4   Conclusion

In this paper, we presented a new algorithm to compute the canonical direct unit basis by exploiting the notion of minimal dual transversal. These minimal dual transversals are computed by means of a merge process from minimal generators of subcontexts.

This point of view allows the computation of the canonical direct unit basis in a parallel and distributed way, since all the subcontexts can be considered simultaneously. More precisely, this algorithm can be translated into a MapReduce job in which the Map step consists in computing the CDB of each subcontext while the Reduce step consists in merging the generated CDBs from all subcontexts.

We plan to conduct an empirical study of different implementations of our algorithm against an iterative approach in order to assess the performance and scalability of our solution. Very first experiments with a parallelized algorithm seem to show that the execution time of the merging process is too high to get an efficient computation. Further distributed experiments using MapReduce will be conducted.

Our algorithm can also be generalized to association rules. Indeed, rules that are valid in a subcontext having n_1 rows over the whole set of n objects have a confidence greater than or equal to $\frac{n_1}{n}$. So the merge process can be adapted to get all the rules with a confidence strictly over $1 - \min(\frac{n_1}{n})$.

Acknowledgment. One of the authors acknowledges the financial support of the Natural Sciences and Engineering Research Council of Canada (NSERC) and all the authors would like to warmly thank referees for their comments and suggestions that helped improve the quality of this paper.

References

1. Adaricheva, K., et al.: Measuring the implications of the D-basis in analysis of data in biomedical studies. In: Baixeries, J., Sacarea, C., Ojeda-Aciego, M. (eds.) ICFCA 2015. LNCS, vol. 9113, pp. 39–57. Springer, Cham (2015). doi:10.1007/978-3-319-19545-2_3
2. Adaricheva, K.V., Nation, J.B., Rand, R.: Ordered direct implicational basis of a finite closure system. In: ISAIM (2012)
3. Armstrong, W.W., Deobel, C.: Decompositions and functional dependencies in relations. ACM Trans. Datab. Syst. (TODS) **5**(4), 404–430 (1980)

4. Bertet, K., Monjardet, B.: The multiple facets of the canonical direct unit implicational basis. Theoret. Comput. Sci. **411**(22–24), 2155–2166 (2010)
5. Bertet, K., Demko, C., Viaud, J.F., Guérin, C.: Lattices, closures systems and implication bases: a survey of structural aspects and algorithms. Theoret. Comput. Sci. (2016)
6. Cordero, P., Enciso, M., Mora, A.: Automated reasoning to infer all minimal keys. In: Proceedings of the Twenty-Third International Joint Conference on Artificial Intelligence, IJCAI 2013, pp. 817–823. AAAI Press (2013)
7. Cordero, P., Enciso, M., Mora, A., Ojeda-Aciego, M.N.: Computing minimal generators from implications: a logic-guided approach. In: Szathmary, L., Priss, U. (eds.) CLA. CEUR Workshop Proceedings, vol. 972, pp. 187–198. CEUR-WS.org (2012)
8. Cordero, P., Enciso, M., Mora, A., Ojeda-Aciego, M.N.: Computing left-minimal direct basis of implications. In: CLA, pp. 293–298 (2013)
9. Fu, H., Nguifo, E.: Partitioning large data to scale up lattice-based algorithm. In: Proceedings of 15th IEEE International Conference on Tools with Artificial Intelligence, pp. 537–541, November 2003
10. Ganter, B., Wille, R.: Formal Concept Analysis - Mathematical Foundations. Springer, Heidelberg (1999)
11. Guigues, J.L., Duquenne, V.: Familles minimales d'implications informatives résultant d'un tableau de données binaires. Mathématiques et Sciences Humaines **95**, 5–18 (1986)
12. Krajca, P., Outrata, J., Vychodil, V.: Parallel algorithm for computing fixpoints of Galois connections. Ann. Math. Artif. Intell. **59**(2), 257–272 (2010)
13. Krajca, P., Vychodil, V.: Distributed algorithm for computing formal concepts using map-reduce framework. In: Adams, N., Robardet, C., Siebes, A., Boulicaut, J.F. (eds.) Advances in Intelligent Data Analysis VIII. LNCS, vol. 5772, pp. 333–344. Springer, Heidelberg (2009)
14. Kriegel, F., Borchmann, D.: Nextclosures: parallel computation of the canonical base. In: Proceedings of the Twelfth International Conference on Concept Lattices and Their Applications, Clermont-Ferrand, France, 13–16 October 2015, pp. 181–192 (2015). http://ceur-ws.org/Vol-1466/paper15.pdf
15. Kryszkiewicz, M.: Concise representations of association rules. In: Proceedings of Pattern Detection and Discovery, ESF Exploratory Workshop, London, UK, 16–19 September 2002, pp. 92–109 (2002)
16. Mora, A., Cordero, P., Enciso, M., Fortes, I., Aguilera, G.: Closure via functional dependence simplification. Int. J. Comput. Math. **89**(4), 510–526 (2012)
17. Mora, A., de Guzmán, I.P., Enciso, M., Cordero, P.: Ideal non-deterministic operators as a formal framework to reduce the key finding problem. Int. J. Comput. Math. **88**, 1860–1868 (2011)
18. Obiedkov, S., Duquenne, V.: Attribute-incremental construction of the canonical implication basis. Ann. Math. Artif. Intell. **49**(1–4), 77–99 (2007)
19. Renaud, Y.: Join on closure systems using direct implicational basis representation. In: Le Thi, H.A., Bouvry, P., Pham Dinh, T. (eds.) MCO 2008. CCIS, vol. 14, pp. 450–457. Springer, Heidelberg (2008). doi:10.1007/978-3-540-87477-5_48
20. Rodríguez-Lorenzo, E., Bertet, K.: From implicational systems to direct-optimal bases: a logic-based approach. Appl. Math. Inf. Sci. **9**(305), 305–317 (2015)
21. Rodríguez-Lorenzo, E., Bertet, K., Cordero, P., Enciso, M., Mora, A.: The direct-optimal basis via reductions. In: Proceedings of the Eleventh International Conference on Concept Lattices and Their Applications, Košice, Slovakia, 7–10 October 2014, pp. 145–156 (2014)

22. Ryssel, U., Distel, F., Borchmann, D.: Fast algorithms for implication bases and attribute exploration using proper premises. Ann. Math. Artif. Intell. **70**(1–2), 25–53 (2014)
23. Tsiporkova, E., Boeva, V., Kostadinova, E.: MapReduce and FCA approach for clustering of multiple-experiment data compendium. In: Causmaecker, P.D., Maervoet, J., Messelis, T., Verbeeck, K., Vermeulen, T. (eds.) Proceedings of the 23rd Benelux Conference on Artificial Intelligence (2011)
24. Valtchev, P., Duquenne, V.: Towards scalable divide-and-conquer methods for computing concepts and implications. In: Proceedings of the 4th International Conference Journées de l'Informatique Messine (JIM 2003): Knowledge Discovery and Discrete Mathematics, Metz (FR), pp. 3–6 (2003)
25. Valtchev, P., Duquenne, V.: On the merge of factor canonical bases. In: Medina, R., Obiedkov, S. (eds.) ICFCA 2008. LNCS (LNAI), vol. 4933, pp. 182–198. Springer, Heidelberg (2008). doi:10.1007/978-3-540-78137-0_14
26. Vu, L., Alaghband, G.: Novel parallel method for association rule mining on multi-core shared memory systems. Parallel Comput. **40**(10), 768–785 (2014)
27. Xu, B., Fréin, R., Robson, E., Ó Foghlú, M.: Distributed formal concept analysis algorithms based on an iterative MapReduce framework. In: Domenach, F., Ignatov, D.I., Poelmans, J. (eds.) ICFCA 2012. LNCS (LNAI), vol. 7278, pp. 292–308. Springer, Heidelberg (2012). doi:10.1007/978-3-642-29892-9_26
28. Zaki, M.J., Pan, Y.: Introduction: recent developments in parallel and distributed data mining. Distrib. Parallel Datab. **11**(2), 123–127 (2002)

Author Index

Albano, Alexandre 39
Andrews, Simon 56

Bedek, Michael 3
Bertet, Karell 228
Borchmann, Daniel 23, 72
Bosc, Guillaume 106
Boulicaut, Jean-François 106
Bourneuf, Lucas 89
Braud, Agnès 138

Codocedo, Victor 106

Demko, Christophe 228
Dragoş, Sanda-Maria 122

Ganter, Bernhard 3, 23

Hanika, Tom 72
Heller, Jürgen 3

Kaytoue, Mehdi 106
Kriegel, Francesco 155, 168
Kuznetsov, Sergei O. 184

Le Ber, Florence 138

Makhalova, Tatiana 184
Missaoui, Rokia 228

Napoli, Amedeo 106
Nica, Cristina 138
Nicolas, Jacques 89

Obiedkov, Sergei 72

Peláez-Moreno, Carmen 211
Priss, Uta 198
Prochaska, Juliane 23

Săcărea, Christian 122
Şotropa, Diana-Florina 122
Suck, Reinhard 3

Valverde-Albacete, Francisco José 211
Viaud, Jean-François 228

Wille, Rudolf 23

Printed in the United States
By Bookmasters